Springer Biographies

The books published in the Springer Biographies tell of the life and work of scholars, innovators, and pioneers in all fields of learning and throughout the ages. Prominent scientists and philosophers will feature, but so too will lesser known personalities whose significant contributions deserve greater recognition and whose remarkable life stories will stir and motivate readers. Authored by historians and other academic writers, the volumes describe and analyse the main achievements of their subjects in manner accessible to nonspecialists, interweaving these with salient aspects of the protagonists' personal lives. Autobiographies and memoirs also fall into the scope of the series.

Martin Beech

William Frederick Denning

Grand Amateur and Doyen of British
Meteor Astronomy

 Springer

Martin Beech
Department of Astronomy
Campion College at the University of Regina
Regina, SK, Canada

ISSN 2365-0613   ISSN 2365-0621   (electronic)
Springer Biographies
ISBN 978-3-031-44442-5   ISBN 978-3-031-44443-2   (eBook)
https://doi.org/10.1007/978-3-031-44443-2

This Springer imprint is published by the registered company Springer Nature Switzerland AG
The registered company address is: Gewerbestrasse 11, 6330 Cham, Switzerland

Paper in this product is recyclable.

Prologue

*In the silence of the night when the living have left,
the whispers begin*

The Library, by Rosemary Griebel

I am not sure that I actually like the idea of searching out the final resting places of the long departed (it seems somehow macabre and intrusive), but on an overcast, gray-sky day I boarded a Bristol-bound train at London's Paddington Station in the hope that I might locate the grave site of William Fredrick Denning. I knew from the outset that the graveyard where Denning had been buried was now an urban park, but it seemed worth a visit, just in case some physical traces had survived.

I arrived at Bristol Temple Meads by mid-morning. The sky was still overcast and now threatening rain. From the train station, I had about a one mile walk to Brunswick Square. The roads were heavy with traffic, and the noise seemed almost overpowering. The buildings around me were new, modern, tall, and obscure, giving no indication of what business or trade the people inside might be plying. I walked through Cabot Circus, an aptly named mall alive with commerce and shops. Moving closer toward Brunswick Square, however, the traffic diminished, and a quiet began to descend. I saw fewer and fewer people, and the houses began to look ever more weather-worn, lived-in, and less cared for. A sense of the real city, the bit that people actually live in, as opposed to simply rush through, started to emerge before me. Upon entering

Brunswick Square, a sad feeling of urban decay overpowered me. The houses were old, tall, and clearly built in an era now long past. They were glaringly at odds with the newer buildings that have sprung up around them—the older houses belonged to a changed landscape that their earlier owners would now no longer recognize. The past confronted the modern at Brunswick Square's boundary, and it was clear that the new would soon overpower the old. The grass in the center of the Square was cut, but not especially cared for. The trees seemed subdued, overburdened by their environment, and they looked less than sturdy.

Founded in the 1750s, the Brunswick Square cemetery is no longer in active use (Fig. 1). Indeed, it was turned into a public park in 1988. A white marble stone located close to the entrance informs the visitor that, "burials took place during the period 1768–1963. Many original monuments having been removed [in 1988] to permit the laying out of the site as open space for the benefit of the citizens of Bristol". It is not an overtly pleasant park, and it exuded a sense of rejection and detachment—the public had seemingly not come to enjoy the greened space. The walls around the cemetery were topped with both barbed and razor wire, and bright yellow signs warned that, "the public consumption of alcohol is forbidden". Many of the gravestones and monuments have been moved to the periphery of the park; they are mostly old, weather-worn, and broken. Many of the headstones have been warn smooth by the elements, the names of those commemorated now lost to my gaze. Some of the chiseled text had moss and lichen growing into the cut marks, providing the would-be reader with an embossed living font. I searched for about an hour. There are perhaps only fifty gravestones left in the park, but none bear witness to the resting place of Denning. Indeed, no gravestones revealed his family name at all. Denning's final resting place is now lost to us, and even his headstone has been removed. Later inquiry to the Bristol City Archives revealed that Denning's headstone had not been kept, it being destroyed along with the other headstones after their removal. Furthermore, no specific information concerning the location of grave plots has been kept.

Brunswick Square was quiet during my visit, and I saw just a few people using the park to walk their dogs. It is inherently a peaceful place by day, but not a place in which to dwell at nighttime. Most people walked straight through, neither looking to the left nor the right—the presence of their forebearers being of no apparent interest. One section of the park contains a multi-component sculpture, commissioned in 2010, by artist Hew Locke—it is incongruous and jars somewhat with its surroundings—which was perhaps the point. Indeed, the sculpture had resonance with the park setting since its

Fig. 1 Brunswick Square cemetery, Bristol, Denning's now unmarked resting place. The small building to the middle-right is a Georgian mortuary chapel. Photograph by the author

title is *Ruined* (Fig. 2). A plaque explains that the sculpture was composed in order to reflect the boom-and-bust cycle of Bristolian industry. The rusting of the various iron pieces, symbolic gravestones, seemed poetic, as if the very elements wanted to erase their presence from the park—the slight rain that had begun to fall also added to the general sense of wanting to wash the park clean of this modern-day intrusion. As I leave the park, I am stopped by a young man, an earnest Jehovah's Witness. He hands me a pamphlet with the title *Would you like to know the truth?* It seemed a poignant moment.

The bodily remains of Denning are interned in Brunswick Square cemetery, somewhere below the grassy surface where the living still walk. His resting place is localized, but not locally marked. His presence is recorded, but as a person he resides beyond our enquiring gaze. Indeed, this is a recurring theme that will appear more than once in the pages of this book: Denning, as an individual, is both near and far, muted and aloof. The greater quantity of his life is lost to enquiry, the records destroyed by war and entropy, or misplaced, or just plain never made. Nearly a century after his death, however, his name, influence, and presence can still be witnessed, and those facts and records that do remain make for a storied life. For all this, Denning remains an enigmatic figure—a figure from a world now long past. In life, he attained national and international fame, and his death was recorded in headlines published across the world. While Denning's publication legacy will live on as long as there are astronomical records, the record of his life, and

Fig. 2 Corner of Brunswick Square cemetery, showing (at center) the industrial-style, iron headstones of *Ruined* by Hew Locke. Photograph by the author

the measure of the man has passed into shadow. The house in which he lived during the final years of his life, 44 Egerton Road, still stands—the bricks and mortar having held up against the ravages of time, war, and endless British rain. The remembrance plaque that was erected at the house, however, with due reverence and respect, was removed sometime in the past and is now lost. Even the church where Denning's funeral service took place is no more—the Church of St. Michael and all Angels was demolished in 1997.

And so, as the day began to close, I found myself on the train back to London: Through the window, the trees and green fields are motion blurred—there is a hypnotic passing of the nearby landscape. The sky had been overcast all day; looming and overarching. The weather enhanced the mood and abstraction. It has been a contemplative, out of focus, and far away day. Denning, the person, it seems is a dissipated entity, a shadow-being, his life obscured by lost records; an uncertain birth place, an uncertain childhood, an obscure personal life, a lost graveside, a removed headstone, and a lost commemorative plaque. Perhaps, I wondered, as the train hurried away from Bristol, this was how the story should end, even though it left me with a deep sense of non-closure. Denning, the man, has passed into an unreachable realm, and our only fixed point, indeed the focus of our initial gaze, is that of his many contributions to the astronomical literature and to his seemingly endless enthusiasm to observer the heavens.

The theme of this book is twofold. Firstly, it will present a biography of Denning, considering his life, achievements, writings, and awards. The reader is reminded, however, that all biography, at some level, is provisional and subject to change. Letters, analysis, and understanding continue to evolve, and what were once gaps in our knowledge of a person's life are sometimes bridged. I believe that this present biography makes for a near complete, in terms of presently available data, review of Denning's life, but it is to be hoped that more details will eventually become available—especially in connection to his formative teenage years. As it is, we present, in Chap. 1, a temporal, year-by-year, account of Denning's life and times. There we attempt to find the storied man himself. Chapters 2 and 3 turn more or less away from biography, and I concentrate on Denning's contributions to science, presenting a detailed summary of his astronomical discoveries and his science writings. In Chap. 4, I turn to the second of the themes to be followed in this book and that is the rise of the amateur astronomer in Britain during the late nineteenth century. In this manner, I shall explore the role that Denning played in promoting and motivating amateur astronomy. Chapter 5 again turns away from the direct study of Denning and has as its main focus the rise of meteor astronomy in the UK from circa 1830 to circa 1930. In this chapter, Denning certainly features as a major player, but my aim is to determine how British researchers set out, starting in the early decades of the nineteenth century, to understand, codify, and arrange the details of meteor astronomy. Such developments, of course, did not occur solely in England, and the many fundamental contributions by other global players will also be examined. Indeed, one of the major issues, upon which Denning was very much on the wrong side (as it turned out) of the argument, was that of stationary radiants. Here Denning came into direct conflict with the new meteor astronomy being developed by many practitioners, but generally spearheaded by Dr. Charles Olivier of the American Meteor Society. Chapters 6 and 7 explore additional aspects of Denning as a motivating force. Chapter 6 concerns the appearance of the first astronomy journal, the *Astronomical Register*, produced with the amateur astronomer directly in mind. In addition to reviewing the importance of this publication, and Denning's contributions to it, I shall also attempt to discern who the typical amateur astronomer was in late nineteenth-century Britain, and I shall ask what sort of instruments, and observing projects it was that such amateurs were both using and interested in performing. Chapter 7 concerns Denning directly in that he was one of the key instigators in forming the Observing Astronomical Society—a dynamic, if not eclectic, group of amateur astronomers interested in performing scientifically useful work. Although short-lived, the Observing Astronomical Society paved the way for the later formation of important provincial and national amateur astronomical societies.

Contents

1

A Man of Parts

It was a man of many parts
  Who in his coffer mind
  Had stored the Classics and the Arts
  And sciences combined
  (James Whitcomb Riley)

Each chapter of human history is highlighted by its greater thinkers, the doers and the achievers. Be it in the political, literary or scientific realm, history has repeatedly shown that it requires just a few special individuals to significantly change the direction of human thought and understanding. Not least among the sciences, astronomy has recorded its fair share of luminaries; the intellectual giants who outstripped their contemporaries in terms of analytic skill or far-reaching observations.

For obvious reasons historians of science have tended to direct their attention towards the most distinguished scientists of an era. While such selection of key individuals does allow for a clearer understanding of the major advancements that were made, such an approach often overlooks the contributions made by the lesser but equally important and interesting workers in the field. This study turns to one such lesser luminary; lesser, that is, only in degree rather than kind. We shall present in this study what is essentially an emerging profile of a grand amateur astronomer. Indeed, a grand amateur astronomer whose influence has often been acknowledged but whose life has never been deeply explored. Our study is concerned with the life, influence, and works of William Frederick Denning (Fig. 1.1).

© The Author(s), under exclusive license to Springer Nature
Switzerland AG 2023
M. Beech, *William Frederick Denning*, Springer Biographies,
https://doi.org/10.1007/978-3-031-44443-2_1

Fig. 1.1 William Frederick Denning at his home of 44 Egerton Road in Bristol. Wearing an observer's greatcoat and felt cap, Denning, meteor-wand in his right hand, contemplates his Carey Celestial Globe. The photograph is dated April 1926, at which time Denning would have been 78 years old. Image from the author's collection

Although considered an amateur astronomer, Denning made significant contributions to the development of meteor astronomy. Not solely content with observing meteors, however, Denning also studied the planets, and was an avid sweeper of comets. In practice his field of interest was wide, and his enthusiasm for observing inexhaustible. Denning was an acknowledged inspiration to many of his contemporaries, both professional and amateur, and his contributions to astronomy have been long-lasting and influential. For all this, he was also, on occasion, a person involved in scientific controversy.

Throughout most of Denning's long life, there was no clear or recognizable distinction that could be drawn between the professional and amateur astronomer. The professional, as such, was typically an Oxbridge educated scholar, who may or may not have been affiliated with some university teaching position. Certainly, the idea and embodiment of a research scientist did not exist in Denning's lifetime—indeed, such aspirations are very much a modern, largely post second World War II, desire and possibility. Many of the most celebrated scientists of the eighteenth and nineteenth centuries were independently wealthy practitioners. William Herschel, for example, the most celebrated astronomer in eighteenth century England was not allied to any university or government office—although he did, in later life, receive a Royal stipend. A similar story applies to his son John who followed an illustrious and highly celebrated career, but who never once taught a university course, or, for that matter, collected a wage from any employer. Astronomy as a profession, that anyone might aspire to and additionally, thereby, earn a living, only began to develop with the dawn of the twentieth century. When Denning began exploring his astronomical interests, the only professional and government-paid astronomer was the Astronomer Royal. For all of the relatively recent, twentieth century, development of 'scientist' as a reasonable career choice, learned societies, such as the Royal Astronomical Society, were to begin forming in the early to mid-nineteenth century. These prestigious societies, however, did not cater to the needs or wants of the amateur, or to those simply interested in studying the heavens. These societies were about formality, and the establishment of authority. Indeed, it was not until the later decades of the nineteenth century that the amateur astronomer (as a recognizable person) might be said to have appeared on the scene. Such amateurs plied their interests as they chose, and generally had sufficient free-time, and funds to indulge their interests—they were effectively evolved practitioners from the burgeoning, and increasingly influential middle-class citizenry of Britain. More than playing the mere dilettante, however, the late nineteenth century amateur astronomers strove to find meaning in their activities. The desire was to observer, contemplate, and organize, and to use observing time to further their own intellectual development. Furthermore, the adversity inherent to their efforts was something to be worn as a badge of honor. Indeed, there was a certain romantic aura that surrounded the stalwart amateur, toiling away at their chosen subject, in the late night, post-workday, hours. Novelist Thomas Hardy tapped into this romantic conceit through the character Swithin St. Cleeve—the star-crossed, and hapless young astronomer described in *Two on a Tower* (published in 1882).

Denning described the outlook of the Victorian amateur astronomer in volume 3 of *Science for all* (edited by Robert Brown) published in 1897. Writing on what astronomy might be performed with the naked eye, Denning informs the reader that, "The chief thing necessary to success is a great love for the subject, for this is required to sustain the enthusiast through nightly vigils". To this he added, "unremunerative labour tires most men, but the true student of science will make any sacrifice of time and labour in the development of his observations or theories". Here, the key ideas are enthusiasm, graft, and personal sacrifice. These values, Victorian to the very core, are a repeating motif throughout Denning's life and legacy. Unlike the wealthy and financially independent grand amateurs, such as William Lassell (owner of a successful brewing company), or William Parsons (3rd Earl of Rosse—born into a titled position among the landed gentry), Denning was relatively low-born, and never financially well-off. To perform his astronomical works, he sacrificed career, comfort and worldly security. For all this, he was supported by his family, and personified the Marxian ideal of plebian scientist and intellectual.

William Frederick Denning was born in the small Somerset village of Wellow on 25 November, 1848. At that time his father, Isaac Poyntz Denning (1820–1895), was identified as being manager and accountant of the Braysdown Collieries. Four years earlier, however, on the day of his second marriage, Isaac had described himself as a schoolmaster. Denning's [I shall generally use Denning to refer to W. F. Denning] mother, Lydia Padfield, was the seamstress daughter of Richard Padfield, a wagon loader at Coal Barton Colliery. Isaac Poyntz Denning's father, Isaac Denning, is identified in 1844 as a retired Sergeant Major. Indeed, Isaac Denning served in the 53rd Shropshire Regiment of Foot (now the Kings Shropshire Light Infantry), and he saw action in the Peninsular Wars—the wars that helped establish the (so-called) Great Peace that stretched from 1815 to 1914, and the onset of the First World War.

Essentially nothing is known of Denning's early childhood. When he was just eight years old, however, his family moved from the rural settings of Somerset to the prosperous port-side city of Bristol. In the same year that the family moved to Bristol (in 1856) Isaac P. Denning established the accountancy partnership of Denning, Smith and Co., and indeed, he was to work within this partnership until his death in 1884. Isaac, and Lydia Denning were to sire a total of nine children: Isabella (born 1841), Ellen Louisa (born 1846), William Frederick (born 1848), Frederick (born 1850), Charles Poyntz (born 1852), Emma Elizabeth (born 1855), Mary Eveline Poyntz (born 1856), Alice Angelina (born 1857), and Francis (born 1860). Furthermore, the Denning family had many relatives in the Bristol area, and Denning was to live within the city limits for the remainder of his life, indeed, it would appear that he only rarely left its confines. The Denning family prospered in Bristol, but few details of Denning's formative years have survived. It seems reasonable to assume, however, that he received a sound education, some of which, according to Victor Plarr in his *Men and Women of the Time: a dictionary of cotemporaries* (published in 1895 by George Routledge & Sons, London), was conducted by his mother. Certainly, it was the case that Denning had a healthy and athletic adolescence. Fellow meteor observer and long-time friend, J. P. M. Prentice (see Chap. 5) commented that, "in his younger days he [Denning] was somewhat of an athlete, delighting in running and playing hockey, and a keen and skillful cricketer" [1]. Denning's skill as a cricketer was described in several of his obituary accounts, and T. E. R. Phillips recounts in particular that, "he once told the writer that W. G. had invited him to keep wicket for Gloustershire" [2]. "W. G.", of course, was none other than Dr. William Gilbert Grace who captained the renowned Gloustershire Cricket Club from 1871 to 1898. Then, as today, such an offer would have only been extended to players of some distinction. For reasons that we may only guess at Denning declined Grace's invitation, but, one suspects, the offer was probably declined because the young Denning had already set his sights on a living related to science writing and astronomy.[1] At the time that the offer to play for Gloustershire was made Denning would have been in his early to mid-twenties.

Denning had a life-long interest in natural history (see Chap. 3), an interest fostered at an early age by his mother, but according to his own words, he began to specialize in astronomy at 17 years of age [3]. In his *Men and Women of the time,* Plarr indicates that Denning's first telescope, a 3-inch refractor, was a gift from his father (presented circa 1864/5). His father subsequently brought him a 4 ½ -inch refractor. With these humble, but typical for the time, telescopes (see Chap. 6) Denning began to study the heavens and hone his observing skills. It is probably significant, however, that within a few years of his decision to pursue astronomy, Denning was fortunate to witness the spectacular Leonid meteor storm of 1866 (see Chap. 5), and the flight of an awe-inspiring fireball on the night of November 6th, 1869 (Fig. 1.2). It seems reasonable to suppose that it was the observation of these two impressive celestial events—both unexpected and visually spectacular—that turned Denning's interests towards meteor astronomy, a science that was then in its early stages of development. The 1869 fireball event was particularly significant for Denning since it initiated a long-running correspondence with Alexander Stewart Herschel (see Chap. 2) who was then one of the leading experts on meteor astronomy in Great Britian [4].

---

[1] Denning in fact used his knowledge of cricket on at least one occasion [3] to illustrate his frustration at the manner in which some observatories set about using untrained observers to make meteor observations. He noted, "it seems to be the fashion at certain observatories to set a number of observers (some of whom have perhaps never registered a meteor path before) watching and recording meteors, and then to investigate their results as though they could be thoroughly depended upon. It is similar to placing a man, who has never played in a cricket match before, as wicket-keeper to fast bowlers like Mold, Richardson and Woods, and expect his performance to be creditable". The analogy is in fact a good one, and indeed, Denning was often critical of the many poor observational accounts that found their way into the literature.

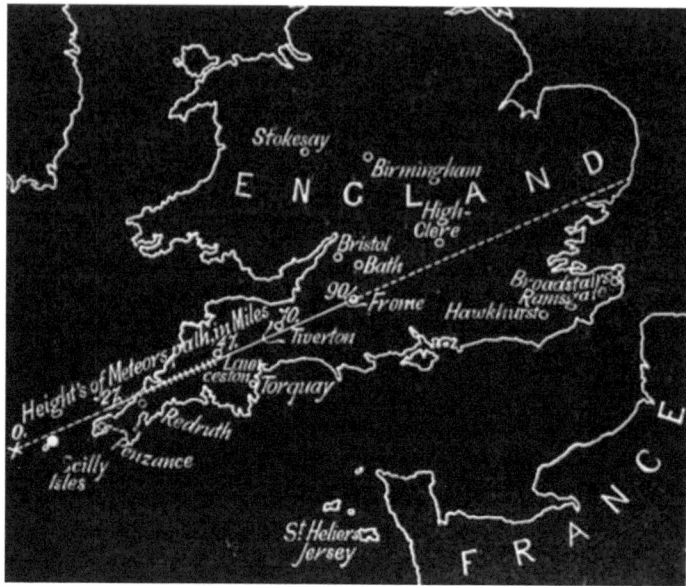

Fig. 1.2 The path of the November 6, 1869 fireball. By triangulation (see Chap. 5) methods, A. S. Herschel determined that the beginning and end heights of the fire-ball were 47 km and 27 km respectively, with the luminous trail length being some 170 km. The estimated duration was 5 s, with the speed being determined as 34 km/ s. From Bristol, the fireball would have been a truly spectacular sight. The luminous trail left in the fireball's path persisted for some 50 min. Image from *The Story of the Heavens,* by Robert Ball

While meteor observing was to dominate Denning's later research inter-ests, his early observational projects were many and varied. With respect to astronomy, however, Denning built his first telescope, a refractor using two pre-shaped lenses in 1866. But unlike many of his contemporaries, Denning was not a skilled optician and/or lens grinder, and all his future instruments were purchased items. As testament to his all-round enthusiasm, however, among Denning's first published notices are accounts relating to Sunspot groupings, the timings of Jovian satellite transits, and a record of the transit of Mercury as seen on 5 November, 1868 [5]. All these early publications appeared in the *Astronomical Register* (see Chap. 6) [6]. Denning became a regular contributor to the *Register*, and he used its pages to describe, and initiate a series of observational campaigns. Details relating to the first obser-vational campaign that Denning organized can be found in a Letter to the Editor of the *Register* dated March 16th, 1869. Indeed, Denning, along with sixteen other observers proposed to continuously monitor the daytime

Sun between March 14th and April 14th (1869), "with the view of re-discovering the suspected intra-Mercurial planet Vulcan" [7, 8].[2] It is not clear how, or when, the members involved in this study were first assembled, but it was probably initiated through the *Astronomical Register*, with the organizing correspondence taking place exterior to its pages. While the search for Vulcan was unsuccessful (we, of course, now know that any such search would have to be unsuccessful), Denning's enthusiasm for organized observing clearly remained high, and on July 1st, 1869 he became one of the founding members, as well as Honorary Treasurer and Secretary, of the Observational Astronomical Society (OAS—see Chap. 7). This short-lived society was in many ways the forerunner of the present-day Liverpool Astronomical Society (founded 1881), and the British Astronomical Association (founded 1890). The initial OAS membership was stated to be about fifty observers [9]. Denning compiled quarterly reports on the observations collected by OAS members, and these were published in *The Astronomical Register* and the journal *Nature*. Denning was closely involved with several additional campaigns in search of planet Vulcan, and (more fruitfully) in a series of studies of planet Venus. Importantly for his future research work, in 1871, Denning purchased a 10-inch With-Browning reflecting telescope. This telescope became his primary research instrument, and he used it to great effect in his subsequent planetary work (Fig. 1.3). The location from which the image in Fig. 1.3 was taken is not clear (see Table 1.1, however), other than it was within the boundaries of Bristol. Furthermore, the image indicates that the telescope was not housed within an observatory. Indeed, there is no indication that Denning ever constructed, or had use of an observatory shed or dome—his observing was invariable *en plein air*.

In late 1871 Denning published, through Wymann and Sons of London, his first astronomy almanac. Titled *Astronomical Phenomena in 1872* this text was written with the amateur astronomer in mind, and it was probably intended as a handbook for OAS members. The pamphlet, however, was reviewed in the influential journal *Nature* with the anonymous reviewer giving it a decidedly poor evaluation [10]. The reviewer, possible missing the point concerning its intended audience, was to write that the, "general remarks on astronomical observing … are addressed to the simplest tyro, and are so meagre as to give the impression of a want of accurate knowledge". The

---

[2] The first supposed sighting of Vulcan was made on March 29th, 1859 by the French country doctor Edmond Lescarbault. Denning was later to write of Lescarbault that he, "obviously lacks the experience and caution necessary to command credit". These comments followed in the wake of Lescarbault's announcement that he had discovered a new star in Leo on the night of January 11th, 1891. Incredibly, this new star was not a nova at all, but the planet Saturn, and indeed, Denning's comments seem apt.

Fig. 1.3 Denning by the side of his 10-inch aperture With-Browning telescope. Notice the telescope's somewhat unusual altitude-azimuth mounting. Image circa 1871, and courtesy of the Royal Astronomical Society

Table 1.1 Known addresses for Denning from 1868 to the time of his death in 1931

Year	Address
1868	2 Sussex Place, Ashley Road, Bristol
1872	HollywoodLodge, Cotham Park, Bristol
1876	Tyndale House, Ashley Down, Bristol
1890	17 Berkley Road, Ashley Down, Bristol
1891	Bex Villa, Morley Square, Bishopston
1892	Shannon Court, Corn Street, Bristol
1895	102 City Road, Bristol
1896	51 Brynland Avenue, Bishopston
1898	102 City Road, Bristol
1899	51 Brynland Avenue, Bishopston
1906–1931	44 Egerton Road, Bristol

Clearly Denning did not lead a particularly settled life, but the times spent at Tyndale House, and later 44 Egerton Road were apparently the most stable review concluded, "altogether the book is a very weak production". The editor of the *Astronomical Register*, Sandford Gorton (see Chap. 6) was equally as dismissive of the text, and somewhat arrogantly, chastised the text for its price, and for its occasional incorrect use of Latin terminology. One assumes that such a review would have disappointed Denning, and certainly no subsequent handbooks were produced. It was presumably a further disappointment to Denning that the OAS ceased activity circa 1872 [11]. On a more encouraging note, however, it was in December 1872 that Denning published his first observational results [12] in the *Monthly Notices of the Royal Astronomical Society* (MNRAS). At about the same time, Denning proffered his candidacy for fellowship to the RAS. He was, however, unsuccessful in this first attempt, and was not, in fact, elected a Fellow until 1877. The RAS does not keep records outlining the reasons for non-election of candidates, but one might speculate that the demise of the OAS, of which Denning was the prominent public figure, and the poor reviews of his first text were contributing factors.

Undaunted by the souring events of 1872, Denning continued to pursue his astronomical observations and interests. Indeed, throughout the 1870's the majority of Denning's published notes were concerned with the observation of meteors, and the reduction of meteor radiants (see Chap. 5). His first radiant catalogue was published in the *Monthly Notices* in 1876 [13], and was based upon observations collected between 1872 and 1876. An interesting point concerning Denning's 1876 radiant catalogue is that it was communicated to the Society by Robert P. Greg. Along with A. S. Herschel, Greg was one of the foremost authorities on meteor astronomy in England at that time. Greg's endorsement of Denning's meteor work was an important factor in determining Denning's eventual election as a Fellow of the RAS. From a few surviving fragments, it is clear that Denning and Greg must have exchanged a good number of letters and postcards concerning their meteor observations, and this approach is indicative of Denning's preferred mode of communication (see Chap. 2).

Denning made his first significant contribution to meteor astronomy in 1877. It was in that year that he was able to demonstrate a steady night by night movement in the location of the Perseid meteor shower radiant (see Chap. 2). These observations essentially confirmed, for the first time, a long-postulated expectation. More significantly, however, in the following year (1878) Denning published his first research paper suggesting that some meteor radiants were in fact stationary (with respect the background stars) in the sky for weeks to months on end [14]. The stationary radiant hypothesis, as Denning's claim came to be known, was problematic, however, since it was clear from the outset that the origin of such meteoroid streams (and their associated radiant positions) could not be explained through an association with cometary decay [15].[3] The issues surrounding the existence of stationary radiants took many years to resolve, and even at the time of his death, Denning still firmly believed in their existence (see Chap. 2).

With the close of the 1870s Denning began to publish an increasing number of observational notices on the planets. Specifically, with the aid of his 10-inch With-Browning reflector, he embarked on a series of studies to determine planetary rotation rates. While he regularly observed Mercury, Venus, Mars, and Saturn, the greater amount of his attention was directed towards Jupiter. Indeed, great interest had been excited towards Jupiter after what is now known as the Great Red Spot came in to prominence in 1878. Denning made a detailed historical study of the Red Spot's appearance (published in 1899), and in his lifetime recorded many hundreds of transits (see Chap. 3).

Parallel to his increased interest in planetary observing, Denning also began to search for comets in the late 1870s. His efforts were rewarded in 1881, when on the morning of October 4th, he discovered his first comet. This find turned out to be a short period comet, and the story of its discovery was often used by Denning as an example of why an, "observer should never hesitate" [16]. He noted, "on July 11, 1881, just before daylight, I stood contemplating Auriga, and the idea occurred to me to sweep the region with my comet eye-piece, but I hesitated, thinking the prospect not sufficiently inviting. Three nights later [John] Schäberle at Ann Arbor, U. S. A., discovered a bright telescopic comet in Auriga! Before sunrise on October 4 the same year I had been observing Jupiter, and again hesitated as to the utility of comet-seeking, but, remembering the little episode in my past experience, I instantly set to work, and almost at the first sweep alighted upon a suspicious object which afterwards proved itself a comet of short period". From all indications, Denning took this missed cometary episode to heart, and thereafter never hesitated at the thought of making any observation. Indeed, on the comet-seeking front Denning continued to have a good measure of success, discovering three more comets in 1890, 1892, and 1894. For each of these discoveries he was awarded Donohoe Medals by the Astronomical Society of the Pacific. Denning was also co-discoverer, with Edward E. Barnard, in America, of a comet in 1891.

---

[3] That meteoroid streams could be produced through cometary decay was realized in the mid-1860s. The Italian astronomer Giovanni Schiaparelli first demonstrated this in 1866 when he found that the Perseid meteoroid stream had orbital parameters similar to those of periodic comet 109P/Swift—Tuttle (see Chap. 5). A more detailed understanding of the cometary—meteoroid stream formation process did not become available, however, until well into the last century, with the pioneering work being performed by Fred Whipple in the 1950s.

By the mid-1880s Denning's publication rate had risen to about 20 articles and observing notices per year (see Chap. 2), and his reputation as a writer and skilled observer was becoming widely recognized. This recognition resulted in his being elected an honorary member of the Liverpool Astronomical Society (LAS) in 1882. His high-standing among amateur astronomers being further recognized between 1887 and 1888 when he served as President to the LAS. The LAS, which still thrives to this day, had been founded in 1881, and at the time of Denning's Presidency it boasted some 440 members world-wide. In addition to being President, Denning was also elected to the Directorship of the meteor, and comet-seeking section, as well as the planetary (Jupiter) section.

Denning's return to organized amateur astronomy seems to have been largely successful. During his yearlong Presidency, the LAS continued to prosper [17], with its membership increasing to 641. The society held monthly meetings in several locations throughout England, and it appears that Denning was a regular attendant at the meetings held in London, at which he read papers on meteors, comets, and planetary observing. He also wrote a collection of articles on telescopes, and telescopic work for the society's journal. These articles were later collected and expanded in book form. Denning's new work, entitled *Telescopic Work for Starlight Evenings*, was published by Taylor & Francis (London) in 1891 (Fig. 1.4) and, just like his first pamphlet, it was reviewed in the journal *Nature*. This time, however, the text received whole hearted praise [18]. The anonymous reviewer wrote, "As might be expected from such an experienced and enthusiastic observer as Mr. Denning, this book is thoroughly practical. He is not content with describing the beauties of the skies, but gives invaluable information as to how to see them better". With similar praise, the reviewer concluded, "Everyone who uses a telescope, or who intends to use one, of whatever dimension, should read Mr. Denning's book". Indeed, glowing reviews poured in: the *Philosophical Magazine* reviewer writing, "all telescopists should procure a copy without delay"; the *Observatory* magazine adding, "written in a straightforward and earnest fashion that must not only make difficulties clear to his readers but inspire them with something of his own determination and enthusiasm"; the *Weekly Mail* indicated that, "Mr. Denning is one of the keenest, most unwearying observers in England, and the book is worthy of his reputation", and the *Saturday Review* noted that, "it deserves to be read with respect and attention by every amateur astronomer". All high praise indeed, and the reviews clearly indicate that by the early 1890s Denning was a recognized, respected, and celebrated amateur astronomer, if not a burgeoning public figure.

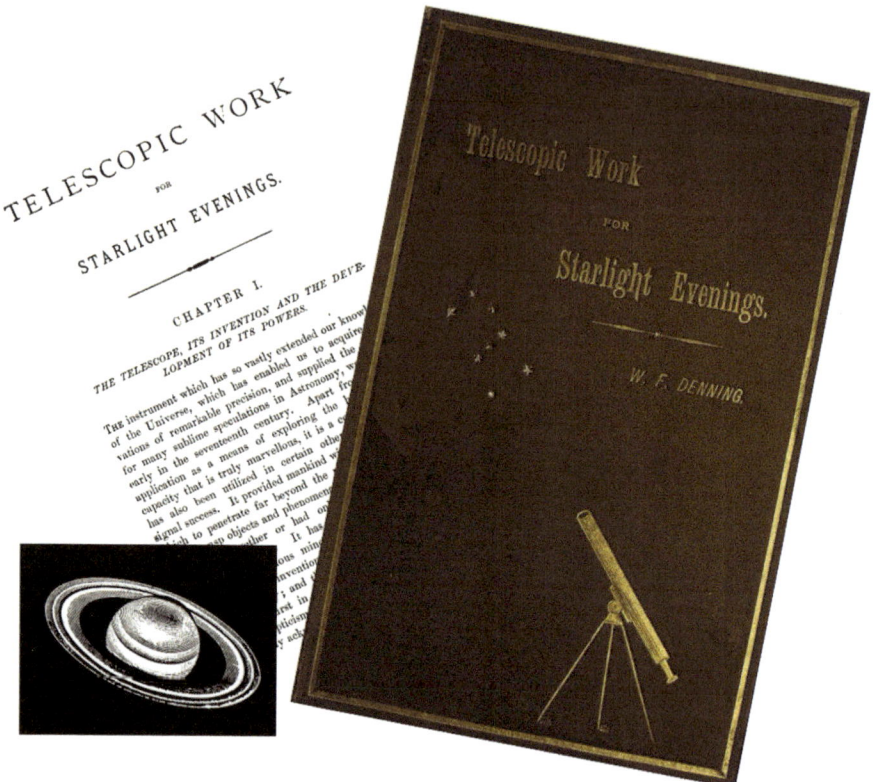

Fig. 1.4 Front cover and insets from Denning's *Telescopic Work for Starlight Evenings*. The drawing of Saturn is by Denning, as seen through his 10-inch aperture With-Browning telescope (power 252) on 28 December, 1885

It must be acknowledged that *Telescopic Work for Starlight Evenings* is an important book in the history of amateur astronomy. Indeed, it remains in print, and largely relevant even to this very day. Within the pages of *Telescopic Work*, Denning sets out to get the amateur astronomer organized and orientated towards the goal of performing scientifically useful work. It exudes practical knowledge, and it is clearly based upon the experiences of a long-time observer. The first 4 chapters of *Telescopic Work* are concerned with the telescope—that is, how the various optical designs and configurations work, the relative merits of small and large apertures, and how to use them effectively. Chapters 5–13 work their way through the solar system, from the Sun to Uranus and Neptune, including the main belt asteroids. Chapter 14 is about comets and comet-seeking, while Chap. 15 deals with meteors and meteorites. The final two chapters are concerned with stars, stellar clusters and nebulae. The text is liberally illustrated, often with drawings made by

Denning while at the telescope eyepiece, and the writing style is largely non-technical, but always encouraging.

Following his term as President to the LAS, Denning served as Vice President to the society through 1888. After this time, he was to continue his involvement with organized amateur astronomy through the British Astronomical Association (BAA). The announcement to establish what was to become the BAA appeared in the *English Mechanic* magazine on July 18th, 1890. Interestingly, while we know Denning read this magazine, he made no attempt to become involved with the initial formation of the Association. One wonders if memories of the short-lived OAS tempered his outlook on the prospects for this new amateur body. If this was the case, he had changed his mind by June 1891, since at that time, having been described by the Association's President as, "an earnest and successful comet-seeker" Denning was elected a BAA member. He was also invited to be the first Director of the Association's comet section [19]. The chief objectives of this Section were later outlined by Denning as, "comprising the discovery of new comets, nebulae, and recording telescopic meteors" [20]. At its inception, next to Denning there were three other observers in the comet section. This was to rise to seven observers by the time Denning retired his Directorship in 1893. Denning cited poor health as the reason for stepping down as Section Director, and it would appear that the general state of his constitution deteriorated from about 1890 onwards.

In addition to being elected Director of the BAA's cometary section, Denning was elected [21] a Corresponding Fellow of the Astronomical and Physical Society of Toronto, Canada, in 1891. This society, later to become the Royal Astronomical Society of Canada (RASC), had approached Denning with reference to his work as an observer, and upon his being a prolific writer on astronomical matters. Indeed, Denning became a regular contributor to the Journal of the RASC, and within its pages was to publish numerous articles popularizing meteor astronomy [22].

Denning's high standing as an observer, both within the British Isles and internationally, was recognized in the 1890s through the award of several prestigious medals. The French Académie des Science was to bestow their Valz Prize on Denning in 1895 in recognition of his meteoric studies [23]. The RAS also honored Denning with the award of their highest medal, the Gold Medal, in 1898 [24]. This latter honor, bestowed in his fiftieth year, marked the zenith of Denning's career. His skill as a planetary observer, and recorder of meteors was now universally recognized. Not just in the scientific domain, however, Denning's authority extended even to the literary world. Writing in his new book, *The War of the Worlds* (published 1898), the renowned novelist H. G. Wells was to open Chap. 2;

Then came the night of the first falling star. It was seen early in the morning rushing over Winchester eastward, a line of flame, high in the atmosphere. Hundreds must have seen it, and described it as leaving a greenish streak behind it that glowed for some seconds. Denning, our greatest authority on meteorites, stated that the height of its first appearance was about ninety or one hundred miles. It seemed to him that it fell about one hundred miles east of him [25].

Writing a few years before Wells, in his science-fiction-catastrophe book *Omega: the last days of the world* (published in 1894), Camille Flammarion considered the consequences of a cometary impact upon the Earth, and while his story was set in the twenty-fifth century, he writes that all of the classic treatises on comets were consulted, "chapters on comets written by Newton, Halley, Maupertuis, Lalande, Laplace, Arago, Faye, Newcomb, Holden, Denning, Robert Ball and their successors, had been re-read." Here we find Denning cast amongst august company; set amidst a veritable pantheon of celebrated scientists and mathematicians, all of who, it being taken for granted, were readily known to the general public and the casual reader.

During the last few years of the 1890s Denning was to publish several important studies on meteor astronomy. In 1897 Denning's third book, *The Great Meteoric Shower of November* appeared [26].[4] Solely concerned with the Leonid meteor shower, this slim volume was based upon a collection of articles previously published in the astronomical literature (mostly in the *Observatory* magazine). The book was a grand synthesis of virtually all that was known about the Leonid shower at that time (see Chaps. 2 and 5).

In 1899 Denning published what was to become one of his most important works [27], the *General Catalogue of the Radiant Points of Meteoric Showers and of Fireballs and Shooting Stars Observed at more that one Station*. In brief, here, this catalogue was a general survey of observational work, not just by Denning, but many other observers, concerning the determination of meteor radiants. The *General Catalogue* was appropriately published in the same year that Denning held the Directorship of the BAA's Meteor Section. Denning was to hold this latter office for just one year, and again was to cite poor health for retiring from the post. It is not entirely clear what form of ill

---

[4] Denning was to later produce another short text that was based on collected *Observatory* magazine articles. This book, *The Planets Mercury and Venus*, appeared in 1916.

health Denning was suffering from at this time, but it is noteworthy that he had also cited health problems as the reason for not attending the ceremony at which his RAS Gold Medal was awarded [24] in 1898.

In 1904, at the age of 56 years, Denning was awarded a Civil List Pension by the British Government [28]. This pension, which amounted to £150 per annum, was presented "in consideration of his services to the Sciences of Astronomy, whereby his health has become seriously impaired, and of his straitened circumstances". In 1906 Denning's health took a sufficiently bad turn that he elected to abandon planetary observing altogether, and he thereafter concentrated his efforts towards naked-eye astronomy, and the reduction of meteor radiants.

The first decade of the twentieth century delineates a clear transitional period in Denning's life. Not only was his health deteriorating, but he became an increasingly reclusive figure. From circa 1900 onwards he apparently saw few people, and only rarely left his home. Certainly, it would seem that he did not travel beyond the confines of Bristol in spite of prestigious invitations to do so. During the last thirty years of his life Denning's only contact with the astronomical community was through an extensive correspondence. It is a sad misfortune to the historian that only a small fraction of this correspondence seems to have survived to the present day. One of the largest surviving fragments of correspondence, however, is that with A. S. Herschel (see Chap. 2). But even this collection is sadly one sided, with only the letters from Herschel to Denning (Fig. 1.5) surviving. The Herschel-Denning correspondence that has survived dates from the period August 31, 1871 to September 12, 1900. The letters and postcards cover all aspects of meteor astronomy, and offer a few snippets of personal information, and activities. Most of the correspondence, however, deals with the exchange of observational data.

Alexander Herschel died in 1907, and Denning was to write his obituary [29] for the journal *Nature*. In his account Denning explained, "it is not too much to say that without the deep interest incited by Prof. Herschel's letters the meteoric observations obtained at Bristol during the past thirty-five years may never have been made". Indeed, Denning lost an important friend and confidant when Herschel died. Not only had Herschel openly encouraged Denning in his meteoric work, but he was also a strong supporter of the stationary radiant concept. The stationary radiant issue had by circa 1910 become critically controversial, and Herschel's death left Denning with virtually no supporters from within the ranks of professional astronomers.

It is probably safe to suggest that some of Denning's growing reclusiveness was due to the increasing number of attacks on the stationary radiant

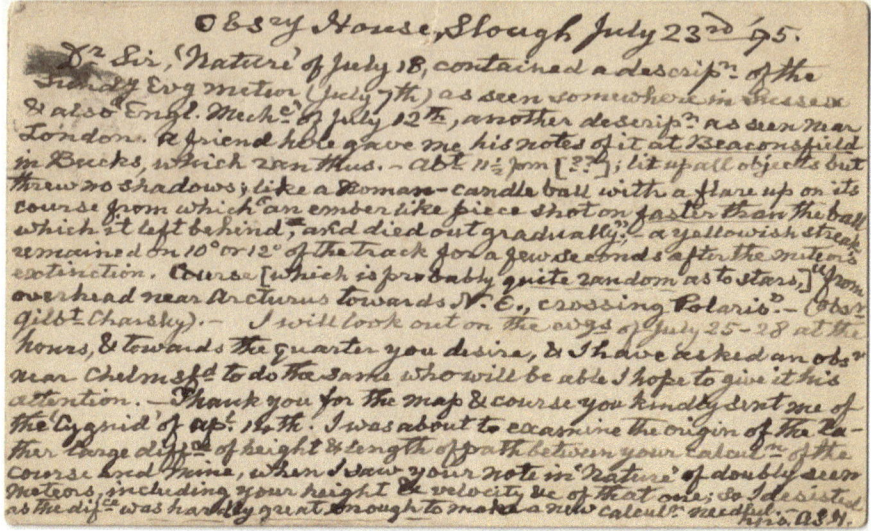

Fig. 1.5 Postcard to Denning from Alexander Herschel, Observatory House, Slough—dated 23 July, 1895. The dense writing discusses the appearance of a bright fireball seen on July 12th, "lit up all objects but threw no shadows, like a Roman candle ball with a flare up on its course". Card size: 12.2 cm by 7.5 cm. Image from the author's collection

concept. Some hint of his bitterness towards this controversy can be found as early as 1891 when he was to write, "as a rule, amateurs should avoid controversy, because it very rarely clears up a contested point … it wastes time, and often destroys that good feeling which should subsist amongst astronomers of every class and nationality … competition and rivalry in good spirit increases enthusiasm, but there is little occasion for the bitterness and spleen sometimes exhibited in scientific journals" [30]. For the moment it is simply observed that the existence of stationary radiants was an issue that dominated the last thirty years of Denning's life. It is also worth noting that Denning's continued belief in the existence of stationary radiants saw him become increasingly alienated from the main-stream of astronomical thought. In spite of such matters, he quietly continued his work on the reduction of meteor radiants. Denning was to return to the astronomical limelight, however, in 1920 when on the night of August 20th, he noticed a new star (nova) in the constellation of Cygnus [31–33] (see Chap. 3). This latter discovery offers clear testament to Denning's tenacity as an observer, and to the great acuity of his then septuagenarian eyes.

In 1922 Denning's high-standing and respect as a meteor observer was further acknowledged by an invitation to become the first President of the Commission des e'toiles Filantes in the then newly formed International

Astronomical Union (IAU). Denning had in fact been approached with this offer in 1919, and some measure of his dire circumstances at that time can be found in one of the few surviving letters that he wrote to his niece Christine Gravely (daughter to his sister Ellen Louisa).[5] In a letter dated 1919, November 25th Denning comments,

> [I have] been placed at the head of a committee of the International Astronomical Union appointed to study and advance our knowledge of meteors and meteoric phenomena. I hope to manage it all by correspondence. The first meeting does not take place until 1922 in Belgium, and it is quite a strain that someone will have to be chairman in my absence.

Denning continued in the same letter to his niece,

> My night-watching's have been few lately - I find it rather trying to be sitting out in the garden for hours on these damp cold nights. If I could take solid food and use a hamper all could be easy but things are different with me now to what they were 40 or 50 years ago when I found it quite pleasant to be out in the frost all night long!

Understandably, advancing age and ill-health were having their effect on Denning's ability to observe and actively participate in astronomical research. It would also appear that his spirits were at a low-ebb. Writing again to his niece on September 4th, 1923 Denning explained,

> I am sorry not to have answered your letter before, but I have done very little writing indeed for some weeks having been suffering more that usual. Your bright and interesting letter was very entertaining and was read by me with great pleasure at a dull time.
>
> I am glad you still find enjoyment in natural history subjects. If I could only get into the fields, I should watch the birds more and make notes of what I saw and heard, but I am not often out in my own garden now and I have made hardly any observations during the last two years.
>
> This has been detrimental to me in various ways. I used to have the expectation of a night's successful observation and after that the discussion of my results and comparison with details obtained before proved engrossing. It passed the time agreeable and occupied my thoughts - whereas now I have no new work to think about or fresh discoveries to look forward to. Day and night succeed each other with dull monotony.

---

[5] I am indebted to Maurice Brain for making copies of these letters available.

You were kind enough to inquire about the articles I had been writing lately for a serial now being issued entitled The Splendour of the Heavens. It is a grandly illustrated work. I have written rather long chapters in part 6 and 9 and another chapter on shooting stars will be printed in part 11 or 12. The work is full of most attractive pictures, and I believe they will be appreciated by everyone fond of astronomy.

The text that Denning refers to in this letter, *The Splendour of the Heavens*, was edited by W. H. Steavenson, and was destined to become a popular, and widely read astronomy text (see Chap. 3). Steavenson recalled[6] a visit to Denning's home, circa 1922, and remembered being greeted by a subdued figure draped in an old overcoat warming himself by a fire in surroundings of dank poverty. It is also at this time, the story is recounted, that the young children on Denning's home street (Egerton Road) used to taunt and shout at him whenever he left the house. It would seem that Denning's reclusive, poverty-stricken life-style, along with his interest in the stars had become the cruel butt of school boy humor.

In 1927 Denning was awarded the degree of Master of Science, *honoris causa*, by the University of Bristol. Denning's health at this time had so deteriorated that he was unable to attend the award ceremony, even though the University was just a few kilometres away from his home. Indeed, he even felt unable to receive a deputation from the University. His degree was eventually conferred in absentia.[7] It is fitting that Denning's final academic award for services to astronomy was bestowed by his home-town University, and the degree represents the well-deserved recognition of a life dedicated to the furtherance of meteor science. Right up to the very last Denning was recording, and collating meteor data, and his final observing note, published in the *Observatory* magazine, appeared just a few weeks before he passed away [34].

Ironically, in terms of known facts, it is Denning's death that provides us with the most detailed information about his life, family and friends. Denning died of atrial fibrillation on Wednesday 9th June in 1931—he was 83 years old. His death was announced in the *Times of London* newspaper for June 10th under the headline "famous astronomer dead", to which was added, "Mr. William Frederick Denning, the world-famous astronomer and discoverer of five new comets died at Bristol last night". The *Times* ran a special article on Denning the following day, June 11th, and included a picture of Denning with the caption, "the well-known amateur astronomer,

---

[6] James Muirden, personal communication 1989.

[7] Wright, E. C., Registrar and Secretary, University of Bristol, personal communication 1987.

MR. W. F. DENNING, the
well-known amateur astronomer,
discoverer of the new star in
Cygnus, whose death is announced

Fig. 1.6 The portrait picture of an aged Denning carried by the *Times of London* on 11 June, 1931

discoverer of the new star in Cygnus whose death is announced" (Fig. 1.6). Denning's death was announced in the *New York Times* on June 10. Run as a "special cable to the NYT", the article had the cover line, "discoverer of five comets and new nebulae succumbs in England at 82 [sic]". The announcement of Denning's death was placed on page 25 of the NYT, while the front page of the same issue carried a story about the work of Albert Einstein—again, as a special cable, this time from Berlin. The leading column begins, "Einstein tells of advance in his field theory in unifying gravity and electromagnetism". This juxtaposition of locations, Einstein on the front page, Denning on the 25th seems to summarise the fate of Denning's life and works. By 1931, he was an observer from the past, and the interests of scientists (and the public) was moving into areas completely undreamed of when Denning started observing.

Denning's interment took place on the 13th of June, and the *Western Daily Press* newspaper tells us that, "the service was of a simple nature indicated by his retiring mode of life and held at St. Michael and Angels". The service was conducted by Rev. D. H. Hall, assisted by Denning's nephew Rev. A. A. Cockle, rector of Aston Sommerville. It was the Rev. Cockle who took the committal at the graveside in Brunswick Square cemetery. The principal mourners were, "Mrs. Willetts, Norman Denning, Mrs. C. F. Denning, Mr.

Fig. 1.7 Plaque unveiling ceremony held at 44 Egerton Road on 18 December 1931. To the right is Dr. Knox Shaw, President of the Royal Astronomical Society. To the left (checking his notes) is Normal Langley Denning, closest to the window, in a rain coat, is Professor Lennard–Jones of Bristol University, and next to him is Christine Gravely (daughter of Denning's older sister Ellen Louisa). The gentleman still wearing his hat is Frederick Denning (Denning's younger brother). The gentleman to the far left (facing camera) is Ernst Edward Denning, and the woman facing the camera is Mary Willetts (née Denning). From the author's collection

and Mrs. Ernest Denning, Mrs. Katie Norman, Mr. James Norman, Miss Daisy Denning, Mr. Vicars Webb, Miss Graveley, Mr. Willets, Mr. and Mrs. Norman, Miss Denning and Mr. H. W. Cockle". We are also informed that the Royal Astronomical Society sent a wreath. In unison with the astronomical community, the City of Bristol also honored Denning's achievements by erecting a memorial tablet at 44 Egerton Road on 18 December 1931 (Fig. 1.7). At this ceremony Dr. Knox Shaw, President of the Royal Astronomical Society, remarked that, "Mr. Denning was an amateur in the true sense of the word. He studied the heavens not in the hope of gaining fame or renown, but because he could not help it" [35].

Awards made during a person's lifetime provide one measure of their standing according to fellow peers. Posthumous awards, however, reflect a more enduring, indeed, historical recognition of a person's contributions. In this respect, main-belt asteroid 71,885, discovered on 16 November 2000 by the Spacewatch team of observers at Kitt Peak, has been officially named after

Fig. 1.8 Far-side lunar Crater Denning (the large crater to the upper left of the image). Apollo 15 image courtesy of NASA

Denning. Appropriately, the discovery date for this asteroid was the same day (but 109 years after events) that Denning was elected a corresponding Fellow of the Royal Astronomical Society of Canada. The designation information for asteroid 71,885 Denning reads, "British amateur astronomer William Frederick Denning (1848–1931) was renowned for his visual study of the heights and velocities of meteors and for his catalogues of meteor radiants. He also maintained an interest in Jupiter's red spot and discovered five comets, two of them of short period". While Denning did not dedicate much time to observations of the Moon, at the General Assembly of the International Astronomical Union held in Brighton, England in 1970, a lunar crater was named in his honour. It does seem somehow apropos, however, that crater Denning (Fig. 1.8) is on the far-side of the Moon: remote, un-seen by casual human gaze, and yet close to us. A Martian crater (see Chap. 3) was also named in Denning's honour at the IAU General assembly in Sydney, Australia, in 1973.

We have now accomplished the main chronological description of Denning's life. There are, however, a few points that should be addressed before moving-on to discuss his work on meteors. One issue that has proved difficult to fully determine, is how Denning made a living—that is, how did Denning acquire money for his every day expenses? Victor Plarr indicates in his *Men and Women of the Time: a dictionary of cotemporaries* that, in late

1863 Denning was employed as, "a clerk to a manufacturing firm in Bristol", although which firm is not indicated. Several of his obituary accounts [1, 2] suggest that Denning had trained as an accountant, but there is, in fact, no evidence to verify this claim. Certainly, Denning's father (Isaac Poyntz Denning) was the leading partner in the accountancy firm Denning, Smith and Co., but there is no evidence that Denning himself was ever on the payroll. The partnership of Denning, Smith and Co., is described in the 1868 *Trade Directory for Bristol* as being a, "public and private accountants, auctioneers, bankruptcy, insolvency, and general agents, valuers, etc." Later trade directories indicate that Denning's brother, Frederick Denning, became a partner in the company, and that circa 1890 the partnership was to become Denning, and Co. While Denning, and other family members are listed in the various name directories for Bristol throughout the 1880s and 1890s at no time is Denning described as being an accountant. It would seem that the obituary accounts are confused. Confused, that is, in the sense that there is no indication that Denning was ever a fully trained, registered and practicing accountant. Further confusion is also evident from the observation that Denning's father, Isaac Poyntz Denning, was referred to in several obituary accounts as having been "Borough Accountant of Bristol". This was never the case. Certainly, it is possible that Isaac may have undertaken occasional work for the city, but not through the auspices of any official civic office.

Writing in 1905, Hector Macpherson noted that Denning followed a journalistic line of work [36]. This suggestion makes more sense when one considers Denning's life style, since it would have been difficult to hold down a full-time day-job, and spend as much time observing as Denning evidently did. Certainly journalism, and popular writing were two ways that Denning would have made some money. Many newspaper articles, and popular accounts by Denning are in existence (see Chap. 4), and one can find general astronomy articles by him in, for example, the *Boys Own Magazine, Scientific American, Knowledge*, and *Encyclopedia Britannica*. It has also been suggested that Denning received occasional monies from Queen Victoria.[8] Again, however, it has not been possible to confirm this, although such gifts were indeed conferred to other people on occasion.[9] The *Times* obituary of June 11th [37] also notes that Denning's income had, "been augmented in recent years [circa 1920?–1930] by subscription among his brother astronomers who have thereby shown their appreciation of his quality".

---

[8] Maurice Brain personal communication, 1989.

[9] Derrett, A., Assistant Registrar to the Royal Archives, Windsor Castle. Personal communication, 1989.

The only regular source of income that can be attributed to Denning is that of his Civil List Pension, which was awarded in 1904. He was, however, 56 years old when this was first paid out. Prior to this, all that can be said is that he may have performed some accountancy work, and supplemented his income through writing and journalism. On a few occasions Denning's income was supplemented by prize money. The Valz Prize (awarded by the French Academie des Sciences), for example, came complete with a cash award of fr. 460. The indications, therefore, that Denning led a reserved and frugal life would certainly appear to be true. Some further indication that Denning was unable to fully support himself financially is offered by the fact that he was listed as living at the same address as his parents in the 1880 and 1890 *Bristol Name Directories*, at which time he would have been respectively 32 and 42 years of age. That Denning lived in rented houses seems likely given his regular change of address (see Table 1.1). All of the known addresses are within the confines of Bristol, ranging, however, from the Montpelier district to Bishopston and Ashley Down (the later being close to the County Cricket Ground). Denning moved to his last, and probably best-known address, 44 Egerton Road, in 1906. That he was able to support himself there in the long-term was, no-doubt, due to the financial security afforded by his Civil List Pension. Although he moved on a regular basis, the post office seems to have had little problem in forwarding letters to him. There exists one letter (which found its destination), for example, dated 24 February, 1928, to Denning from the artist Arthur Huish Webber, in which the address was simply given as, "in or near Bristol".

As testament to his all-round interests, it is worth pointing out that not all of Denning's writings were concerned with astronomical research. Indeed, Denning would on occasion express his thoughts and feelings through prose and poetry. There is no direct evidence to support the notion that Denning was a deeply religious man, but it is clear that he rejoiced in the study of nature, and in the observation of the heavenly cycles. Being largely self-taught, and with no formal (that is university level) scientific training, Denning relied purely on what he saw with his own eyes when formulating his ideas, making no speculative suggestions above or beyond what he recorded. In this sense Denning was a truly empirical or Baconian scientist. Indeed, Denning lived at a time during which philosophers struggled to not only define what a scientist was, but what the very principles of science, and what makes it work, were. At question was literally the problem of extracting truth from nature. How is it that progress is made, and how can one use the knowledge so-gained to establish universal rules. Francis Bacon, working in the early to mid-seventeenth century, was among the first new practitioners to take-on such questions, and move away from the practices of the ancient Greek philosophers—he also gave us the well-known dictate that 'knowledge is power'. Indeed, he was particularly interested in determining how (what would now be called) scientific knowledge was obtained. Bacon recognized that knowledge, and especially systematically derived knowledge, was the underpinning of technology, and at issue was how such technology related to human progress. Furthermore, Bacon strove to place the attainment of knowledge on a new and firm foundation. His concern was to move away from the practices encapsulated in Aristotelian methodology, where one begins from a set of broad assumptions and statements concerning the working of the world, and then narrowed down these premises to explain specific situations. On this basis, it was argued, provided that the starting assumptions and statements were actually true, then what was logically deduced from them must also be true. Here we have reasoning such as the following at play: given the statements (1) Steven Hawking is a man, and (2) all men and mortal, so it must be the case that (3) Steven Hawking is mortal. Result (3) was (sadly) verified empirically on 14 March, 2018.

Bacon rejected the Aristotelian approach by effectively reversing its methodology. Rather than using a deductive approach, Bacon argued for an inductive one—that is, he reasoned that one should start by understanding the small and the locally observed phenomena, and slowly build-up to a bigger and broader understanding of how nature works—if you like, the premises and axioms come last in the process, after all the important observing has been done. Through this process of the gradual, step-wise accumulation of facts, based upon empirically verifiable statements, Bacon felt that (scientific) knowledge could be placed upon a sure and solid foundation. Scottish philosopher David Hume, in the eighteenth century, was to criticized Bacon's approach of induction, however, arguing that the problem with induction was that it extrapolated from the known (the observations) to the unknown (that is, the theoretic description) with no clear guarantee that any newly observed phenomena must be explained by the same principles and causes that worked in earlier situations. Philosophers John Stewart Mill and William Whewell, in the nineteenth century, made further attempts to improve upon Bacon's notion of induction, but again, both (ultimately) failed to clarify the picture. Whewell felt that all scientific investigations should start with some hypothesis ("happy guesses", as he called them[10]), with the hypothesis dictating where and what to look for. The hypothesis under review

---

[10] Whewell also used the wonderful phrase, "felicitous and inexplicable strokes of inventive talent".

was then put to the test by confronting it with the data appropriately gathered. In contrast Mill argued that the causes of observed phenomena should be discovered through the successive elimination of unsuitable hypothesis (a process he called, eliminative induction[11]). For Mill, there are echoes of Sherlock Holmes's famous maxim of, "once you illuminate the impossible, what ever remains, no matter how improbable, must be the truth". For all the philosophical posturing, however, it is now accepted that science progresses, and scientific knowledge grows, by both inductive, and deductive methods, supported by an almost limitless number of ad hoc methodologies. For Denning, scientific knowledge was entirely empirical, and if the theoreticians had problems finding an explanation for what he observed, then that was their problem. Certainly, for so it would seem, Denning was driven to make his observations according to a quest for truth, understanding, and scientific progress, and he was to write, for example, "the work of observations must go on continuously. It is like a river which runs endlessly along the shores of time connecting the past with the present and the present with the past".[12] For Denning the act of observing was part and parcel a quest for understanding, and a contribution to the reservoir of knowledge. Denning's feelings towards observing are further expressed in his comments concerning the supposed observation (in 1900) of markings seen on Saturn's disk, "it resolves itself into a question of ethics. There are men who will report nothing but what they are absolutely certain is presented to their eyes, and are unbending in their regard for the truth" [38]. Denning's outlook on life and science was seemingly guided by high principle, and it would seem that truth, commitment, and personal integrity were the underlying ethics that Denning employed in his studies.

There is no specific evidence to support the idea that Denning thought himself a serious poet. Rather it would seem that he occasionally used poetic language to express his enjoyment at participating in the scientific process. What poetic verse Denning did write may have been fanciful and romantic, but it was not romantic in the poetic sense. The romantic poetry of Wordsworth and Keats, after all, was essentially a reactionary backlash against the critical rationalism of science. In this manner Denning's few surviving poems have more in common with the works of Mark Akenside and James Thomson, poets who rejoiced at the wealth of knowledge that scientific study brought forwards, rather than with those of Alfred Tennyson and

---

[11] Mill actually introduced five methods of induction, but they need not all be considered here.

[12] This reference is taken from an unidentified newspaper cutting from among the Denning archives held by the British Astronomical Association's Meteor Section. The article was seemingly written circa 1900.

Thomas Hardy, who were more inclined to see scientific study in a darker and more ominous light [39, 40]. One example where Denning used both prosaic language and poetry to express his feelings can be found in an article[13] written circa 1895. In this account Denning rejoices at the imminent arrival of spring, and he expounds,

> Oh, Spring! Dear Spring! Thou more dost bring
> Than birds, or bees, or flowers -
> The good old times, the holy prime
> Of Easter's solemn hours;
> Prayer's offer'd up and anthems sung
> Beneath the old church towers.

Following this triumphant outpouring Denning continues in prosaic tones:

> The opening of the snowdrop and crocus tells us that spring is near, the bloom of the primrose and violet brings us the realization. March, though it has its keen winds and sometimes wears a wintry frown, yet proves that the dark days are past, and towards the end of the month gently introduces us to the summer's advent in the person of her younger and sweeter sister spring.

There is a clear romantic sensitivity in Denning's writing, and indeed, one can sense that he has a deep respect for the workings of nature. Clearly, it would seem, Denning experienced a heart-felt joy in observing the seasonal changes.

While smatterings of poetic verse from Virgil, Homer, Shakespeare, and Milton, can be found in *Telescopic Work for Starlight Evenings*, Denning was not above adding-in in his own lyrical musings. Concerning the observation of Mercury, Denning writes,

> Come, let us view the glowing west,
> Not far from the fallen Sun;
> For Mercury is sparkling there,
> And his race will soon be run.
> With aspect pale, and wav'ring beam,
> He is quick to steal away,
> And veils his face in curling mists, -
> Let us watch him while we may.

---

[13] This material is taken from an unidentified newspaper cutting in the archives held by the British Astronomical Association's Meteor Section. The article was written circa 1895.

And in similar tones he was to write of the ringed planet Saturn,

Muse, raise thy voice, mysterious truth to sing,
How o'er the copious orb a lucid ring,
Opaque and broad, is seen its arch to spread
Round the big globe, at stated periods led.

Interestingly, while the study of meteors was one of Denning's major pre-occupations only one poem on this topic seems to have seen print. This poem, simply called *Falling Stars*, appeared in the *Journal of the Royal Astronomical Society of Canada* in 1915, and was written just two days after his 66th birthday [41]. The poem seems reflective and offers some insight into Denning's personal thoughts and inner feelings:

Bright falling stars I greet you with a smile,
While you beguile,
My loneliness, with pleasure pure and sweet
In moments fleet.
In coloured beauty and in lustre dressed,
Never at rest,
You span the sky and guild the heav'nly way
With sparkling ray.
I only know the moments of your birth,
Above the earth;
As she performs her yearly round in space
You run your race
And pierce the blue just as a flashing blade
Too quick to fade.
Along your flight the burning embers sow
An after-glow,
To mark your path amid the stars of night,
With guiding light.
I never know the instant when you will
Disturb the still
Of Heaven's stars and speed athwart the sky
All silently.
Nor can I tell in Nature's open book,
Just where to look,
To watch your coruscations wax and fade
Amid night's shade.
Adown the east or west your fiery ball
May headlong fall,
Or, slowly, stream along the starry height

In graceful flight.
Whene'er you come you bring a joyous thrill
My soul to fill.
Oh, messengers from distant worlds! I yearn
Your tale to learn,
And I await, amid earth's frosted dews,
Celestial news.

This poem is clearly one of celebration. It celebrates the visual appearance of shooting stars, and it celebrates the joy related to their study. It is also a personal poem that hints at loneliness, and uncertainty. Indeed, there is a feeling of the sublime in Denning's lines. He seems to be echoing German philosopher Emanual Kant's notion of the sublime being generated by human reason as it searches for absolute ideas located beyond the grasp of the senses and from any form of tactile exploration. At various times, during the 1890s, Denning submitted articles to the magazine *Great Thoughts from Master Minds* (edited by Robert Colville), and here he would often let his best poetical thoughts run free—in one article concerning fireball meteors, written in 1895, he atones,

> The fugitive torch falling in the dark blue ether, from Heaven to Earth, and dying amid the blaze of its coruscating glories, creates an effect which altogether baffles description. Like a beautiful sunset and some other of Nature's attractions, it must be witnessed to be appreciated in its grandest effect, for no one can paint it either in colours or in words.

In this same article Denning describes the scene that he witnessed during the 1866 Leonid meteor shower, "they [the shooting stars] fell in showers as thickly as the particles of sand in the winds of the Libyan desert". Clearly, the influence of Calliope was running high in Denning on this particular occasion.

Denning's reclusiveness became a well-known, later-life characteristic, and he seems to have had few close or personal friends. In a letter to Grace Cook (Fig. 1.9), however, Denning wrote of the fatigue that he experienced after a long observing session [42],

> I fancy it does me good intellectually and physically to be at work exercising my patience in this way. Anyone who really loves the stars for their own sake need never despair of finding, sooner or later and whatever troubles may oppress him, not only a solace but a supreme happiness in contemplating them.

Fig. 1.9 Alice Grace Cook seen here seated in her meteor observing deckchair. In her right hand and across her lap can be seen a "meteor wand"—a visual aid for fixing the direction and position of meteor trails upon the sky. Image from the *Daily Mail* newspaper for June 19th, 1918

He also wrote to Cook on another occasion commenting that once the "spirit of the night" had appeared to him after a long observing session. Upon being asked if he [that is Denning] would see her again, the spirit replied, "at some hour when you feel weary with your labors and the night is far spent, I will come to cheer you".

Alice Grace Cook was in many ways kindred spirit to Denning. She joined the British Astronomical Association in 1911, and was Director of the BAA Meteor Section from 1921 to 1923—this term being held after serving as joint temporary-director of the section, along with Fiammetta Wilson (born Hellen Francis Worthington—Fig. 1.10), during the years of the First World War (see Chap. 5). Cook was also one of the first women to become a Fellow of the Royal Astronomical Society (elected in 1916 and, indeed, her candidacy was proffered by Denning). Wilson was a trained musician of international renown, but upon becoming interested in astronomy joined the BAA in 1910 and then, along with Cook, a Fellow of the Royal Astronomical Society in 1916. She published numerous notices on meteors, comets, aurora and the zodiacal light, and was the only person to co-author a research paper with Denning—this being in the *Monthly Notices* of the RAS, and concerned

Fig. 1.10 Fiammetta Wilson. Image from *Knowledge and Illustrated Scientific News,* vol. 12 (1915)

with double station observations of the Quadrantid (now so-called) meteor shower of 4 January, 1918.

More will be said about the meteor-related work of Cook and Wilson in Chap. 5, but it was both of these ladies who helped form, and promote, the somewhat obscure Chaldæan Society. Indeed, Wilson was directly involved with the founding of the London-based Chaldæans in November 1916, at which time the aim of the society was stated as being the promotion of naked-eye astronomy. Wilson was appointed the Society's Director of Observations in 1919, but died within a year of taking-on the office. In October 1921 James Hargreaves took over as Society President,[14] and pushed for the development of regional branches [43]. Soon thereafter local Observing Correspondents were identified in Luton, Tottenham, Hertfordshire, Warwick, Ipswich, Lethchworth, the Isles of Wight and in Scotland. Alice Cook was one of the founding members of the Ipswich branch of the Chaldæan Society (established in August of 1921), and she soon established sub-branches in her home town of Stowmarket, and in the village of Yoxford in Suffolk. At the October 1921 meeting of the Society, the

---

[14] James Hargreaves (October 1899–November 1985) became interested in astronomy at an early age, and was educated at the University of Cambridge. An engineer by training, he moved to Canada in the early 1930s, first to raise pure-bred cattle in Quebec, and later taking-up an astronomy teaching post at the University of Ottawa. He was a great traveler and toured the world in order to study numerous solar eclipses.

Reverend David Ross Fotheringham outlined areas in which society members might work[15]—including observing the first appearances of planets emerging from conjunction, the appearance of Vesta (the only asteroid visible to the naked-eye), the appearance of lunar and solar haloes, and the monitoring of naked-eye variable stars such as Mira Ceti and Algol. These suggested projects confirming the Society's aims of promoting naked-eye astronomy.

The Society published a quarterly journal, *The Chaldæan*, from 1918 to 1927, and this was edited by various society members, including Fotheringham, Sidney Mattey,[16] James Hargreaves, and Fiammetta Wilson. It is not clear if Denning was ever a member of the Chaldæan Society (there was no Bristol branch), but he did publish articles in society's journal *The Chaldæan*, and he was consulting editor to the spring 1918 issue of the journal, which contained a supplement on meteor astronomy. The Ipswich branch of the Chaldæan Society resolved to close, due to lack of members, on 29 September 1924. The London Chaldæan Society continued to operate for two more decades, but formally disbanded in 1944, with all remaining regional sections following suit soon there after. The Society seems to have been particularly active with respect to observing the solar eclipses on 8 April 1921 [44].

Wilson's death on 21 July, 1920, not only robbed the Chaldæan Society of a stalwart member, but so to did it rob astronomy of a pioneering amateur and tireless practitioner. Indeed, Wilson had been awarded the Edward C. Pickering Fellowship for Women in 1920. This award, the first presented outside of America, consisted of $500 US, and was engineered through the efforts of Herbert H. Turner (University of Oxford) in 1919. Writing in

---

[15] David Ross Fotheringham (1872–1939) had a life-long interest in astronomy, and kept up an extensive correspondence with his brother, J. K. Fotheringham who was a reader in Ancient Astronomy and Chronology at Oxford University. Educated at the City of London School, and Queen's College, Cambridge, Fotheringham graduated in 1895, and was ordained in 1896. He eventually accepted a curacy in Camden Town (London)—a post that he held until 1912. After this time, Fotheringham moved to Charing in Kent, adopting the duties of country clergyman. Fotheringham became a Fellow of the Royal Astronomical Society on 9 December 1910. He published many books and pamphlets on biblical chronology (some still available in reprint form from *Amazon* to this very day), and the history of ancient Greece, and in 1928 published the well-received pamphlet *The Date of Easter and other Christian Festivals*. During the last several decades of his life, Fotheringham became profoundly deaf, but continued to conduct his clerical duties and editorial work for the Chaldæan Society. He died on 30 June, 1939.

[16] Sidney Batt Mattey (16 April, 1886–22 December, 1940) was born in London, and after taking his BSc. at University College in 1908, he became science master at Shooters Hill School, Woolwich. He was a keen meteor observer, and offered adult classes in popular astronomy. He was elected a Fellow of the Royal Astronomical Society on 14 February, 1919.

the 18th Annual Report (1920) of the Nantucket Maria Mitchell Association (NMMA), Annie Jump Cannon[17] reported that Turner had inquired
if the astronomical fellowship for women might be awarded to an "English
woman", arguing that it would be, "a friendly act at the present moment of
great events and important new departures [i.e., the end of WWI] …. To
the advantage which such action would have both in encouraging women's
work generally, in cementing friendly relations between two nations, and in
creating a new form of recognition". While Wilson had been the initial candi-
date for the Pickering award, here death resulted in its transferal to Alice
Cook. Writing in the 1921 Annual Report of the NMMA, Cannon introduced Cook as, "an experienced observer of meteors", and presented a vivid
account of how observations were made: "Seated in a deck chair, with a wide
expanse of the sky before her, armed with a stop-watch and a very straight,
thin wand about five feet in length [recall Fig. 1.8], she observers hour after
hour". Cannon then quotes from a letter by Cook, "the mind must be kept
on the alert and the body poised for instant attack", further adding, "as soon
as a meteor is seen, the wand is held along its path, the stop-watch is clicked
for duration of the flight, and the magnitude is estimated by comparison
with adjacent stars or planets. The position of the beginning and end points
are then estimated. Cook indicated that the whole logging process could be,
"noted down in a record book in thirty seconds from the first sight of the
meteor". Cannon also explains in the 1921 Annual Report that, "on very
clear nights she [that is Cook] searches for comets with a small telescope and
looks for novae in the Milky Way". Furthermore, it is noted by Cannon,
that Cook has used the funds from the award, to purchase reference books,
meteoric star maps by T. W. Backhouse (see Chaps. 5 and 7), and a portable
telescope to be used for comet hunting and observing telescopic meteors.
Cook reported on the meteor observations made during her fellowship in the
January 1921 issue of *Popular Astronomy*. Amongst the basic results presented
there, we also learn that Cook observed, "keeping strict watch in the direction of Bristol [basically towards the west from Stowmarket] and the south of
England, where the other observers are mostly stationed". The aim here was
to gather data on especially bright meteors that might be doubly observed,
and accordingly have their true atmospheric paths determined (see the next
chapter). Clearly, by looking towards Bristol, Cook was singling out Denning
as a key observational partner. In the Annual Report to the NMMA, for
1922, Cannon records that Cook had set-up a "Davidson Meteoroscope

---

[17] Cannon lived a storied life, but is perhaps best remembered today as both a pioneer astronomer
(and suffragist), and for the development of the Harvard spectral classification scheme for stars. She
was also the first female recipient of an honorary doctorate from Oxford University in 1925.

[described in Chap. 5] … in her garden and used [it] for locating the radiant points of the meteors she observed". In the early 1920's meteor photography was still in its infancy (see Chap. 5), and it is noteworthy that Cook was a pioneer of applying instrumental techniques to the study of meteor trails, and specifically to the determination of meteor radiants—something that Denning, as a purly naked-eye observer of meteors, never did. As will be seen in enfolding chapters, the decade beginning circa 1920 was a time of great change in meteor astronomy. Not only did observational practices change, with new instrumental techniques being developed, but so to did the reduction methodologies for the determination of shower radiants undergo a dramatic re-thinking.

In this chapter it has been revealed that Denning was a complex, sensitive, genuine, and passionate man. Not only did he dedicate the majority of his life, at the cost of great personal poverty, to the detailed study of nature, he also contemplated the universe in all its beauty and bounty. Denning, for so it would seem, literally found wonderment wherever he looked, and he unashamedly indulged his eyes and appetite. He was a celebrated amateur observer during his lifetime, and he was a fundamental player in the development of meteor astronomy (see Chap. 5), and a leader in the promotion of amateur astronomy in Britain (see Chaps. 4 and 7).

Bibliography

1 Prentice, J. P. M. (1931). *Journal of the British Astronomical Association, 42,* 36–40.

2 Denning, W. F. (1895). *Nature, 51,* 320–321.

3 A self-made English astronomer. *North British and Ladies Journal*, April 4th, 1904.

4 Beech, M. (1992). The Herschel—Denning correspondence. *Vistas in Astronomy, 34,* 425–447.

5 Denning, W. F. (1868). *Astronomical Register, 6,* 92.

6 Johnson, P. (1990). The astronomical register. *Journal of the British Astronomical Association, 100,* 62–66.

7 Denning, W. F. (1869). The supposed new planet Vulcan. *The Astronomical Register, 7,* 89.

8 Beech, M., & Peltier, L. (2015). Vulcanoid asteroids and sun-grazing comets—Past encounters and possible outcomes. *American Journal of Astronomy and Astrophysics, 3*(2), 26–36.

9 Newall, H. F. (1987). *History of the royal astronomical society: 1820–1920* (p. 135). Blackwell Scientific Publications.

10 Anonymous report (1872). *Nature, 4,* 261–262.

11 Denning, W. F. (1972). *Nature, 6*, 94.

12 Denning, W. F. (1872). Observations of luminous meteors. *Monthly Notices of the Royal Astronomical Society, 33*, 93–95.

13 Denning, W. F. (1876). Radiant—Points of shooting stars. *Monthly Notices of the Royal Astronomical Society, 37*, 282–284.

14 Denning, W. F. (1878). Suspected repetition, or secondary outbursts from radiant points; and the long duration of meteor showers. *Monthly Notices of the Royal Astronomical Society, 38*, 111–114.

15 Beech, M. (2016). *The wayward comet: A descriptive history of cometary orbits, Kepler's problem and the cometarium.* Universal Publishers.

16 Denning, W. F. (1891). *Telescopic work for starlight evenings* (p. 79). Taylor and Francis.

17 Liverpool Astronomical Society (1888). *Observatory, 11*, 309–311.

18 Anonymous review (1891). *Nature, 44*, 467.

19 Prentice, J. P. M. (1948). *Memoirs of the British Astronomical Association, 36*(2), 10. See also chapter 5.

20 Denning, W. F. (1891). *Journal of the British Astronomical Association, 1*, 490.

21 Council Report (1891). *Transactions of the Physical Society of Toronto, 2*, 45.

22 Beech, M. (1990). William Frederick Denning: In quest of meteors. *Journal of the Royal Astronomical Society of Canada, 64*(6), 383–396.

23 Denning, W. F. (1896). *Nature, 53*, 215.

24 Council Report (1898). *Monthly Notices of the Royal Astronomical Society, 63*, 242–253.

25 Wells, H. G. (1975). *The war of the worlds.* Pan Books Ltd.

26 Anonymous Review (1897). *Nature, 57*, 7.

27 Denning, W. F. (1899). *Memoirs of the Royal Astronomical Society, 53*, 203–292.

28 House of Commons paper 201 (1905). The British Government Archives.

29 Denning, W. F. (1907). *Nature, 76*, 202.

30 Denning, W. F. (1891). *Telescopic work for starlight evenings* (p. 56). Taylor and Francis.

31 Denning, W. F. (1920). *Nature, 105*, 838.

32 Luyten, W. J. (1920). *Monthly Notices of the Royal Astronomical Society, 81*, 61–65.

33 Beech, M. (1993). Denning on Novae. *Journal of the British Astronomical Association, 103*, 130.

34 Denning, W. F. (1931). Autumnal meteors. *Observatory, 54*, 271–272.

35 Newspaper clipping. *Western Daily Press*, December 19, 1931.

36 Macpherson, H. (1905). *In astronomers of today* (pp. 172–178). Gall and Inglis.

37 Maddocks, H. (1931). Mr. W. F. Denning: Doyen of amateur astronomy, in the Times newspaper, Thursday, June 11th, page 16, column b.

38 Denning, W. F. (1900). Notes on Saturn and his markings. *Nature, 67*, 237.

39 Beech, M. (1989). Meteor imagery in English poetry. *New Comparison, 7*, 99–112.

40 Beech, M. (2021). *A cabinet of curiosities: The myth, magic and measure of meteorites*. World Scientific.

41 Denning, W. F. (1915). The claims of meteoric astronomy. *Journal of the Royal Astronomical Society of Canada, 9*, 57–60.

42 Cook, A. G. (1931). *Journal of the British Astronomical Association, 42*, 1931.

43 Hargreaves, J. (1921). *Nature, 108*, 288.

44 Hargreaves, J. (1921). Nature, *106*, 830.

2

In Quest of Meteors

In the course of its orbit around the Sun the Earth intersects a multitude of meteoroid streams. About a dozen of these streams produce strong and reliable meteor showers, each year, at the present epoch. When a meteor shower is active the meteor rate projected from a small region of the sky, called the radiant (Fig. 2.1), is enhanced over that expected from the so-called sporadic background. The sporadic background is (broadly speaking) the average number of meteors that a visual observer might see on any given night of the year irrespective of shower activity. The sporadic meteor rate does vary according to time of day and the season, but for visual observers, the rate is typically about eight to ten meteors per hour on a clear moonless night [1].

We have all seen meteors, usually purely by chance, and the situation would have been no different for our distant ancestors. For all this, however, the actual study of meteors is a relative newcomer to the pantheon of astronomical topics. Indeed, meteor astronomy essentially saw its birth in the nineteenth century, this being the time when practitioners first began to discard ancient ideas, and to develop dedicated observing methodologies (see Chap. 5 for further details). Discussing the state of meteor astronomy in 1864 the Reverend Charles Pritchard (former Headmaster of Clapham Grammar School in London, and soon to be President of the Royal Astronomical Society) commented [2] that,

> it is to amateurs in astronomy especially that we must look for assistance in this interesting branch of celestial mechanics: these mineral fragments, these celestial rockets, this fiery dust from the lathe of the Omnipotent Worker, will furnish to him the correlative to that which the naturalist so fondly traces in

© The Author(s), under exclusive license to Springer Nature Switzerland AG 2023
M. Beech, *William Frederick Denning*, Springer Biographies,
https://doi.org/10.1007/978-3-031-44443-2_2

Fig. 2.1 An artist's illustration of the radiant associated with an active meteor shower. A perspective effect (see Chap. 5) gives the impression that the meteors from the shower radiate from a specific point in the sky. Illustration from *Star-Land—being talks with young people about the wonder of the heavens,* by Robert Ball (Cassell and Co, London, 1893)

> the organic regions of creation - all space teeming with life - beauty, order, scattered on all sides with a lavish hand - yet everywhere, and in all things, amenable to the control of law.

in this wonderful passage, Pritchard accurately captured the zeitgeist of the era, and his call to the amateur astronomer was correctly placed. Meteor astronomy in the mid-nineteenth century was a topic primed, ready, and waiting for the appearance of an enthusiastic and dedicated observer to carry its cause forward—Denning was destined to be that key figure. Indeed, in an interview Denning gave to *Tit Bits* magazine in 1895 [3], he commented

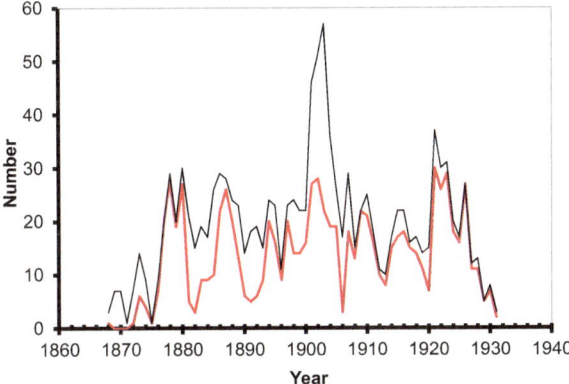

Fig. 2.2 Denning's yearly publication numbers. The upper (black) curve represents the total yearly count, while the lower (red) curve indicates the number of meteor-specific articles—see text for details

that he first turned his full-time attention to astronomy in the mid-1860s, but in the early 1870s developed a more intensive and directed interest towards meteors and meteor showers. It was also at this time that Denning began to publish extensive numbers of reports and notices relating to his observations. Figure 2.2 is a summary plot of Denning's publications per year from 1873 to 1932. The data displayed in the figure have been gleaned from many sources including:

- The *Astronomical Register*
- The *Observatory*
- *Nature*
- *Monthly Notices of the Royal Astronomical Society*
- *Memoirs of the Royal Astronomical Society*
- *Astronomische Nachrichten*
- *Astronomie*
- *Publications of the Astronomical Society of the Pacific*
- *Journal of the British Astronomical Association*
- *Journal of the Royal Astronomical Society of Canada*
- *Symons Meteorological Journal*
- *Quarterly Journal of the Royal Meteorological Society*
- *The English Mechanic and World of Science*
- *Knowledge and Science News*
- *Sidereal Messenger* (after 1892: *Astronomy and Astro-Physics*)
- *Popular Science Review*
- *Popular Astronomy*

- *Scientific American.*

In compiling the numbers plotted in Fig. 2.2, along with journal index searches, use was also made of Poggendorff's *Biographisch*, the *Royal Society Catalog of Scientific Papers*, and the *International Index of Periodicals*. Articles and notices sent to newspapers have not been included in the yearly counts, and accordingly the numbers shown in Fig. 2.2 represent Denning's minimum yearly output, rather than the maximum. Indeed, many letters by Denning appeared in the *Times of London, The Daily Mail*, and *The Western Morning News*, along with numerous letters to smaller-circulation, provincial newspapers. The latter letters tending to be concerned with specific fireball events, and were placed in order to explain what had been seen, and to ask readers to send him additional details. The number of journals and magazines that Denning sent his articles and letters to makes for an impressive catalog, and it illustrates a broad reach in readership, including both professional and amateur astronomers, as well as the general public. Indeed, Dennings diverse use of journals from across the UK, France, Germany, and the United States, helped to establish his name as an internationally recognized figure.

In total, 1263 research articles, letters, communications, and commentaries were identified in our survey of Denning's publications, with his peak output being recorded in 1903 when he published at least 57 notices and articles. There is an intriguing, about 10-year period variation apparent in the data portrayed in Fig. 2.2, and this presumably relates to interest in certain topics peaking and waning, and, in later life, to bouts of ill health. The two curves shown in Fig. 2.2 are an attempt to separate-out Denning's meteor work, from his other astronomical interests. Where the total output curve is most noticeably different from his meteor-work curve, between 1900 and 1906, the difference is largely due to Denning's interest in measuring the rotation rate of Jupiter, and with studies relating to the dynamics of that planet's red spot (see Chap. 3). Of Denning's collected communications, 871 relate to meteors, meteors showers, and fireball observations and their analyses (that is 70% of his total output), and his average yearly output in this area was about 17 articles per year.

The Astrophysical Data System (SDS), maintained by the Smithsonian Astrophysical Observatory (under a grant from NASA), allows for the identification of Denning's most referenced works, and the top one-dozen of these are summarized in Table 2.1. The top ranked works were published in the time interval between 1879 and 1923. All but one of the top-ranked works relate to meteor observations, with the exception (ranked number 2) being his book *Telescopic Work for Starlight Evenings* (published in 1891). The top

Table 2.1 The top 12 most-referenced works by Denning

No	Year	Title	Comments
1	1899	General Catalogue of the radiant points of meteoric showers and fireballs and shooting stars observed at more than one station	*MNRAS*
2	1891	*Telescopic Work for Starlight Evenings*	Book (see Chap. 4)
3	1916	Remarkable meteor shower on [1916] June 28	*MNRAS* Boötid meteor shower
4	1923	Radiant points of shooting stars observed at Bristol chiefly from 1912 to 1922	*MNRAS*
5	1926	A new cometary meteor shower (1926, October 9)	*MNRAS* Draconid meteor shower
6	1893	The August Meteors, 1893	*The Observatory* κ Cygnid meteor shower
7	1923	Stationary Meteors and meteors near stationary observed at Bristol since 1879	*MNRAS*
8	1899	Ephemeris of the radiant point of Perseid Meteors	*Astronomische Nachrichten*
9	1916	The remarkable meteors of February 9, 1913	*Nature* (See Chap. 3)
10	1912	Real paths of 429 fireballs and shooting stars observed in the British Isles during 15 years 1897 to 1911 inclusive	*MNRAS* (See Chap. 3)
11	1879	Catalogue of 222 stationary meteors	*MNRAS*
12	1886	Distribution of meteor streams	*MNRAS*. A study on radiant distributions

Data from SAO/NASA ADS website. Column 1 provides the ranking, column 2 indicates the year of publication, and column 3 indicates the title. Column 4 provides a few brief comments about the article, and indicates in which journal it appeared, and where further discussed in this text

ranked reference work is that of his *General Catalogue of radiant points*, and this will be discussed in detail below. Three of the works (ranked 3, 5 and 6) relate to the discovery of new periodic meteor showers, while one (ranked 9) relates to a remarkable fireball procession, seen from central Canada, the eastern seaboard of the United States, and ultimately by ships in the Atlantic Ocean, east of Brazil.

Denning's first radiant catalogue was published [4] in the *Monthly Notices of the Royal Astronomical Society* (MNRAS) in 1876. This catalogue was based

Fig. 2.3 Denning seated next to his Carey Celestial Globe, at his home in Egerton Road, Bristol, in August 1924. The globe now resides in the RAS archive. Image from the authors collection

on "nearly 900 shooting stars" observed from Bristol between 1872 and 76. A total of 27 meteor radiant locations being derived from the data set by, "penciling the courses [of the meteors] on a Cary's 18-inch globe (Fig. 2.3), and prolonging them backwards, the average places of convergence being selected as the approximate areas of radiation". (We shall have more to say on Denning's reduction methods below). A second article, also published [5] in 1876, provides some indication of Denning's dedication to observing at that time. Indeed, he writes that, "between October 13 and November 28, watching for forty-nine hours, I observed 367 shooting stars, 306 of which were well seen and their paths registered". Given the typically poor weather in Britain during the autumnal months, Denning must have been observing at virtually every opportunity to amass so many hours of observations.

Denning's observational intentions were made clear in an article published in 1879, where he explained, "In March, 1876, I commenced a series of watches for shooting stars, and have continued them to the present time; the result of the two years' work being that I have observed 3,749 of these bodies in 368 hours of work. My chief object all through has been to discover as many new systems as possible and to get the radiant points with accuracy" [6]. This approach was typical of Denning's observing philosophy, and we find

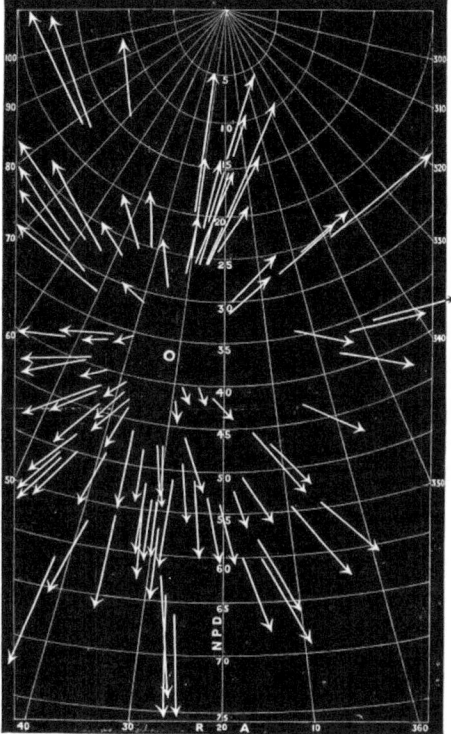

Fig. 2.4 Perseid meteor paths (for 38 July to 1 August 1878) plotted according to a gnomic projection (this projection shows all great circle arcs as straight lines). The radiant point is indicated by the small central circle. Note that the meteor trail lengths increase the further the beginning point is from the radiant. The north celestial pole is located at the center-top of the figure

echoes of this ideology appearing in his *Telescopic Work for Starlight Evenings*. There he comments, "Nearly all the most successful observers have been men of method. The work they took in hand had been followed persistently and with certain definite ends in view. They recognized that there should be a purpose in every observation" [7]. He further noted, "It need hardly be said, however, that every difficulty may be surmounted by perseverance, and that a man's enthusiasm is often the measure of his success, and success is rarely denied to him whose heart is in his work". The benefits of adopting such an approach to observing soon paid dividends for Denning, and his first genuinely new contribution to meteor astronomy was announced in 1877, in a short article [8] published in the journal *Nature*. Denning's important announcement was concerned with observations on the radiant drift of the Perseid meteor shower (Fig. 2.4).

The Perseid meteoroid stream produces a rich annual meteor shower that is active each August, and the stream produces a steady display of meteors over several weeks [1]. For such a long-lived meteor shower it is expected that the radiant location should show a night-by-night migration in its position with respect to the background stars. This shift being due to the Earth's movement through the stream (see Chap. 5 for further details). Through his accurate and near continuous observation of meteors in the summer of 1877, Denning was able to not only show that the Perseid meteor shower was active from early July to late August, but that its radiant point moved across the sky exactly as expected.

Denning continued to monitor the Perseid stream for many more years,[1] and in 1884 published [9] a complete, and detailed review of his observations (Fig. 2.5). In 1899 Denning published in the January issue of *Astronomische Nachrichten* (AN) an ephemeris for the radiant of the Perseid shower, showing that the typical hourly rate at which meteors were produced varied from as low as 0.2 to a maximum of 50 (Fig. 2.6). The data analysed in his AN article were collected between 1869 and 1898, during which time Denning notes that some 6479 meteors, including 2409 Perseids had been observed. In total some 73 Perseid radiant point locations were determined, and the activity profile (Fig. 2.6) captures the distinctive asymmetry of Perseid shower activity surrounding the time of its maximum, with a slow rise to maximum followed by a fairly rapid decline. This activity profile is in contrast to that deduced for the Delta Aquarids, which has a more symmetrical rise and fall about the time of maximum activity.

In addition to collecting his own meteor path data, Denning also worked with data obtained by many other observers. In 1877, for example, he published [10] a radiant catalogue from meteor observations collected by Captain George Lyon Tupman (Fig. 2.7) between 1869 and 1871. Tupman certainly qualifies as a grand amateur in the sense that he did not make a living through his scientific endeavors. From an early age Tupman was interested in astronomy, and became a Fellow of the RAS, at age 25, in 1863. For all this, Tupman made a carrier as an officer in the Royal Marine Artillery. His first observational reports were published in the *Astronomical Register* starting in 1865, and his first paper on meteors appeared in the November 1867 issue. Indeed, this first article on meteors covered the basics of two-station observations, and he called upon observers to send him observational data in order that detailed atmospheric paths, velocities, radiants, and orbits might

---

[1] Indeed, it was probably equivalent to something like an annual pilgrimage. The author, like many other meteor astronomers, always makes an effort to see the Perseids even if no specific research activity is being undertaken.

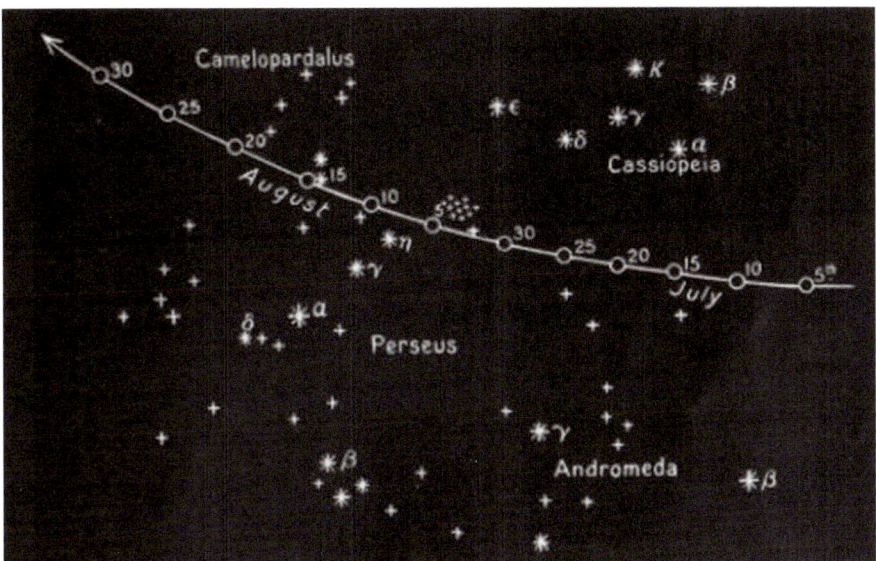

Fig. 2.5 Drawing by Denning to illustrate the radiant drift of the Perseid meteor shower radiant between 5 July and 20 August. The shower undergoes its strongest activity between the 10th and 13th of August

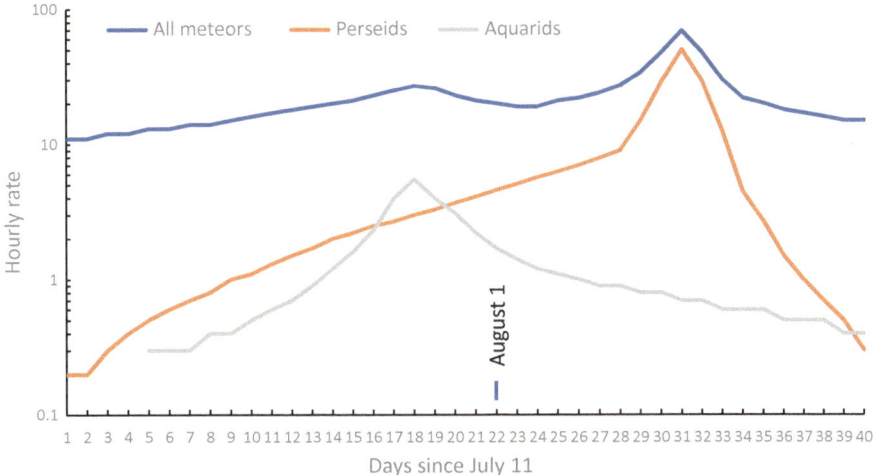

Fig. 2.6 Meteor activity profiles (average hourly rate, versus time), as derived by Denning, for the Perseid, and Delta Aquarid showers, along with the total number (including sporadic meteors) as seen in the 29-year time interval from 1869 to 1898. The time scale axis is based upon the number of days since July 11

be calculated. In this latter respect his first detailed results appeared in the appendix of the December 1876 issue of the *Register*, and concerned three very bright fireballs seen from across southern England on the 3rd, 7th and 14th of September 1875. Meanwhile, while stationed in the Mediterranean between 1869 and 1871, Tupman, at the suggestion of Julius Schmidt,[2] began observing meteors. Keeping detailed records, Tupman, collected information on some 3800 meteors over 180 clear nights. Of these about 2000 were registered on star charts resulting in 102 radiants being deduced. It was from this data set, in 1870, that Tupman discovered the Delta Aquarid meteor shower[3] (see Fig. 2.6). In 1872 Tupman was assigned as an assistant to an expedition to study the 1874 transit of Venus. This he dually did from the Sandwich Islands, and upon returning to England spent several years (as an unpaid assistant) at the Greenwich Observatory reducing the data obtained. Between 1873 and 1880 he served on the Council of the RAS, and was the Society's treasurer from 1884 to 1889. From his home in Harrow, he set up Hillfoot Observatory in the early 1880s, and equipped it with a transit circle and an 18-inch aperture reflector by George Calver (see Chap. 6). In later life he turned his attention towards observing occultation events, along with the appearance and motion of new cometary apparitions. He also developed a keen interest in the then new technology of wireless telegraphy.

In 1878 Denning published [11] another radiant catalogue. This time the catalogue was based upon observations collected by Italian observers in 1872. Denning projected 4143 meteor trails from these observations, and found 315 radiant points. A measure of Denning's remarkable commitment to radiant reductions can be found in an 1897 article that was published [12] in the *Bristol Naturalists Society Proceedings*, where he comments, "The number of meteors actually projected by me on star charts, including those observed and those selected from published catalogues, reaches over 10,000; but, in addition to this, the paths of fully 20,000 others were examined".[4] Denning believed that his observations and reductions indicated the existence of at least several hundred annual meteor showers. This number is an overestimate, in modern terms, of the actual number of annual meteor showers by

---

[2] Johann Friedrich Julius Schmidt was director of the National Observatory of Athens in Greece. He was an avid lunar observer, and in his early carrier studied meteors with Johann Benzenberg in Bonn (see Chap. 5).

[3] The delta Aquarid shower actually has some complicated radiant structure, and is typically divided into northern and southern components, each of which having different radiant locations, activity rates and peak activity nights.

[4] It is one of those great and sad ironies that essentially all of Denning's hard work on meteor trail projections is of little contemporary value. It is also to be regretted that the vast bulk of Denning's original trail projections, notebooks, and correspondence have not survived to the modern era.

Fig. 2.7 George Lyon Tupman seen here preparing to observe the 1874 transit of Venus

a factor of about 4 or 5. We shall see later, however, that Denning's estimate of the number of active meteor showers became even more extreme. At the present time it is believed that there is good evidence to support the existence of some 50–60 annual meteor showers. It is important therefore to explore the reasons why Denning believed that so many meteor showers existed.

Denning's strong (indeed, unshakable) belief in the existence of many hundreds (and later, circa 1900, many thousands) of meteor radiants is an example of what might be called the philosophical parallel [13]. That is, Denning was led to an erroneous conclusion by the unquestioned acceptance of a theoretical paradigm that was, in reality, untrue. The modern-day meteor astronomer now knows that probably only 10–20% of visually observed meteors actually belong to well defined meteoroid streams [14]. What this means is that the vast majority of observed meteors cannot, in fact, be traced to some well-defined group radiant. This principle while clear to modern astronomers was not, however, obvious to Denning or any of his contemporaries. They believed, in contrast, that all meteors could be traced to a group radiant, and that each radiant point could probably be associated with a parent comet. The paradigm under which Denning operated was that all meteors could be traced to a radiant point belonging to an active meteor shower. Under modern practices a meteor shower is deemed to be active if at least four meteors can be unambiguously associated with a common radiant during the course of one night's observing session (i.e., within a time span of some six to eight hours) by a single observer. Denning would on occasion deem a shower to be active on the basis of observing just one specific meteor per night.

The term sporadic meteor was apparently coined by the pioneering meteor astronomer Edmund Heiss (see Chap. 4). Working in the late 1840s through to the early 1860s, Heiss and co-workers distinguished between the periodic meteors, which they believed returned in yearly showers, and those which fell outside of the times of yearly activity [15]. More extensive, long-term surveys by, for example, A. S. Herschel, and Denning began to reveal, however, that more, and more periodic meteor showers existed, and accordingly it was generally believed that only a few meteors were actually sporadic. Denning, and Herschel found few sporadic meteors, of course, because they believed that all meteors could be traced to some active group radiant. In 1885, Denning was to write [16] of sporadic meteors, "the term hardly seems to me a commendable one, though undoubtedly useful to cancel our ignorance of the contemporary streams supplying meteors unconformable to any special display that may be under observation". Norman Lockyer (see Chap. 4) also commented [17] on sporadic meteors in his book, *The Meteoric Hypothesis* (published 1890—see Chap. 5), and exclaimed that, "the term sporadic [is] simply a measure of our ignorance". To this he added, "with every new radiant thus established the number of sporadic meteors naturally become less and less". The reason that Denning and his contemporaries found so many radiants, therefore, is explained on the basis that every meteor was believed to issue from an active meteor shower radiant.

The crowning achievement to Denning's study of meteor radiants was the publication, in 1899, of his *General Catalogue*. This catalogue was not only a summary of his own observations, but a summary of the observations collected by many other observers including Robert Greg, Henry Corder, Edward Heis, George Tupman, Giovanni Schiaparelli, and Alexander Herschel. The *General Catalogue* contains information on 43,647 radiants, and Denning commented that the total number of projected meteor-paths, from which the radiant points were determined, was, "approximately 120,000" [18]. Denning's *General Catalogue* was a monumental work, but even so he did not claim completeness, rather, he merely believed that it, "includes the bulk of such radiants as have hitherto been published".

Denning believed that his catalogue demonstrated that, "there are considerably more than 50 showers in play on any and every night of the year, and, moreover, certain (in fact the great majority) of these are not confined to limited periods, but extend their activity over several weeks, and in some cases over several months". As explained earlier the belief in the existence of many shower radiants was an artifact of the early plotting methods, and virtually all of the radiants listed in Denning's catalogue, and those of his contemporaries, are spurious. Spurious, however, only in the sense of not being related to some active meteor shower. This is not to say, however, that the early radiant catalogues are of no contemporary value. Certainly, all the major, and lesser meteor showers are included within their various pages, and it remains likely that there are some weak showers in the catalogues that do exist, but have not been confirmed, as such, to date. Furthermore, the catalogs do provide a clear indication (in modern terms) of structure within the sporadic background of meteors [19].

Although sporadic meteors can enter the Earth's atmosphere from any location on the celestial sphere, there is, in fact, a distinct clustering of eight diffuse sporadic background radiants (Fig. 2.8). These are the Helion, Antihelion, North apex, South apex, along with their antiapex counterparts, and the North toroidal and Southern toroidal regions. Meteors associated with the Helion and Antihelion regions are derived from meteoroids travelling along elliptical orbits about the Sun in the plane of the ecliptic, and moving in the same (prograde) direction as the planets. Antihelion meteoroids encounter the Earth from directions located 180° away from the Sun on the sky. Helion meteoroids encounter the Earth from the same direction as the Sun on the sky (i.e., during the daytime) and are only detectable via radar and radio reflection techniques. The Helion and Antihelion populations are taken to be derived (via gravitational scattering and orbital migration) from meteoroid streams associated with short-period comets. The North and South Apex populations, in contrast are thought to be derived from long-period comets traveling along retrograde orbits, which enter the inner solar system at high inclinations to the ecliptic. These meteoroids encounter the Earth in the direction of its motion about the Sun (that is, the Apex of the Earth's way). The antiapex sources, in contrast, a related to meteoroids that encounter the Earth from a direction opposite to its direction of motion. The apparent gap between the north and south apex regions is attributed to the Earth clearing-out meteoroids moving directly in the plane of the ecliptic. The north and south toroidal source regions are thought to be derived from material derived from high inclination short-period comets.

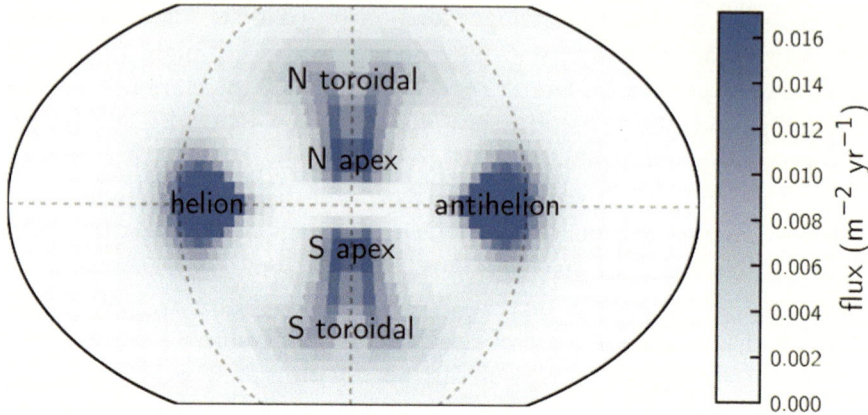

Fig. 2.8 Location map for the primary sporadic background radiants. The map is an all-sky projection in heliocentric coordinates (the Earth's orbital plane, the ecliptic, is the central horizonal line). The Sun is located at the point where the zero-longitude great circle intercepts the ecliptic (to the left in the diagram). The apex of the Earth's way (its direction of motion) is situated 90° away from the Sun, and is located at the center of the map. To the right in the diagram is a colour scale showing the meteoroid flux in units of meteoroids per square meter per year. Note, the flux from the various radiant locations is seasonal, and not constant over the course of the year. Image from Aletha Moorhead (NASA Meteoroid Environment Office). http:// ntrs.nasa.gov/api/citations/20190032387/downloads/20190032387.pdf

Figure 2.9 is a plot of the deduced sky distribution of meteor shower radiants, in terms of number per 10° interval of declination on the sky. The curves are based upon data extracted from three catalogs: Denning's *General Catalog*, the radiant catalog produced by Ernst Opik from the meteor paths recorded during the Arizona Meteor Expedition from 1931 and 1933 (see Chap. 5), and the radiant catalog compiled by Ronald McIntosh from observations gathered by members of the New Zealand Astronomical Society between 1927 and 1935. All three of the catalogs considered are based upon visual observations of meteor trails. The catalog by Denning is the most extensive in terms of radiant numbers (43,647 as seen earlier), Opik deduced 279 (high probability) radiants, while McIntosh found 320 radiants. In each catalog the radiants were derived by projecting the observed meteor trails to common intersection points on the sky. Both Denning and Opik pooled data over many nights to bring-out 'weak' radiants. Interestingly, McIntosh wanted to apply more rigorous reduction methods, using only those meteors observed on a single night to deduce radiant locations, but found that the practice produced very few results. Rather than derive effectively no radiant points, McIntosh noted that, "a little elasticity justified the inclusion of radiants based on observations spread over several nights". In addition to showing the

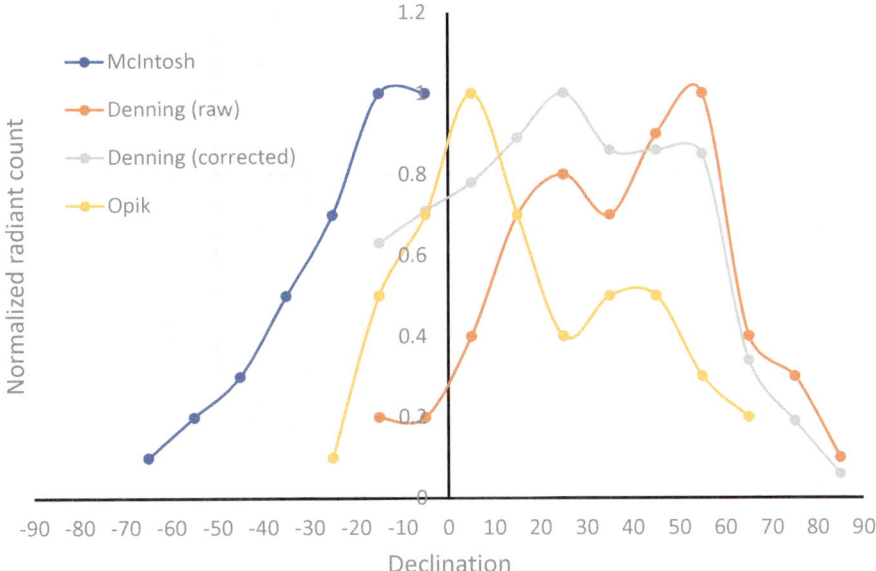

Fig. 2.9 Normalized number of radiants per 10° interval in declination on the sky. The normalization constant for the data presented by McIntosh is 83, that for Opik's catalog it is 57, while those for Denning's raw and corrected catalog numbers are 654 and 988 respectively. See text for details

number of radiants, per 10° declination interval, as given by Denning in his *General Catalog* [labeled Denning (raw) in Fig. 2.9], a second, 'corrected', curve is also shown [labeled "Denning (corrected)" in Fig. 2.9] and this corrects for the amount of time that a given region of the sky is visible at the latitude of Bristol, UK [19]. Since the more southerly regions of the sky spend less time above the horizon than the more northerly ones, the correction term acts to enhance the distribution of radiants in the region of the celestial equator.

The curves shown in Fig. 2.9 are revealing in two ways. With respect to Denning's (raw data) *General Catalog* distribution, the peak number of radiants occurs at a declination of about 50°, and this corresponds to the latitude of Bristol, from where he was observing. Likewise, the smaller peak in the data given by Opik is at about 40°, and this is close to the latitude of Flagstaff, Arizona from where the Arizona Meteor Expedition was conducted. The New Zealand data was predominantly collected from the North Island with a latitude of approximately − 35°. What these peaks in the radiant distribution data indicate is that most of the radiants are located close to the declination corresponding to the latitude of the observer—in other words this is a selection effect. Since most meteors are observed as they move downwards across the sky, there will be a tendency to deduce radiants in the region of the observer's zenith. The other important point to take away from Fig. 2.9 is that, after the local—zenith related—peaks, by far the greatest number of the radiants are situated in the declination range between − 25 and 25°. This is the region of the sky that includes the ecliptic—in other words, the majority of deduced radiants are situated in the plane of the Earth's orbit about the Sun. In more contemporary terms this is the same as saying that the vast majority of sporadic meteors are moving around the Sun in the same plane as the planets (recall Fig. 2.8 and [19]).

2.1   The Stationary Radiant Debate

The idea that the radiant location of a meteor shower might remain fixed on the sky, with respect to the background stars, and remain active for days, weeks, or even months on end was something that Denning felt he had proved beyond any doubt. He was mistaken. Importantly, however, he was honestly mistaken, and the controversy concerning the existence verses the non-existence of such stationary radiants behoves us to focus attention upon Denning's observing methods, his observing philosophy, and his seeming mistrust of theoretical argument. Firstly, Denning had a strong philosophical outlook, and insisted that theory must be guided by, and fully based upon, the observations, and not the other way around. For Denning, what he saw with his eyes (that is meteor trails), and what was recorded in his data (that is meteor speeds, directions and colors) was the true measure of what was going on in the sky. This philosophy is (within limits) a reasonable approach to observing, but to unflinchingly follow entrenched methods of data reduction can, and indeed, often does, result in confusion and even delusion. In the stationary radiant debate, the problem was not the observing methodology. Rather it was the manner in which the meteor trail data was analyzed. Denning refused to reconsider his analysis methodology, partly, one suspects, through his lack of formal scientific training, and partly because he simply stuck with what he knew. In short, he stuck with what he had learned during his formative years from the leading observers. The absence of critical questioning, and no doubt some good old stubbornness, restricted Denning's ability to adjust his reduction methodology, and his doggedness, although admirable in itself, reflected a blind trust in the ideas and methods of his teachers.

Denning began his meteor work at a time when the primary occupation of all observers was to simply record meteor trails on the sky, and then from these trails deduce radiants. He also began his meteor work at a time when it was generally believed that there was no such thing as a sporadic meteor, and, in Britain at least, astronomers such as Norman Lockyer still strongly suspected that not all meteors were derived from cometary bodies orbiting about the Sun along closed orbits. Indeed, the meteoritic streams of Giovanni Schiaparelli and Lockyer were cosmic in nature, and, it is the case that any meteors derived from an actual cosmic stream (given they actually existed) would, indeed, appear to be derived from a radiant point fixed amid the background stars. To achieve this, however, they would need to be travelling with so-called hyperbolic velocities amounting to at least a hundred kilometers per second. Accordingly, Denning had reason enough to believe in the possibility of fixed radiants, but he did not have the critical reasoning skills (or, perhaps, he chose not to apply them) to deeply question alternative interpretations to those that he was finding. Following an approach of logical positivism, as strongly espoused by physicist Ernst Mach (1838–1916), Denning effectively adopted the attitude that empirical data must always trump theory-based deductions. Such an outlook, of course, is part and parcel of the scientific method, but it can be problematic, and Mach's opposition, for example, to those researchers that tried to establish an atomic theory of physics, on the grounds that no person had ever seen an atom, seems archaic and blinkered today. Yes, Denning accepted that some, if not all, periodic meteor showers were associated with, and derived from cometary streams, but he saw absolutely no reason to suppose that all meteoroids had to be derived from comets.

George Tupman signaled problems with the idea of stationary radiants as early as 1878, noting in an article published in the January issue of the *Monthly Notices* that such radiants were most likely a consequence of pooling data over multiple nights. To this, however, he added a potential loophole, commenting that:

> there is a special case in which the Earth can remain in a stream for several weeks with a nearly fixed radiant. The orbit must nearly coincide with the plane of the ecliptic, the perihelion distance of the central portion be a little less than unity and the motion be direct.

If the dynamical and geometrical conditions so outlined did not apply, then, Tupman argued, the apparent reappearance, or seeming continued activity over long intervals of time, must indicate a coincidence in the activity of two or more distinct meteor showers which by chance had radiants in the same region of the sky. In prescient terms, Tupman further noted that the pooling of meteor paths, over many nights, to determine radiant positions, as begun by "Heiss, and used by Greg, Herschel and Schiaparelli", should be relaxed, and that, "radiants proper to each day of the year [should] be deduced separately". It would appear that Tupman's advice concerning radiant reduction protocols fell on deaf ears, and it was to be some 40 years before they were generally heeded.

With the above being said, what was it that Denning was actually claiming with respect to stationary radiants, and how was it that he, and other observers, could readily find data to support such claims? The claim itself is easy enough to state, the radiant point does not show any night-by-night drift—such as that deduced for the Perseids (recall Fig. 2.5)—and it remains active, that is meteors could be traced to the radiant point each night over time intervals amounting to many weeks and even months on end. As such this is simply a claim that Denning made according to his reduction methods—it was simply what he found to be the case within his reduced data set. There was no inherent analysis error in his reduction methodology—he recorded a meteor's path on the sky, he noted its swiftness and color, and then he wrote these data down in a notebook. Later he would take the collected meteor trail data and mark the beginning and end points on a globe or sky map. This is all fine and good, and indeed, given Denning's experience as an observer, his data was probably as good (that is accurate to within a fraction of a degree on the sky) as naked-eye observations of meteors can be. It was at the next step in his analysis that problems began to emerge. And, the problems emerged because Denning stuck to the idea that all meteors must be derived from an active group radiant, and to locate a radiant one has to project at least two meteor trails backwards on the surface of a globe or star map to find an intersection point (see Chap. 5). Accordingly, Denning projected all his paths backward until some intersection point emerged. Critically, as well, he was prepared to use and project meteor trails on his various maps that had been obtained over many nights of observing, or even on the same nights over multiple years. From such practices, stationary radiants will emerge simply as false-positives in the random background of apparent intersection points.

Not only were practitioners, such as Tupman, beginning to question the process of pooling meteor trail data over many nights 'to bring out radiants' in the late 1870s, the physical origin of meteors was also being questioned. As the nineteenth century ended so the concept of cosmic streams, as advocated for by Schiaparelli and Lockyer, fell into disfavor. Indeed, the majority of practitioners began to adopt the stance that meteoroids must have an origin from within the solar system, and in particular that they were derived through the process of cometary decay. The details surrounding this latter development will be described in Chap. 5, for the moment, however, it suffices to say that, dynamically speaking, apart from the unlikely geometrical circum-stance outlined by Tupman in 1878, if meteors are derived from comets, then stationary radiants are an impossibility. The impossibility being dictated by the indomitable weight of celestial mechanics, and the stalwart equations of Newtonian dynamics.

For all of their unexpected existence, Denning did have influential supporters who were openly prepared to accept his claims on stationary radiants. Richard A. Proctor, in 1885, for example, argued that stationary radiants could generally be explained in terms of drawn-out meteoroid streams ejected from near-by stars. Alexander Herschel, further attempted to explain stationary radiants as relics of past collisions of cosmic streams with a ring of meteoritic material in orbit about the Sun ("perhaps like the annulus of matter round the planet Saturn"). By way of analogy, Herschel explained, "as a tree is robbed of many leaves by a strong gust of wind sweeping through it, …it is here suggested now that it is the debris of such gusts, not the gusts themselves [that is the cosmic streams], which we long see afterward". Writing in the *Monthly Notices* for January 1899 Herbert H. Turner, Director of the Radcliffe Observatory in Oxford, outlined a repeated stream perturbation scenario for the origin of stationary radiants—Turner's analysis, however, was restricted to those streams located in the ecliptic. Likewise, William Henry Pickering, in America, argued in 1909, that stationary radiants were entirely possible, and proceeded to expand upon the ideas outlined by Turner. From among these theoretical possibilities, even if they were somewhat vague and poorly constrained, Denning could at least take some comfort in the fact that not every theoretician rejected, outright and out-of-hand, the possibility of such long-lived radiants from existing.

Claiming that stationary radiants exist is one thing, but having confi-dence that naked-eye observations are accurate enough to truly deduce such results is another. Certainly, Denning's long experience would suggest that his observational skills were about as good as any human-being might attain, but where they accurate enough. The young British astronomer Bryan Cookson took-on the question of eye-observation accuracy in the March 1901 issue of the *Monthly Notices*, and after much mathematical discussion concerning error analysis, concluded that Denning's observations concerning radiant positions could be relied upon to within ¾ of a degree. Further-more, Cookson argued, eye-observations were of sufficient accuracy to reveal the existence of stationary radiants. For all this, however, Cookson, also noted that the whole issue of stationary radiants rested upon the validity of combining meteor trails to form a radiant. Mathematician H. Wallis Chapman further considered the question of radiant reduction in the January

1905 issue of the *Monthly Notices*. Specifically, Chapman noted that many radiants were deduced on the intersection of just 3 or 4 meteor trails, and he questioned if this was a reasonable procedure. In a highly mathematical study, Chapman argued that the probability of finding a false positive radiant from 4 or less meteor trails, projected at random on the celestial sphere, was, in fact, remarkably high. In this manner, Chapman drew attention to the problem of co-adding meteor trails over many nights. Denning quickly responded to Chapman's paper, and writing in the April 1905 issue of the *Monthly Notices*, he, "offer[ed] a few remarks from an observer's point of view". Key hear is Denning's use of the term "an observer's point of view". Certainly, Denning would not have appreciated the detailed mathematical analysis presented by Chapman,[5] but he did agree with Chapman that radiants should be deduced on the basis of as many meteor trails as possible. This being said, Denning argued that, "much depends upon the experience, judgement and method of the observer", and he side-stepped the issue of pooling data over many nights.

Circumstances concerning stationary radiants changed dramatically for Denning when, in 1912, Charles Pollard Olivier (Fig. 2.10), a 28-year-old American astronomer from Virginia, published the results of his Doctoral thesis in the *Transactions of the American Philosophical Society*. Entitled, *175 Parabolic Orbits and Other Results from Over 6200 Meteors*, Olivier's work was both detailed and highly mathematical. It was also highly critical of Denning's reduction methods (that is in the pooling of data over many nights), and the very observational existence of stationary radiants was brought into question. Olivier specifically called-out Denning for using a pooling methodology, but in fact, he could have named any number of internationally established meteor observers. In over three decades of observing and producing radiant catalogs, however, Denning had never been so openly challenged or criticized.

From an early age Olivier spent many hours observing meteors, and by 1912 he was a young university graduate looking to make his name, and secure a career in astronomy. Indeed, in 1914 he accepted a position at the Leander McCormick Observatory at the University of Virginia, and later, in 1928, he became Director of the Flower Observatory at the University of Pennsylvania. Olivier's pedigree as a meteor observer, and organizer of other enthusiasts, was established in 1911, when he founded the American Meteor Society, a society that he directed and nurtured for the remainder of his life, and which still thrives to this day. Importantly for meteor astronomy, Olivier developed his ideas and practices from outside, and independently of the British community of meteor observers—a community largely accepting of

---

[5] Indeed, only a very few readers of Chapman's article would have appreciated the 16 journal pages of dense algebra and differential calculus presented.

Fig. 2.10 Charles Olivier (1884–1975). Founder of the American Meteor Society, and principal driver of the new reduction methodology for meteor radiant identification

Denning's leadership. In this manner Olivier had the freedom and opportunity to reassess the practices and ideas pertaining to meteor astronomy, most of which had been developed in the early to mid-1800s, and to bring more modern practices to bear.

For all of his sound academic credentials, and evident observational skill, Denning did not take the criticisms promulgated by Olivier lying down. Indeed, Denning replied with some passion, writing that Olivier, "has adopted a method of reduction which, instead of exhibiting what the data really teach us, veils their actual meaning, and at the same time unduly complicates and multiplies results". Furthermore, Denning complained that his (that is Olivier's), "errors of observations appear to be very large generally". The gloves were off, and Denning's response to Olivier's work ran along the lines that, if other (younger and less experienced) observers did not find the same results as he did, then they must be in error. Olivier, however, was just as frosty in his response to Denning, noting in a 1913 publication that, "the word error as used by Mr. Denning simply means my results differ from his own. The more my conclusions differ the greater the error". Olivier's rebuttal to Denning's bluster was formal and carefully constructed, and his denial of stationary radiants was built upon both theoretical (that is on dynamical) grounds, and Denning's reduction protocols. On the theoretical side, Olivier rejected outright the possibility of cosmic streams, and insisted that all meteoroids must be derived through the process of cometary decay (see Chap. 5). While, at this time, it was still far from clear what comets actually were, it was generally believed that they were typically bound to the Sun, and accordingly they, and any material ejected from them, must be explainable in terms of well-developed Newtonian techniques. For all this, the question concerning the origin of comets was entirely open circa 1912, it certainly was not clear, then, that they formed at the same time as the Sun, and or the planets, or, indeed, whether they had been gathered, from interstellar space, long after the solar system had formed.

In spite of Olivier's criticisms, Denning continued in a non-plussed manner, and confidently announced in the August 1913 issue of the *Observatory* that, with respect to the identification of stationary radiants, "it is a thing that lives indelibly in the sky, and though previous attempts to account for it are not generally accepted, no observations have succeeded or will succeed in obliterating it from the firmament". Denning, in this response, is trying to separate theoretical complications from the observations, arguing that observations must always triumph over theory. Writing in the March 1914 issue of the *Journal of the British Astronomical Association* (JBAA), Denning reiterated his stance concerning the possibility that stationary radiants might be composed of distinct showers that chanced to share a common radiant location on the sky. In the process he also criticized Olivier's argument that radiants should only be derived on the basis of data gathered during a single night's observing. He writes,

> if I see a shower on 10 nights from the same apparent point amongst the stars, it is to be presumed that a similar point of radiation would have been seen in the daytime had conditions permitted. As an observer I call this one shower. If it is called 10 different showers, you may as well disconnect a shower seen during the whole of one night into 12 different showers, for theory imposes apparent changes in the radiant for every hour of the night as well as for successive nights. Observers cannot be trammelled by any arbitrary premises based on the requirements of theory. They must state just what they see when absolutely sure of the accuracy of their observations.

Within the above comments, Denning makes several entirely valid and quite fundamental points, but his arguments fail to satisfy the problem at hand—by which, in the latter statement, it is required that the scientific understanding of any physical phenomenon must, ultimately, have some theoretical underpinning. In his March 1914 comments, Denning wants the stationary radiant issue to be pursued along purely observational lines. Indeed, in the same March letter quoted above, Denning continues that,

"I must confess that I am tired of enforcing this feature, and tired too of waiting for an observer to corroborate it and a mathematician to explain it". While Denning wanted to turn the stationary radiant debate into an observational, and recording issue, he still did not address the issue of pooling data over many nights. It would appear that from an observational standpoint, Denning's *idee fixe* was that no matter when a meteor was observed, if its path could be traced back to some already identified radiant point, then it was by default part of the meteor shower associated with that specific radiant. In this manner, Denning de-coupled the name of a meteor shower and its radiant point from the dynamical origins of its component meteoroids. On purely classification grounds, there is nothing specifically wrong with this procedure, it organizes the sky in terms of those location points from which meteors appear to radiate from. Unfortunately, any point on the sky could, under such a scheme, correspond to a radiant, and it ignores the notion that each annual meteor shower is associated with a radiant related to a specific meteoroid stream that has, in turn, been derived from a single, parent cometary nucleus.

Denning was not alone in criticizing Olivier's animadversions, and, for example, the Reverend Martin Davidson (Director of the BAA's Meteor Section from 1911 to 1921), in the May 1914 issue of the JBAA, noted that, "I for one am more disposed to believe in the results of Mr. Denning's work". For all this, while Davidson rejected the proposed mathematical solutions to the problem, as being impractical, he did indicate that, "the problem, to my mind, turns on the question of the probability of a number of meteor streams existing with the proper elements to give a stationary radiant". In some sense, this is a statement that Denning had been too successful in finding, and then filling the sky with myriad radiants—no matter where or when a meteor was seen in the sky it could likely be traced back to some radiant already identified in Denning's *General Catalog*.

Henry Crozier Plummer, at the time Director of the Dunsink Observatory in Ireland, picked-up on the need to find a theoretical explanation for stationary radiants in the December 1920 issue of the *Monthly Notices*, seeing the problem as being both "obscure and controversial". Restricting his analysis to stationary radiants located close to the plain of the ecliptic, Plummer set about calculating the required structure of a meteoroid stream on the basis that it was required to produce a stationary radiant. In this way he estimated the orbital parameters that the stream meteoroids would have to display (for more details of this see Sect. 2.2.1). The end result of Plummer's analysis, however, indicated that under highly artificial circumstances stationary radiants might be explainable, arguing that, "the failure to find a physical explanation cannot be set against the observations. Contrary to a view often expressed, the difficulty appears to be one rather of practical improbability than of theoretical impossibility". Although the Reverend Davidson echoed the very same point in 1921, suggesting that, "there is no theoretical impossibility in stationary radiants under certain circumstances", the situation really was untenable. The requirements of a stream's structure, that is the variation in the orbital parameters of its constituent meteoroids, required so as to produce a long-lived, active, stationary radiant, were so convoluted that it was simply impossible to accept that they were the result of a natural processes (e.g., through gravitational perturbations and cometary decay). Olivier echoed this point in a paper published in the December 1922 issue of the *Monthly Notices*. Performing an analysis identical to that by Plummer in 1920, Olivier simply noted of the end result that he had, "not the slightest belief that any such highly artificial and complicated system exists in nature". One notes, however, that Olivier's analysis was developed under the pretext of a fixed radiant point delivering meteors on every day of the year—something that Denning never actually suggested was happening. Furthermore, and in light of his analysis revealing the highly contrived orbital structure needed to produce a stationary radiant (see Sect. 2.2.1), Olivier attacked (one could perhaps write, ridiculed) Denning for his suggestion, made in the June 1919 issue of the *Observatory*, that a stationary radiant might be produced through the decay of a single short-period comet. To Olivier, it was the dynamical theory that triumphed over the observations.

Olivier's last words on the topic of stationary radiants appeared in his book *Meteors* (published in 1925). Indeed, one can argue that, apart from Denning's later writings, the publication of *Meteors* was the final word on stationary radiants. The book is a masterpiece, and it ushered in a new era of meteor astronomy. Over the course of its 24 chapters, Olivier recounts the history of meteor showers, their association with comets, and the computation of their orbits. Two chapters were dedicated to stationary radiants, and step-by-step, Olivier dismantled and exposed the observational and theoretical problems associated with their existence. Olivier pulled no punches in *Meteors*, and he roundly criticized the practice of data reduction then being employed, especially by British observers (by which he meant Denning), to derive shower radiants. By being much more conservative with respect to using observations derived on a single night, Olivier dramatically reduced the number of active meteor shower radiants, and reasoned that virtually all of the 43,647 radiants presented by Denning in his *General Catalog*, were spurious and associated with a sporadic background of meteors. To say that *Meteors*

changed meteor astronomy is to make an under statement of the circumstances. Indeed, Olivier's book effectively re-set observational practices, and
it established a clear theoretical foundation for the understanding of meteor
showers and meteoroid streams.

Olivier was a member of the new, and growing population of establishment astronomers. Not only was science and scientific research in America
in its ascendency during the early decades of the twentieth century, so to was
the confidence of its practitioners. Olivier not only began to wrestle control
of meteor astronomy from the amateur observers, he set about making it
a professional field of study. The amateur, of course, was still fundamental
to the data gathering process, but with Olivier's pushing (and the formation of the AMS) the gathering of meteor data and the analysis process
was to become tightly controlled. Indeed, moving towards the mid-twentieth
century, the theory and analysis side of meteor astronomy had largely passed
out of the hands of the amateur observer. Furthermore, from circa 1930
onward, meteor astronomy, as a *bona fide* research field, had largely passed
beyond the technical reach of most amateur astronomers. Not only was
Olivier ushering in new methodologies with respect to data analysis in the
mid-1920s, but the data itself was increasingly being gathered by auto-
mated instruments. The amateur astronomer was not ruled out, but the
role and relevance of the amateur was increasingly subservient to that of the
professional astronomers, and their government-funded, research-orientated
practices.

While by mid-1920s most of the astronomical community had serious
doubts about the existence of stationary radiants, and reasonable questions
were being asked about the reduction techniques being applied to derive
them, Denning stuck to his ideals. In one surviving letter from this time,
written on 20 June, 1920 (see Fig. 2.11), Denning writes:

> I have long been hoping that some really good observer would apply himself
> energetically to the task of practically investigating stationary radiants and some
> other features in meteoric astronomy - But no one has done so up to the
> present time - It could only be accomplished by a man of thorough capacity,
> perseverance and sound unprejudiced judgement. I believe firmly that were
> such a research undertaken it would lead to the absolute corroboration of my
> results and disarm those critics who object to my observations as being very
> difficult to explain, if not directly inconsistent with theory

These are clearly the worlds of an embittered observer, and an elderly
gentleman (he was at this time 72 years old), who could see his life's work
being called into question. By "critics" it is reasonable to assume that Denning

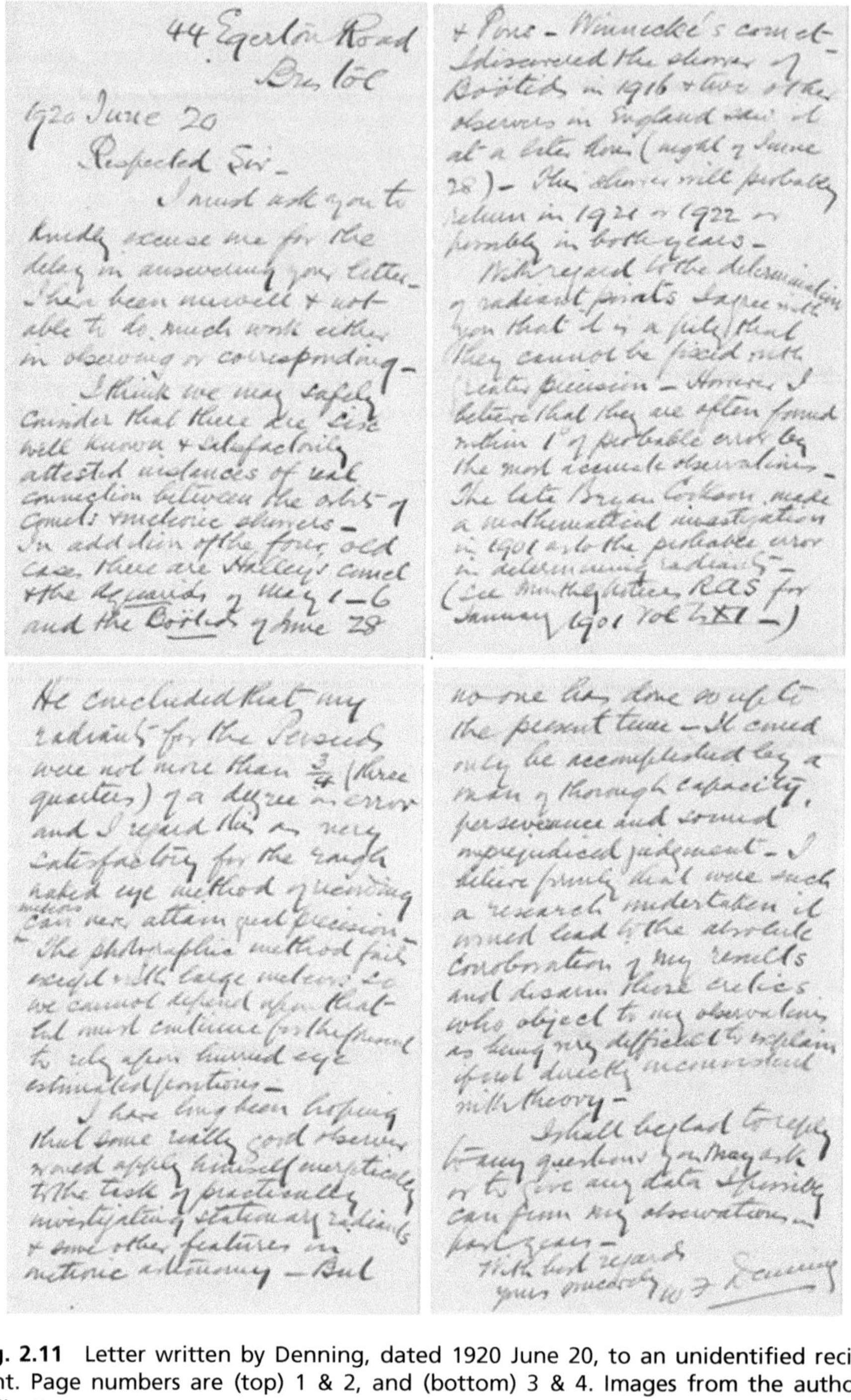

Fig. 2.11 Letter written by Denning, dated 1920 June 20, to an unidentified recipient. Page numbers are (top) 1 & 2, and (bottom) 3 & 4. Images from the authors collection

primarily means Olivier, and it is clear that rather than seek mutual ground, and exchange ideas amicably, Denning had dug-in his heals. Indeed, in Denning's eye's Olivier, and all the other researchers that were critical of stationary radiants and his reduction methods, were committing a personal attack, and casting aspersions on his ability to observe. To Denning, his life-long contributions to meteor science and amateur astronomy, were being called into question. Rather, and in spite of any good intentions, offence had been taken, and Denning felt betrayed. For all this, while Denning might have felt his observational skill was being called in to question, this was not actually the case. It was his reduction methodology that was being questioned, not his actual observations.

Denning's letter of 20 June 1920 is additionally interesting since it indicates a sense of frustration at finding what might be called his successor. Certainly, in the 1920s there were many good observers energetically involved in meteor astronomy. Indeed, moving into the 1920's the rising star of British meteor astronomy was John Phillip Manning Prentice (see Chap. 5). Destined to take over the Directorship of the BAA's Meteor Section in 1923, Prentice had at one time (at least partially) supported Denning's claims concerning stationary radiants. Prentice, however, accepted the reduction methods championed by Olivier, and called for radiants to be derived on the basis of only those meteors observed on a single specific night. By adopting this methodology, Prentice rejected the past practice of pooling data over large intervals of time, and as a result found that stationary radiants simply disappeared from the sky. Writing in the exterior sense of self, Prentice wrote in 1942 that, "Prentice, although he learned his meteor observing from Denning and Miss Cook, was influenced more by the writings of Dr. C. P. Olivier in the U.S.A. and it is to the latter's championship of sound principle that the introduction of reliable methods in the *reduction* of the Section's [the Meteor Section of the BAA] work is ultimately due" (my italics) [20].

Not only did Prentice reject the pooling methods advocated by Denning and earlier researchers, so to did Alice Grace Cook. Indeed, writing in the Report of the BAA Meteor Section for 1922, Cook explains that with respect to determining radiants, "In the past, criticism has been directed to the practice of combining paths seen over a large interval of time to form a single radiant. Care has been taken to avoid this". This being said, Cook additionally commented that with respect to the radiant catalog being presented, "the plan of grouping radiants seen at widely different dates under the heading of the same shower has been followed"—this statement being further qualified as a "convenience of reference". In this manner, while allowing for the notion that stationary radiants can be objected to, "on theoretical grounds", Cook is not fully ruling-out the possibility of their existence on a sky map.[6]

From the early 1920s onwards meteor astronomers began to shift away from the reduction methodologies of early pioneers. Denning, now in his 70's, was no longer the dominant force that he had once been, and, by becoming increasingly entrenched in his beliefs, his influence on the outlook of other, younger, observers began to wain. For all this, Denning's very last publication, a letter written to the *Observatory* on 14 January, 1931 (just 5 months before his death), was both defiant and dictatorial. He claimed, for example, to have "obtained conclusive evidence" that the October Orionids formed part of a stationary radiant that was active throughout November and December. To his detractors, Denning wrote, "let every observer make his observations and derivations as accurate as he possibly can, and then, assured of his unassailable position, fearlessly uphold them". Once again, however, Denning was seemingly missing the point—the issue was not related to his observational skills, but rather with his data reduction procedure. What Denning was unable to accept was that stationary radiants were, at best, the appearance of distinct meteor showers that chanced to have the same radiant point, combined with false-positive associations derived from the predominant number of meteors coming from sporadic background.

The stationary radiant debate, as such, ended with Denning's death, and it seems fare to say that he was the final hold-out. Indeed, by 1931, essentially everyone but Denning had abandoned the concept as illusory, and a spurious result of the data reduction methods being applied.

2.1.1   The Epsilon Arietids: A Case Study

Denning identified a radiant in the region close to the star epsilon Arietid as early as 1886, and indicated in his 1899 *General Catalog* that Weiss had recorded a similarly located radiant as early as 1869. Furthermore, Denning also notes in his *General Catalog* that the epsilon Arietid radiant (radiant XXXVII) is active, "both in August and October. Does not appear to have been recognized in September". The meteors are described as being bright, swift leaving streaks and occasional trains. In the time interval from late July to the end of November, during the years between 1869 and 1898, Denning lists some 19 instances in his *General Catalog* when various observers had identified meteors from the epsilon Arietid radiant, although he also noted

---

[6] Cook's softening of comments concerning reduction methodology are probably based on her loyalty and long friendship with Denning.

that, "in the autumn months a large number of minor radiants are clustered in the region of Aries and Taurus, and in some cases their centers lie so near together that it is difficult to distinguish them individually". To this he adds, "a thorough review of these systems, based on fresh and numerous observations of precision, would afford results of considerable interest".

Denning presented a detailed review of the "Showers of Meteors from near ε Arietis" in the June 1919 issue of the *Observatory* magazine. In this article he identifies some 39 instances of the radiant being identified, and active from mid-July through to late January. The radiant being determined by numerous observers, including E. Weiss, R. P. Greg, A. S. Herschel, H. Corder, and D. Booth. In addition, Denning notes that, "apparently there is no cometary orbit with suggestive resemblance to this meteoric stream, but it may have been, and almost undoubtedly was, derived from a periodical comet". This, perhaps, was the most unfortunate statement Denning made, and an absolute gift to Olivier.

Denning joined forces with Alice Grace Cook and J. P. M. Prentice in 1921 to perform an extensive study the epsilon Arietid radiant. The results of this combined effort were presented in the March, 1922 issue of the *Monthly Notices*, with the radiant being identified on 25 occasions between 10 July and 31 December. Cook as the lead author of the study writes that the determinations made, "afford striking conformation of the stationary character of some meteor showers, which was announced by Mr. Denning many years ago". Cook also drew attention to the fact that the radiants were deduced from, "a good number of paths each night" and not through combined paths, "recorded over a long period". It was a defiant claim, but not one destined to stand. Indeed, Olivier, in his reply published in the December 1922 issue of the *Monthly Notices*, did not question the manner in which the radiants had been derived, other than noting that no apparent corrections had been made for the altitude of the radiant (the so-called zenith correction) during the course of the year. This effect is generally small, but as Olivier noted, not irrelevant. While unable to critique the radiant reduction process further, Olivier changed tack and questioned the reproducibility of the results. That is, he used the American Meteor Society data (now amounting to some 31,000 meteors recorded over 20 years) and commented that there was no trace of any radiant being found near the star ε Arietis at any time of the year. While not fully rejecting the data presented, Olivier did cast considerable doubt upon its veracity, and argued that proof of the radiant's existence (at any time of the year) was still wanting. This was presumably a directed dig at Denning who had published numerous accounts of other observers reporting a radiant close to ε Arietis.

While Olivier cast doubt upon the observational data supporting a radiant near ε Arietis, by far the most damming part of his article in the December 1922 *Monthly Notices*, concerned Denning's comments about the shower being derived from a single periodic comet. In effect, Olivier replicated the analysis of H. H. Turner and Henry Pickering, and asked, if a stream is to produce a fixed radiant point close to ε Arietis during the course of one year, then what would its orbital characteristics have to be. The results of Olivier's calculations are shown in Fig. 2.12. The orbital characteristics plotted are those indicating the perihelion distance, the orbital inclination, and the orbit orientation angles (see inset diagram to upper right). For a typical meteoroid-stream-producing comet the orbital parameters are not going to change appreciably during the course of one year, or even over one orbital period (there are exceptions of course). In other words, the orbital properties of a meteoroid stream required to produce a stationary radiant near ε Arietis are decidedly none comet-like. One specifically damming requirement is that the stream inclination would change from being direct to retrograde in June (after day 163) and back again in October (after day 286). At the same time that the inclination 'switched', the perihelion distance also approached zero, and accordingly any stream meteoroids would be either destroyed in the Sun's atmosphere or undergo a gravitational perturbation that would destroy stream structure.

Olivier's reasons for highlighting the bizarre (indeed, monstrous) orbital requirements needed to produce a stationary radiant close to ε Arietis are straightforward, and he notes:

> one of the principle reasons for publishing this figure is to bring to the attention of the observers themselves just what their results, if we accept then as given, would force us to believe to be the structure of the resulting meteoric system

Olivier's argument, of course, would fall on deaf ears as far as Denning was concerned, but now Olivier was really addressing other observers. Likewise, his arguments may not have convinced Alice Cook, but it certainly swayed J. P. M. Prentice, and for that matter, they convinced the entire corpus of professional astronomers. Indeed, while there is the scientific imperative to let the observations be the arbiter over theoretical prediction, there are also limits to what the observer can expect the theoretician to believe and then account for. What Olivier was pointing out, and what virtually all other's by 1920 were ready to accept, or had earlier realized, was that something must be wrong, not with the observations per se, but with the data reduction process. While Olivier's December 1922 article in the *Monthly Notices* effectively ended the

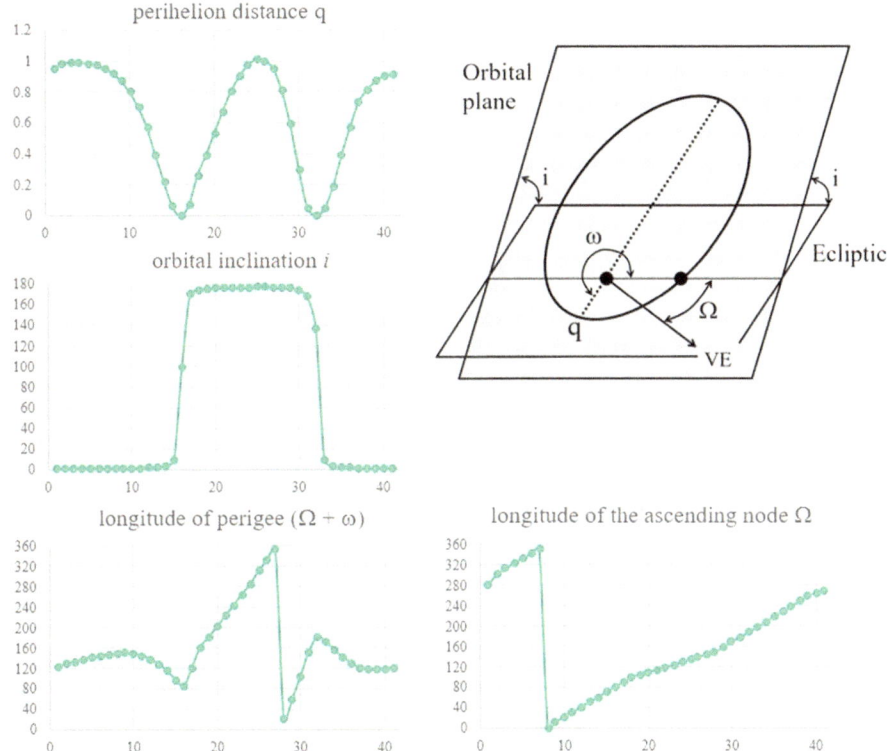

Fig. 2.12 Time variation of the orbital parameters for meteors emerging from the Epsilon Arietid radiant. Data points are calculated at either 5- or 10-day intervals on 41 occasions in the time interval from January 1.5 (first data point) to December 16.5 (last data point). Data points 10, 20 and 30 correspond to April 11.5, July 11.5, and September 11.5 respectively. The data is taken from Olivier's article published in the December 1922 issue of the *Monthly Notices*

stationary radiant debate within the pages of the professional scientific literature, the publication of his book *Meteors* in 1925, essentially wrote its public epitaph. Denning, of course, never let go of the idea, but by this time he was an embittered and embattled player lost in a decisive end game and rout.

While Olivier used AMS observations to question the existence of any shower radiant close to the star ε Arietis, it is perhaps a little ironic that there is a recognized May-time Epsilon Arietid meteor shower—indeed, it is one of the strongest meteor showers of the year. The irony being that it is a daytime meteor shower. The discovery was made with radar equipment at Jodrell Bank (see Chap. 5) in 1947. For all this, there are also low-activity night-time showers with radiant points located within the constellation of Arietis at various times during the year. These include the December delta-Arietids, the November nu-Arietids, the October delta-Arietids, the October sigma-Arietids, and the August zeta-Arietids. Denning's claim of Arietid shower activity in August and October, through to December appears, therefore, to be generally sound, but, of course, it is now recognized that multiple radiants (that is individual showers) are involved.

2.2   The Observer at Work

Having discussed the main achievements, and problems associated with Denning's work on meteor radiants, it is worth saying a few words about where Denning was observing from. We should also discuss his observing methods and philosophy, since how an astronomer observes is just as important as what it is an astronomer sees. It is clear from his many writings that Denning was a man of method. Indeed, he held this principle highly. Writing in his *Telescopic Work for Starlight Evenings* Denning commented, "if the 100 hours of exceptionally good seeing [in England] are to be profitably employed, we must be continually prepared with a scheme of systematic work".[7] Denning's far-ranging systematic program of work will be more fully discussed in Chap. 3, but for the moment, some insight into how Denning collected his meteor-related observations can be gleaned from his meteor catalogue published in 1890. There he explained [21],

> my plan of working may be briefly described as follows: - All the observations were made in the open air and from the garden adjoining the house. Attention was almost invariably given to the eastern sky. In mild weather I sat in a chair with the back inclined at a suitable angle; but on cold, frosty nights I found it expedient to maintain a standing posture, and sometimes to pace to and fro, always, however, keeping the eyes directed towards the firmament in quest of meteors.

Working from an urban, Bristol-back-street garden was probably not as problematic in Denning's time as it would be today. Indeed, Denning often reported seeing faint aurora from Bristol [22]—light pollution levels having now all but made such city-based detection impossible. For all this, Denning's back garden was not an ideal observing site, and one can find occasional references to this fact in his observing journals. He noted on one night, for example, "Depiction bad through smoke from adjoining chimneys", on other

---

[7] Denning's estimate that only 100 h of good seeing is available to British observers follows the earlier remarks on the same subject by Sir William Herschel.

nights he complained about obscuring trees and buildings being in the way of his line of sight. In spite of such drawbacks Denning recounted in one interview [6] that, "I have sometimes watched from my garden for meteors for ten or eleven hours continuously… The hours spent in this way have been intensely enjoyable. Amidst the trees and shrubs in the garden, solitary and with no sheltering canopy but the sky above, I have rarely experienced a feeling of loneliness or nervousness, or have had to make an effort to continue work". As we shall discuss later (Chap. 3), Denning not only spent many nighttime hours observing from his garden, he also spent many daytime hours studying its fauna and flora, and in particular its insect life.

2.2.1 The Observations of Shooting Stars Notebook

Only one of Denning's meteor observations notebooks has survived to the modern era,[8] and this covers the time interval from 1922 to 1924, at this time Denning was in his mid-70s. The journal is a standard-sized notebook and contains snippets of his observing and working practices. Furthermore, the notebook contains the raw data from a number of nighttime observing sessions, as well as newspaper clippings, and hand-copies of letters relating to multiply-observed fireballs. With respect to the meteor observations, Denning was recording data in a standardized and consistent format, providing:

> Month, day, time of event,
> > Magnitude,
> > RA and Dec of starting position,
> > RA and Dec of ending position,
> > Comments.

The right ascension (RA) and declination (Dec.) of the starting and end points of the meteor on the sky would be used to determine radiant positions. Occasionally there is a small diagram indicating how a meteor appeared to move against a set of background stars. Under 'comments' Denning included such information as shower association (e.g., Perseid, or Geminid, etc.), or he recorded the meteor's appearance, indicating whether it was swift, or slow, or if it left a brief train, or had a distinctive color. Counter to modern practice Denning made no attempt to judge the limiting visual magnitude during an observing session, although he did record information such as Moon phase, or

---

[8] My thanks are extended to George Spalding for the loan of this notebook.

clouds, and sometimes he just noted, "a splendid sky". Occasionally he would also add an after-the-event note that some specific meteor had been additionally recorded by another observer. Among the well-known names that can be found in such cases are J. P. M. Prentice, Alice Cook, and William Lockyer (son of J. N. Lockyer).

A study of the entries contained within the observing notebook reveals that 223 meteor trails were observed on a total of 98 nights, with most of the observations being made in the late summer and autumnal months (Fig. 2.13). Observations are particularly numerous around the times of the Perseid and Orionid meteor showers. It is not entirely clear how many hours were spent observing in the time interval covered by the notebook data, but a count based upon those nights when multiple meteors were recorded, we find a minimum of 12½ h observing time in 1922 and 1923, and 13½ h observing in 1924. Based upon these multiple-recorded-meteor sessions a typically observing session lasted for about 2 h, but often 3 and 4 h are indicated. Assuming that a typical observing session lasted 2 h, then Denning was observing for something-like 86 h in 1924, 70 h in 1923 and 40 h (between mid-August and 31 December) in 1922. For a comparison with other leading observers, in 1921, the 44-year-old Alice Grace Cook recorded some 867 meteors over 295-h of observing time, while the 19-year-old, J. P. M. Prentice recorded some 2079 meteors over 349 h of observing time. In terms of hours observing, Denning's yearly totals, while impressive for a septuagenarian, are understandably much lower than the totals maintained in his youth—recall, for example, in 1876, when 28 years of age, he completed some 49 h of observing in just 46 days. Age, of course, does not determine enthusiasm, but it can certainly limit one's ability to observer for long hours on end. Indeed, recall Denning's letter of 25 November 1919 to Christine Gravely (Chap. 1) in which he reminisced about his ability to observe, some 50-years past, when he was in his 20s.

Interspersed between the accounts relating to his meteor observing sessions, the notebook contains many newspaper clippings on the sighting of bright fireballs. One such event is that concerning a very bright fireball seen on Tuesday 11 November, 1924. This section of the notebook contains newspaper clippings, hand copies of letters that he had been loaned, letters he had been sent and one telegram that he had received. Several pages also include note summaries of other letters that he had received (but have not survived) on the topic. The story of this fireball begins with a clipping from the *Daily Mirror* newspaper for 13 November, the clipping being just 5 lines long, but carrying the enticing title of "Fireball Blaze". The notice informs the paper's readers that, "a fireball passed over Belfast early yesterday

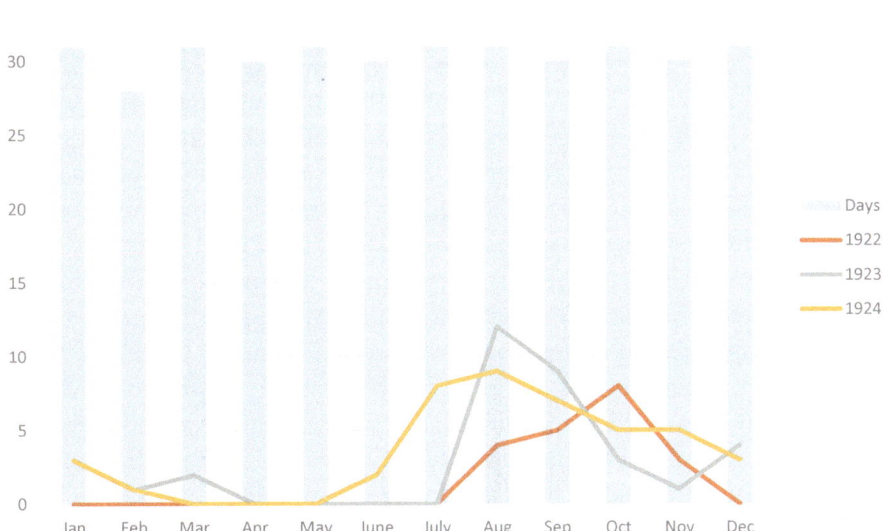

Fig. 2.13 Observing Notebook data: 1922–1924. Vertical bars indicate the number of days in each month, while the colored lines indicate the number of days upon which observations were made in the years 1922, 1923, and 1924

morning and fell into the lough. The meteor was followed by a brilliant train of sparks, and finally exploded". As with most newspaper accounts of fireball sightings, many of the details were wrong. The fireball was seen to the north of Belfast (not over), it did not fall into to the Belfast Lough, and it occurred at 5:40 in the early evening. Another clipping on the same page of the notebook, titled "fireball or Meteor" is from a letter to the *Northern Whig* newspaper by "R.M". In this case we learn that the event occurred in the early evening, and was composed of, "a very bright light white with a long reddish tail which passed across the sky …. slowly … at a height of a few hundred feet". Indeed, "R.M" thought at first that the light was from an aeroplane. The estimated height of the light is entirely wrong, but not an uncommon mistake made (even to this day) by unexpecting observers. By November 15th interest in the event was heating-up, and Denning has cut-out three newspaper clippings from the *Newcastle Journal*, The *Newcastle Chronicle* and the *Daily Mail*. These accounts are more informed and indicate that many hundreds of people had witnessed the event from across northern England and Northern Ireland. The clipping from the *Newcastle Journal* indicates that William Redfern Kelly (one of the founding members of the BAA, and FRAS), who lived in Belfast, "states that if the meteor had struck the

earth the result would have been serious". While this latter claim of devasta-
tion is unlikely, we do learn from the article that the fireball travelled from
east to west, starting close to then rising full Moon, and taking some 20–30 s
to cross the sky, ending below the 'handle' of The Plough (Ursa Major) near
the star Cor Caroli (in Canis Vanatici). The speed of the fireball is remarkably slow, and is indeed suggestive of the fall of a meteorite. The question,
however, is where might the meteorite have landed. The clipping from the
Newcastle Chronicle concerns a letter from F. Sargent of the University Observatory at Durham. This letter calls for eyewitness information on the fireball.
Following these clippings are several pages of hand copied letters and additional newspaper accounts that have been sent to Denning by the Reverend
William Ellison. Ellison was Director of the Armagh Observatory (1819–
1836) and his covering letter indicates that his son had seen the fireball, as
it moved between the stars Capella to Cor Caroli, and that the fireball burst
into fragments at the end of its flight. In his covering letter, Ellison indicates
that he wants to include a discussion of the even in his Astronomical Notes
column in the *Belfast Telegraph*, and would like Denning's opinions on what
had been seen. Denning evidently contacted Ellison, and a clipping from the
2 December issue of the *Belfast Telegraph* is subsequently pasted into the notebook. Following the letter from Ellison the notebook contains several pages
of hand written accounts (from observers located in Ireland). At this stage
Denning must have received a good number of additional letters on the event,
and on one page, along with short clippings from The *Times* of London and
the *Daily Mail*, both for November 21st, he has written out a column of
24 duration estimates, finding an average of some 30 s. The clipping from
the *Daily Mail* indicates that it has received details from Denning and that,
"the fireball had a flight extending over some 490 miles. It was overtaking
the earth, and effected its lengthy course at an apparent speed of 15 miles per
hour. The notebook does not contain any mathematical calculations, and so
the determination of the path length must have been made elsewhere. A few
pages further on in the notebook, a more detailed and revised summary of
results has been penned-in, revealing that the fireball had beginning and end
heights of 55 and 47 miles respectively (indicating a near horizontal atmospheric path), had a trail length of 510 miles, lasted some 33 s and travelled
at some 16 miles per second. Furthermore, the fireball first appeared some
80 miles west of Hartlepool and ended over the Atlantic Ocean at latitude
11.75 W and longitude 56.5 N. Again, no indication of how these numbers
might have been calculated is given in the notebook pages. Two pages over
from the summary table is a clipping from the *Northern Whig* newspaper
for November 27th in which Denning presents a summary of events "to the

Editor". In this column Denning provides a radiant location (not seen in the notebook pages) close to Mira Ceti, and includes a statement concerning the length of fireball's trajectory being some 140° on the sky. With respect to the latter number, we additionally learn from the *Northern Whig* article a snippet of information about Denning's obsessive record keeping: "the writer has been observing meteors since 1866, and has seen about 28,160 meteors, excluding those recorded in the great showers and has never recorded more than two or three meteors with a flight exceeding 120 degrees". Accompanying the *Northern Whig* clipping is another column from the *Times* of London. Headed, "The Recent Display - from a correspondent", this clipping is accompanied by the hand written comment: "WFD <u>Times</u> November". Ostensibly this column is the same as that provided to the *Northern Whig*, although it is a little shorter in length and more matter-of-fact in tone. The final detail in the notebook concerning the November 11 fireball is a diagram, showing the fireballs path inked-in upon a page cut from a pocket atlas. While he clearly spent much time in investigating and making sense of the observational data pertaining to the fireball of November 11th, Denning did not produce a formal publication of his results. A short summary of Denning's *Northern Whig* and *Times* articles was presented in the "Our Astronomical Column" of *Nature* for 6 December 1924, but no other accounts seem to have been published.

2.2.2 Fireballs Galore

In addition to the 11 November 1924 fireball event, Denning's observing notebook contains newspaper clippings, and notes on bright fireballs seen on September 7, 1923, November 2nd, 1924, November 19, 1924, and December 13, 1924. Indeed, it is generally clear that by the early 1920s Denning was both a conduit for receiving information, and the go-to person for explanatory requests from newspaper Editors. Furthermore, it also appears that Denning was a dab-hand at sleuthing-out further details on fireball events. A telling example of this latter skill, and an example of Denning's (international) status as an authority on meteors, concerns a spectacular fireball procession witnessed from across southern Canada, and regions beyond, on 9 February 1913 [23, 24]. The details of this event, often called the Chant Procession, were assembled by Clarence A. Chant (University of Toronto) and published in a number of detailed articles that appeared in the *Journal of the Royal Astronomical Society of Canada* (JRASC). The RASC evolved in 1903 from the Astronomical and Physical Society of Toronto (APST), which in turn had formed on 25 February 1890. The constitution of the

APST called for the election of honorary and corresponding fellows, and Denning had been approached and then elected a corresponding fellow on 16 November 1891, the society having received a "very cordial and gratifying" letter of acceptance. At the time of his election, Denning was 43 years of age, and an established figure among amateur astronomers in England, and it was presumably felt that his association as a correspondent would reflect well of the fledgling Canadian society. It is evident that Denning took his election seriously, and between 1891 and 1928 he published a total of 25 articles in the Society's journal (JRASC). Of these, 17 were concerned with meteor astronomy, 2 with comets, 4 with planetary observations, 1 on newly discovered nebulae, and 1 on observing techniques. Of his meteor related articles, 4 were dedicated to the Chant Procession. The procession was a marvelous parade of numerous meteors which traveled slowly across the sky. The first reports of this spectacular event were obtained from observers in Moose Jaw, Saskatchewan, and the trajectory took the meteors across Manitoba, over Toronto in Ontario, on through New York state, and across to the island of Bermuda. Initial investigations made the ground path of the procession to be at least 4000-km, and many hundreds of eye-witness accounts were collected. Realizing that the end location of the display was only poorly constrained, Denning placed a letter in the *Nautical Magazine* for 4 April 1914, asking for ship borne observations. A year later he received a letter from a Mr. E. Y. Porter, who saw the display from the bridge of the *S.S. Bellucia*[9] when located some 1000-km east of Montserrat, and a further 2000-km south of Bermuda. This ground track length is remarkable enough, but subsequent investigations, made nearly one hundred years after the event, in 2012, of archived ships logs by Donald Olson and Steve Hutcheon discovered records of the fireballs being seen off the coast of Brazil (Fig. 2.14). The estimated ground track length for this spectacular event is currently thought to be of order 15,000-km, and it has been suggested in a research publication by the author and graduate student Mark Comte that what was actually seen was a rare Earth-satellite-capture event involving a multi-meter-sized asteroid [24].

The manner in which Denning recorded his meteor data was straightforward, with his only observing aid being a straight stick, or wand (recall Fig. 1.1). Held at arms length he used the wand to 'fix' the meteor's path upon the sky. Then, by making a mental note of the star fields through which the meteor had passed he marked its corresponding trail on an 18-inch celestial globe. The time, magnitude, appearances, and position of the meteor where then recorded. Again, from Denning's 1890 radiant catalogue [21] we learn,

---

[9] The *Bellucia* was torpedoed and sunk during a U-boat attack off the Cornish coast of England on 1 July 1917.

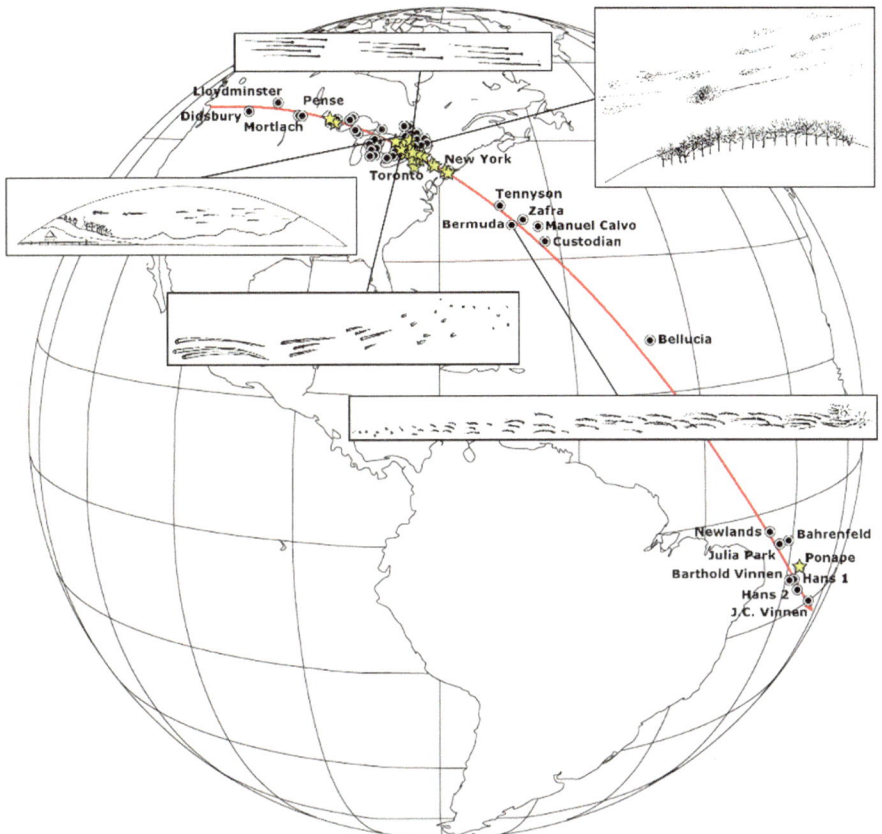

Fig. 2.14 The estimated ground track (red arc) of the 1913 Chant Processions [24]. The path begins over Saskatchewan in western Canada, and proceeds over the Eastern seaboard of the United States, over Bermuda and on past the western most point of Brazil. Inset images show the appearance of the Procession as reported from various locations in Ontario and Bermuda

"at the end of each period of observation I finally discussed the materials collected and deduced the radiants. In some instances, a very definite little shower would be manifest from a single night's work, but I generally found it advisable to combine the paths recorded on several dates in order to obtain satisfactory positions" [25].

With regard to his reduction procedures Denning fell into a situation against which he had warned in his *Telescopic Work for Starlight Evenings*. Specifically, in Chap. 4, Denning remarked that, "a person who relies upon guidance from prior experimentalists will probably make rapid headway… The want of this foreknowledge has often been the main cause of failure, and it has sometimes led to misconceptions and imaginary discoveries" [26]. To this he later added, "let every observer judge for himself to a certain extent and let him follow original plans whenever he regards them as feasible. Let him test preceding results whenever he doubts their accuracy… An observer should take the direction of his labours from previous workers, but be prepared to diverge from acknowledged rules should he feel justified in doing so from his new experiences". Denning did not strongly question the radiant reduction procedure, as discussed earlier, because, as far as he was concerned, it was correctly placing meteors into one radiant or another [27], and because it was the methodology developed by his initial guides and teachers.

2.3   Correspondence

The initial stages of Denning's meteor observing career were nurtured by Alexander Stewart Herschel (1836–1907). Herschel (Fig. 2.15) was one of Britain's foremost authorities on meteor astronomy during the later half of the nineteenth century, and in particular he was a prominent member of the Luminous Meteors Committee which reported to the British Association for the Advancement of Science (see Chap. 5). Alexander was the second son of 12 children born to John and Margaret (née Stewart) Herschel, and ultimately a third-generation astronomer in the William Herschel line. Home schooled for much of his youth he was eventually enrolled at Clapham Grammar School in London, then under the headmastership of the Reverend Charles Pritchard (recall Chap. 1, and see Chap. 4). In 1855 he proceeded to Trinity college, Cambridge and subsequently graduated (as 20th wrangler[10]) in 1859. From 1861 to 1865 he studied meteorology at the Royal School of Mines in London, were he developed his interest in meteors, and specifically meteor spectroscopy (see Chap. 4). In 1866 Herschel was appointed a lecturer at Anderson's College of Glasgow (now Strathclyde University) and worked with the Astronomer Royal of Scotland, Charles Smyth, on developing spectroscopic techniques. It was also in 1866 that Herschel determined an accurate radiant position for the Leonid meteor shower (in that year the shower undergoing a spectacular storm), and indeed, it was Herschel's results that enabled Giovanni Schiaparelli to deduce an orbital identity between the Leonid meteoroid stream and comet 55P/Temple-Tuttle. Herschel also correctly deduced that the Andromedid meteor shower for 1872 might undergo a spectacular outburst of activity—this being precipitated by the break-up of the stream's parent comet, comet 3D/Biela, circa 1844. Herschel

---

[10] A wrangler is a student who attains first class honours in their final year mathematics examinations. The highest scoring student is the senior or first wrangler.

resigned his professorship at Glasgow in 1871, and moved to the University of Durham as professor of physics and experimental philosophy. Between 1874 and 1881 Herschel, working principally with Robert P. Greg, collated vast volumes of data on meteor showers and fireballs for the Luminous Meteors Committee (see Chap. 4). In 1878 Herschel published a landmark paper in the *Monthly Notices* which tabulated some 71 theoretical radiant points for meteors derived from known comets. In this paper he revealed the association between the May-time η Aquarid meteor shower and comet 1P/Halley. Herschel resigned his professorship at Durham in 1886 and shortly thereafter moved to Observatory House (his grandfather's home) in Slough. From his family home Herschel continued observing meteors, concentrating on the determination of meteor heights, and he additionally directed his interests towards photography, and the Society of Arts. He was elected FRAS in 1867, and FRS in 1884.

Denning's first contact with Herschel was prompted by the appearance of a bright fireball seen on the night of November 6th, 1869. At that time Denning was 21 years old, and Herschel 43. A steady correspondence appears to have developed between the two observers, and the surviving

Fig. 2.15 A. S. Herschel (center) with brothers John (left) and William (right). Image taken circa 1870. Image courtesy Strathclyde University Archive

letters clearly indicate that an extensive dialogue on all matters relating to meteor astronomy took place [28]. In the early letters, Herschel was the more experienced observer, and through his office with the Luminous Meteors Committee he encouraged Denning to submit his observations to the yearly reports. Denning acknowledged his great gratitude to Herschel in 1907 when writing his obituary account for the journal *Nature*. Indeed, Denning commented [29] "The writer of this notice will always have reason to be grateful to him [Herschel] for kind encouragement, advice, and instruction in the early years of his observing career".

In total some 230 letters from Herschel to Denning have survived. Written between 1871 and 1900 these letters largely relate to specific meteor observations, but they occasionally reveal a few personal anecdotes.[11] Indeed, the letters were rather formal, and it is amusing from a modern perspective to note that in spite of their long exchange of letters it was to be 28 years before Herschel changed his opening address from "Dear Sir" to "Dear Mr. Denning"—the formality, of course, was a sign of respect rather than one of superiority. There is no recorded evidence to indicate that Herschel and Denning ever met face to face. For all this, in an early letter from Herschel, dated 28 December, 1872, he remarks, "wishing you success in your candidature for the R[oyal] Astronomical Society".

The bulk of the early correspondence between Herschel and Denning deals with radiant reductions, and the exchange of fireball information (days, times seen, and where in the sky—recall Fig. 1.4). The latter information occasionally produced a doubly observed object for which heights could be estimated (see Chap. 5). Interestingly, much of the correspondence between Herschel and Denning focused on information gleaned from newspaper columns. With respect to a bright fireball seen on 14 December, 1890, for example, Herschel informs Denning that:

> I searched all the Daily newspapers for Monday - Wednesday December 15-17 (and some also on Thursday December 18th) that I could find in a City News Room on Thursday last week, and at a literary Club in London, with only moderate success for journal notes of the occurrence. But there were two or three such in some London Daily Newspapers (the Daily Chronicle, Daily News, Evening Standard, and of course Nature) … along with a few more equally vague and scattered notes on the occurrence too, from the Provincial

---

[11] Herschel must have been a keen cyclist, for example, since we learn from one letter dated August 30th 1898, addressed from the Angel Hotel in Ludlow, that Herschel is on a bicycle tour of Herefordshire. Earlier that same year, however, in a letter from June 24th we learn that Herschel had been "floored by sever rheumatism" after a bicycle tour of Oxfordshire. Other letters additionally mention bouts of poor health interfering with observational work.

Weekly Newspapers (of Saturday December 20th) which the same City News Room (172 Fleet Street) enabled me to look through in good abundance.

The great network of newspapers that existed across England in the late nineteenth century, were, in many ways, the equivalent to the internet news services and social media networks of today. From the scattered reports that Herschel had been able to find, however, he, Denning, David Booth and G. L. Tupman attempted to determine the properties of the fireball's atmospheric path. Indeed, it was felt that a meteorite might well have landed near Tilbury, and a letter dated 12 January 1891 indicates that Herschel had made an expedition to search-out possible material:

> I set out for the scene of descent of the fireball on Saturday last and arrived at Tilbury (ferry to Gravesend) that evening to consult the ferry Captains - it had been so foggy on Sunday night 14th that no ferry steamers crossed in the evening hours; and all along the riverside from there to 'grays', where I stopped for the night, the Tilbury dock electric lights were invisible on Dec. 14th 100 yards off; and of course nothing was to be learned there beyond the brilliance of the general illumination and the loudness of the following thunder! … Next morning (Sunday) I walked from 'greys' to Billericay through a vile white Thames fog and fierce frost the whole way …I walked on (eastward) from Billericay to Southern Road, picking up a number more of good measures of position. One or more stones fell from the meteor somewhere between Brent Wood and Southend I don't in the least doubt.

No meteoritic material was found (then or since), but this letter provides a good example of the extraordinary lengths to which Herschel was prepared to go with respect to the investigation of potential meteorite falls. This was not, in fact, Herschel's first investigation into a potential meteorite fall, since he had been a central figure in the recovery of the Middlesborough meteorite[12] that fell of 14 March 1881.

As their correspondence and friendship developed Denning and Herschel began to help each other in directions beyond those of information exchange. Several letters from 1889 and 1898, for example, indicate that Herschel had provided Denning with translations from the German astronomical journal *Astronomich Nachrichten*. Likewise in 1889 Herschel thanks Denning for dealing with a request for information from "General Tennant" (this was James Francis Tennant, President of the RAS: 1890–91). Other letters indicate an exchange of books, letters received from other observers, pamphlets,

---

[12] This is one of the most pleasing of meteorites to look upon, being beautifully sculpted into a conical profile. Herschel described it as being "like an upper oyster shell".

and photographs. Indeed, Herschel's father, John, along with William Fox Talbot, was one of the pioneers of photography during the later part of the nineteenth century. Interestingly, (A. S) Herschel used his photographic skills to make literal "photo-copies" of Denning's *General Catalog of Meteor Radiants*. In a letter dated September 26th, 1897 Herschel expounds on the application of photography to meteor science

> I have also taken a photo-copy of the enclosed meteor list - I dare say this use of photography would do more for hunting out meteor heights and radiant points for years to come yet, than any attempts at photographing sky-views. Photography will hardly steel a march on eye observations, as it can only depict very bright meteors about whose identity eye-observations would scarcely get into any perplexity

Herschel's comments reveal a remarkably lack of foresight with respect to future improvements in photography, and indeed, within a few years of his statement, photography was beginning to yield important results not only in astronomy (see Chap. 5) but many other scientific fields.

Denning, along with his many collaborators (predominantly Herschel, but also in later years, J. P. M. Prentice, Alice Grace Cooke and Fiammetta Wilson) were interested in the determination of meteor heights. There were several reasons for this interest. Firstly, a knowledge of a meteor's beginning and end heights offers important information about the meteoroid ablation process, and second the true path of a meteor can be used to estimate the initial velocity with which it entered the Earth's atmosphere. To determine a meteor's velocity, one needs to know not only the atmospheric path length but also the time it took to traverse the path. Such (human reflex) time estimates are very difficult to make in any accurate and/or consistent manner. Likewise, even simply keeping track of time on a dark night is problematic. Herschel explained to Denning in one letter dated January 10th 1894 that he was, "keeping note of the time by counting alphabets between the half-hour strokes of a neighbouring town-clock, meteors being very scarce, and my hand-lamp and watch having both struck work on the frosty night". In another letter dated December 23rd, 1894, it is evident that Herschel had sent Denning a stop-watch. Indeed, Denning had apparently enquired if more watches were available for other observers. Herschel writes that the stop-watches he had obtained were military surplus (used during the Franko-Prussian War: 1870–71) and obtained by, "Mr. Reid, watch and chronometer maker, Grey Street, Newcastle on Tyne. The price was £2 [each]". This is an interesting example of equipment being shared between members of the meteor observing community, but it would seem that Denning never used his stop-watch for timing meteors (at least in terms of published accounts). Likewise, Herschel appears to have found stop-watch timing impractical, and writing to Denning on December 8th, 1889 he explained an alphabet counting method:

> I am sure enough of my durations, which are really the easiest of all the features of a shooting stars flight to note, because it sticks in the memory through all the puzzling that one goes through to fix the track correctly, and one can settle that, therefore (by the rate of 6 letters of the alphabet repeated clearly = 1 sec), at any time when the rest is decided on. But I always repeat 'abcde' etc instantly after a shooting star has appeared and the letter b, f, I, m or etc where one has to stop to reproduce the time of flight exactly never escapes ones memory afterwards.

For all his confidence in meteor durations, Herschel had little confidence in any deduced velocities. In a letter dated June 16th, 1889, Herschel explains that, "the determination of velocity and durations of flight, are still too inaccurate in all ordinary cases, I'm afraid, to allow us to suppose that we can trust much to any orbit extraction". Indeed, in another letter written on April 13th, 1900 Herschel indicates that he thinks his alphabet counting method provides an over estimate of a meteor's duration. This reflection appears to have been brought about after Herschel had read a "long *AN* paper [by] Von Niessl" who had considered the effects of systematic errors on determining meteor velocities. Indeed, systematic timing errors resulted in the general belief that the majority of meteoroids entered Earth's atmosphere with hyperbolic velocities—that is, with speeds in excess of a hundred kilometers per second. The apparent observation of such very high velocities was significant since it implied that the meteoroids must have an origin from outside of the solar system. It is now known (from more accurate timing data derived from radar and photographic studies—see Chap. 5) that virtually all meteors are produced from meteoroids moving along bound, elliptical orbits, with Earth encounter speeds less than a maximum of 71 km/s—something like 0.1% of all meteors observed might have an origin from interstellar space. It took meteor astronomers many years to understand the hyperbolic velocity problem, but since Denning was not a major player in this particular debate, we do not follow its course here.

Herschel was a strong supporter of Denning's supposed discovery of stationary radiants (recall Sect. 2.1), and it would also appear from their correspondence that Herschel acted to translate for Denning the meaning of some of the early technical (that is, mathematical) papers on the topic. The first discussion on matters relating to stationary radiants is found in two letters, dated April 26th and May 9th, 1892. Specifically in these letters Herschel discusses a research paper by W. H. S. Monck concerning the association between cometary orbits and meteoroid streams. Monck's stance was that there was no physical association between cometary and meteoroid stream orbits, and that any apparent coincidences were purely random. Herschel writes in his May 9th letter

> He blows hot and cold just as he pleases, about the question of stationary radiants, according as he wishes to strengthen or to refute the evidence for cometary accordance … we have yet to learn why stationarily enduring showers exist, and why comets' orbits and radiant points seem to be distinctly associated with them. … he constantly disowns his belief in stationary systems … I prefer, myself, to accept the idea of stationary meteor showers, as a proven feature of radiation.

It is apparent from these two letters in 1892 that Herschel not only supports the idea of stationary radiants, but that he is also prepared to entertain the idea that they have a cometary origin. This being said, as seen in Sect. 2.1, Herschel eventually adopted a model involving cosmic streams, and a primordial ring of meteoritic material in orbit about the Sun. Indeed, in a letter written on January 19th, 1900, Herschel explains that:

> Prof. von Niessl has a theory of his own about these stationary radiant points, depending on assumed hyperbolic meteor velocities and corresponding 'cosmical radiant points' of the meteor streams they are yet at asymptotically immense, or remote and free distances from the sun, which affords some wonderful verifications of stationary and fireball centres.

In this letter, Herschel is referencing Austrian astronomer Gustav Niessl von Mayendorf, who later rejected, at least partially, the idea of stationary radiants in a monograph on *Meteor Orbits in the Solar System* (published in 1907). The term partially rejected is used here since von Niessl argued that while shower meteoroids moved along orbits similar to those of comets, and were accordingly in orbit about the Sun, bright meteorite producing fireballs, in contrast, had an interstellar origin. Herschel returned to the theory of stationary radiants in yet another letter dated April 1st, 1900. In this missive, Herschel was critical of a recent article published by Russian astronomer Fiodor Bredikhine (at the Pulkovo Observatory). Indeed, Bredikhine had both rejected the idea of stationary radiants, and had criticized the perturbed-orbit idea presented by Herbert H. Turner (in 1889). Herschel writes,

Dr. Bredikhine's paper on stationary radiant points was a <u>most</u> absurd one, and Prof. Turner's snubbing of it was well-bestowed and well merited! Dr. Bredikhine is 'behind the age' in trying to disavow stationary points and makes such a mild and timorous attack on <u>them</u> that his paper rather damages the country view which it takes, than supports it very strongly

It is clear from Herschel's letters that he was a strong advocate for the existence of stationary radiants, and that he continuously encouraged Denning to continue his work in this area. Indeed, as argued earlier, when Herschel died in 1907, Denning lost not only a friend but a highly influential ally in the stationary radiant debate.

2.4 Meteor Physics

It was towards the end of the nineteenth century that physicists first began to question the physical mechanism responsible for the appearance of a meteor—that is the light producing phenomenon [30–32]. The first attempt at a detailed physical theory of meteoroid ablation was presented by Frederick Lindemann and Gordon Dobson [33, 34] in 1922 (see Chap. 5). Indeed, since the structure of the Earth's atmosphere was completely unknown in those regions where meteoroid ablation takes place, Lindemann and Dobson used two-station fireball data along with their theoretical calculations to infer the density of the atmosphere at altitudes of order 100-km. This was a new and powerful way of studying the upper atmosphere, and it marks the beginnings of an important transition in application. Increasingly, from the 1920s onwards, meteors were not just objects to study in and of their own right, but they became natural 'tools' through which the physical properties of the upper atmosphere could be probed.

Denning was never truly concerned with the physical details of meteoroid ablation theory, although he did discuss the topic with A. S. Herschel on several occasions. Writing on December 28th, 1872, for example, Herschel explained to Denning that there was, "no possibility of any bolide-looking meteor being of atmospheric origin" (see Chap. 5 for further discussion on this topic). Herschel was entirely correct in this assessment, and Denning employed the argument in one of his few, theory-motivated, articles. Writing in the American journal *Astronomy and Astro-physics* (formerly the *Sideral Messenger*) for May 1892, Denning takes to task an article written by Jesus Muǐoz Tébar published in the 19 November, 1891 issue of the *Publication of the Astronomical Society of the Pacific*. Tébar's essay promoted the idea that all meteors were, in fact, manifestations of ball lightning and that they had no connection to "cometary theory". Denning, in his commentary, laments that Tébar's ideas "carry us back to the times of our forefathers". Interestingly, however, Denning only asserts that, "it is highly probable" that meteors are "small planetary bodies undergoing combustion" in the atmosphere. In a less than confident manner Denning concludes, "many circumstances justify the commonly accepted view that meteors are planetary bodies entering out atmosphere from the outside".

While happily rejecting the idea that meteors were some atmospheric form of ball lightning, Denning was still prepared to entertain a great range of potential source regions for meteoroids. Indeed, in the February 1915 issue of the *Observatory* Denning argued that there are at least 6 source regions—these being:

1. Ejected from terrestrial volcanoes at an early stage in Earth's history
2. Ejected from lunar volcanoes
3. Thrown off the various planets of the solar system
4. Evolved from the Sun
5. Eruptions from stars
6. The "material of comets".

Certainly, all of the source regions mentioned had been entertained at various times before Denning was writing, and it was then far from clear that meteorites had an origin from within the main asteroid belt region between Mars and Jupiter. Accepting that meteoroids might have numerous source regions, Denning then turns the argument on its head, suggesting that, "it is probable that meteoric orbits display similar variety". Here, it would seem, Denning is trying to open-up the idea that meteoroids might not always be derived from comets. He goes on to suggest, for example, that there may be tenuous "invisible" streams in orbit about the Earth, with meteoroids being slowly brought into the Earth's upper atmosphere through interactions with a resisting medium. He also suggests that other planets might "capture" meteoroids, and that planetary perturbations might then act to change the properties of the orbit to become Earth intercepting. He further suggests that cometary tails might, "distribute particles with perihelion distances differing in conformity with the[ir] length". These musing with respect to stream structure and meteoroid origins are all guarded attempts at finding a theoretical (specifically non-cometary related) justification for the origin of stationary radiants

When elected Director of the BAA Comet Section in 1891, Denning indicated that one of the section's goals was to record telescopic meteors. In this respect, writing in the May 1914 issue of the *Observatory*, Denning reported that he had witnessed, "635 [telescopic] meteors during 727 hours of comet-seeking in the years 1881 to 1896", and while he had merely recorded the fact that they had been seen, he argued that they were objects deserving of more study. Having thus introduced his topic, and provided some statistics, the true meaning behind Denning's seemingly sudden interest in telescopic meteors, objects he had formerly paid little attention to in his writings, becomes clear. Indeed, Denning begins to speculate upon what the observations of such meteors might tell us about the Earth's atmosphere, and their potential source regions. Working under the (entirely unjustified) assumption that they had the same pathlengths and speeds as "ordinary" meteors, Denning determined that the observed properties of telescopic meteors (basically he described them as being faint and slow moving) implied that they must be at least "20 times as distant" from us. This led him to claim that at least some telescopic meteors had inferred heights in excess of 1000 miles. Denning is here on dangerous ground, and his lack of expertise in physical theory is telling. Certainly, Denning realized that at these altitudes such bodies could not be undergoing any interaction with the Earth's atmosphere, so he suggested that they might be self-luminous cosmic meteors. Here, again, we find Denning speculating well outside of accepted physical theory, but he is intent on demonstrating that different types, or varieties, of meteors exist, and that meteoroids can be derived from very different source regions and not just comets. In his attempts to justify the existence of stationary radiants, Denning is looking to the past for inspiration, and attempting to resurrect the otherwise discarded idea of cosmic meteoroid streams. For all this, Denning is being consistent in his thinking. If, as Olivier had made clear, stationary radiants could not be produced by any cometary derived meteoroid stream, then to Denning's mind there must be additional source regions for meteoroids, and accordingly why not cosmic streams. There is nothing wrong, of course, with speculation, even wild speculation, but it is reasonably clear that by 1915 Denning was clutching at straws in his stalwart attempts to provide some (even weak) theoretical justification for the existence of stationary radiants.

2.5   Meteorites

In spite of his long-running interest in meteors, Denning did not write extensively on either meteorites or meteorite falls, although he did own at least a few samples from one meteorite. The fragments in Denning's possession were presented to him by Joseph Ward, Director of the Wanganui Observatory and the Wanganui Astronomical Society (WAS), and were from the Mokoia meteorite which fell in New Zealand just after noontime, on November 26th, 1908 (Fig. 2.16). This meteorite is, in fact, a primitive (that is, largely unaltered since it first formed 4.5 billion years ago) CV3 carbonaceous chondrite meteorite, now known to contain complex organic micro-structures [35] and even amino acids [36]. Denning must have had some communications with Ward prior to receiving the meteorite fragments, since, within a scrapbook kept by Denning (now part of the BAA Meteor Section archive) is a newspaper clipping from the *New Zealand Graphic* for 11 July, 1903, showing the Wanganui Observatory (now the Ward Observatory). Ward was born in London, and emigrated to New Zealand circa 1880, and is remembered as an accomplished observer, telescope maker, and astronomy educator. The telescope installed at the Wanganui Observatory was a 9½-inch aperture refractor built by Thomas Cooke and Sons of York (see Chap. 6) in 1857. Purchased from British engineer and safe manufacturer Samuel Chatwood in 1902, the telescope originally belonging to Isaac Fletcher (iron master and Liberal politician) in Cumbria, UK. While the telescope purchased by the society had provenance and history, Ward was to write in a review published in the October 1926 issue of the *Publications of the Astronomical Society of the Pacific* that, "The Wanganui Astronomical Society is distinctly a child of this century". Indeed, this was the case, and as the twentieth century continued to unfold, Denning's influence, in spite of a long and celebrated career as a full-time amateur astronomer, was beginning to fade. Clearly fascinated by the meteorite, however, Denning mused that, "it is interesting, after a person has habitually watched the luminous careers of these bodies during many years, to hold a similar object in one's hand and contemplate it from a much nearer point of view!" [37].[13] In the time interval between 1860 and 1930 (ostensibly Denning's active observing lifetime), 5 meteorites fell in the United Kingdom, these being at Rowton (1876), Middlesborough (1881), Crumlin (1902), Appley Bridge (1914) and Ashdon (1923). Of these, Denning only actively investigated the Appley Bridge fall, and then, only at a distance through newspaper clippings and letters. Writing in the 5 November 1914 issue of *Nature*, Denning summarised the numerous reports that he had received on the fireball, the occurrence of sonic booms associated with the fall, and the recovery of the meteorite.

---

[13] It is not clear what became of the Mokoia sample that Denning received. Inquiries to the City Museum at Bristol have revealed that it was not donated to their collection (Clark, R. D., Assistant Curator, Geology. Personal communication, 1991). A total of about 4.5-kg of material was recovered from the fall. The main mass is located in the Museum at Wanganui, with two small samples being held in the collection of the Natural History Museum in London.

Fig. 2.16 Nicknamed in local newspapers as "the deadly messenger" (although no deaths occurred) the Mokoia meteorite fall was presaged by a bright (daytime) fireball and a loud sonic boom. It was the first witnessed meteorite fall in New Zealand, and within 4 months of its arrival on Earth, fragments were received by Denning in Bristol. Denning was to write that the fragments are, "very dark grey stone or admixture of stone and iron, which has evidently undergone intense heat, and seems of a crumbly nature" [37]. Image courtesy of Whanganui Regional Museum, NZ

Bibliography

1 Hughes, D. W. (1982). The history of meteors and meteor showers. *Vistas in Astronomy, 26,* 325–345.

2 Pritchard, C. (1864). *Monthly Notices of the Royal Astronomical Society, 24,* 139.

3 Denning, W. (1895). *Experiences during thirty years of star gazing.* Tit Bits Magazine, August 31st, 1895.

4 Denning, W. F. (1876). Radiant-points of shooting stars. *Monthly Notices of the Royal Astronomical Society, 36,* 283–285.

5 Denning, W. F. (1876). Radiant points of shooting stars. *Nature, 15,* 158.

6 Denning, W. F. (1879). Shooting stars. *Proceedings of the Bristol Naturalists Society, 2,* 264–278.

7 Denning, W. F. (1891). *Telescopic work for starlight evenings* (p. 78). Taylor and Francis.

8 Denning, W. F. (1877). The radiant centre of the Perseids. *Nature, 16,* 362.

9 Denning, W. F. (1884). The long duration of meteoric radiant points. *Monthly Notices of the Royal Astronomical Society, 45,* 93–116.

10 Denning, W. F. (1877). Radiant points of shooting stars. From captain Tupman's unreduced observations 1869–71. *Monthly Notices of the Royal Astronomical Society, 37,* 349–351.

11 Denning, W. F. (1878). Radiant points deduced from the paths of 4,143 shooting stars observed by the members of the Italian meteoric association in the year 1872. *Monthly Notices of the Royal Astronomical Society, 38,* 315–317.

12 Denning, W. F. (1879). Shooting stars. *Proceedings of the Bristol Naturalists Society, 2,* p.271.

13 Beech, M., & Hughes, D. W. (2000). Seeing the impossible—Meteors in the Moon. *Journal of Astronomical History and Heritage, 3*, 13–22.

14 Hughes, D. W. (1990). *Monthly Notices of the Royal Astronomical Society, 245*, 198–203.

15 Phipson, T. L. (1867). *Meteors, aerolites, and falling stars* (pp. 160–161). Lovell Reeve, and Co.

16 Denning, W. F. (1885). The great shower of Andromedes, November 26, 27, 28, and 30, 1885. *Monthly Notices of the Royal Astronomical Society, 46*, 67.

17 Lockyer, J. N. (1890). *The meteoritic hypothesis* (p. 135). Macmillan and Co.

18 Denning, W. F. (1899). General catalogue of the radiant points of meteoric showers and of fireballs and shooting stars observed at more than one station. *Memoirs of the Royal Astronomical Society, 53*, 203–292.

19 Hawkins, G. S. (1958). Catalogues of meteor radiants. *Smithsonian Contributions to Astrophysics, 3*(2), 7–8.

20 Prentice, J. P. M (1989). The British Astronomical Association—The first fifty years. *BAA Memoirs, 42, 104 - 110.*

21 Denning, W. F. (1890). Catalogue of 918 radiant points of shooting stars observed at Bristol. *Monthly Notices of the Royal Astronomical Society, 50*, 410–467.

22 Denning, W. F. (1885). *Nature, 33*, 152.

23 Chant, C. (1913). An extraordinary meteoric display. *Journal of the Royal Astronomical Society of Canada, 7*, 144–215.

24 Beech, M., & Comte, M. (2018). The chant procession of 1913—Towards a descriptive model. *American Journal of Astronomy and Astrophysics, 6*(2), 31–38.

25 Beech, M. (1991). The stationary radiant debate revisited. *Quarterly Journal of the Royal Astronomical Society, 32*, 245–264.

26 Denning, W. F. (1891). *Telescopic work for starlight evenings* (p. 66). Taylor and Francis.

27 Olivier, C. P. (1925). *Meteors*. Williams and Wilkins.

28 Beech, M. (1992). The Herschel—Denning correspondence. *Vistas in Astronomy, 34*, 425–447.

29 Denning, W. F. (1907). Professor A. S. Herschel, F.R.S. *Nature, 76*, 202–203.

30 Opik, E. (1958). *Physics of meteor flight in the atmosphere*. Interscience Publishers Inc.

31 Bronshten, V. A. (1983). *Physics of Meteoric Phenomena*. D. Reidel Publishing Company.

32 Ryabova, G. O., Asher, D. J., & Campbell-Brown, M. D. (Eds.) (2019). *Meteoroids: Sources of meteors on earth and beyond*. Cambridge University Press.

33 Lindemann, F. A., & Dobson, G. M. B. (1922). A theory of meteors, and the density and temperature of the outer atmosphere to which it leads. *Proceedings of Royal Society, 102*, 411–437.

34 Beech, M. (2021). *A cabinet of curiosities: The myth, magic and measure of meteorites*. World Scientific Publishing.

35 Briggs, M. H., & Kitto, G. B. (1962). Complex organic micro-structures in the Mokoia meteorite. *Nature, 193*, 1126–1127.

36 Cronin, J. R., & Moore, C. B. (1976). Amino acids of the Nogoya and Mokoia carbonaceous chondrites. *Geochimica et Cosmochimica Acta, 40*, 853–857.

37 Denning, W. F. (1909). Fall of an Aerolite in Mokoia, New Zealand, on November 26, 1908. *Nature, 80*, 128.

3

The Heavens Provide

It was revealed in Chap. 1 that Denning had an insatiable appetite for observing, and that his interests were highly catholic. Indeed, for Denning, the heavens provide a great and seemingly limit-less bounty of phenomena to observe. In terms of his published works, Fig. 3.1 shows a percentage breakdown of Denning's observations sorted into several broad research categories. Of the articles that he published 70% relate to meteors, and meteor showers, with some 15% being related to observations of Jupiter and Saturn. Of these two planets, however, it was Jupiter that preoccupied most of his observing time. Just under 8% of Denning's publications are related to comets and comet detection, with the remaining publications (some 7%) being relating to other areas of study, including Mercury, Venus, the Moon, Mars, interstellar nebulae, and novae.

Figure 3.1 clearly indicates that that meteor astronomy dominated Denning's written output, but it is also shows that he spent much observing time, and writing effort in other areas of study. It is these non-meteor areas of study that I will discuss in this chapter. Indeed, it was through these other areas of planetary and cometary studies that Denning became an important voice in the promotion and development of amateur astronomy in the later part of the nineteenth century. Furthermore, it is through these non-meteor related studies that we find Denning working at the telescope, and it is to his thoughts on telescopes, their apertures and usage, that we turn first.

© The Author(s), under exclusive license to Springer Nature Switzerland AG 2023
M. Beech, *William Frederick Denning*, Springer Biographies,
https://doi.org/10.1007/978-3-031-44443-2_3

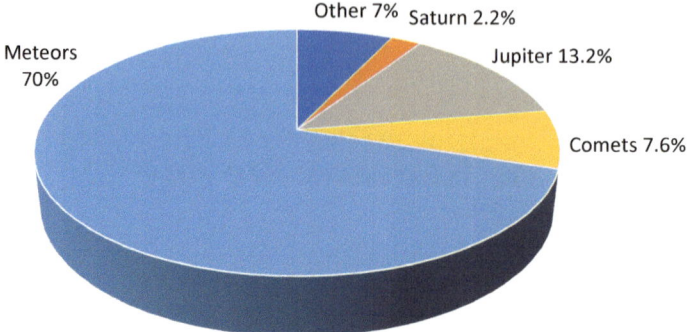

Fig. 3.1 Percentage distribution of Denning's publications (recall also Fig. 2.2) divided into the categories of Meteors, Saturn, Jupiter, Comets and Other areas

3.1 Telescopes: Big Versus Small

Denning held strong opinions as to how telescopic observations should be made. Indeed, he argued that astronomers had an ethical responsibility to seek out the truth, and to report only those facts that were unambiguously observed. In many ways, Denning's perceived quest for *the truth* was a romantic ideal, but an ideal rooted in a strong belief that the pursuit of science was a noble cause. Indeed, his commitment to the pursuit of science was sufficiently resolute that he followed a celibate, and impoverished life-style because of it. Furthermore, although Denning laid great stress on how an astronomer should set-about collecting observations, he additionally expressed strong views as to the type of telescope that should be used in astronomical research. These opinions, on occasion, brought him into disagreement with the outlook of other astronomers. In particular, Denning was critical of the employment of large aperture telescopes in planetary studies, and this belief brought him into direct conflict with professional researchers—especially ones in America.[1] Denning's viewpoint is quite extraordinary from a modern perspective, where the mantra of bigger is better predominates. Once again, however, we find Denning's essential views

---

[1] On the topic of small versus large telescope abilities, Denning had distinctive and sometimes belligerent run-ins with Sherburne W. Burnham (amateur observer and Lick Observatory), Charles A. Young (Princeton University), and G. W. Hough (Dearborn Observatory). There is a definite hint of nationalism and peacockery at play in the letters that were published (on both sides of the argument)—no one came away looking especially pretty. In the late nineteenth century, however, American astronomy was in its ascendency, and large telescopes were being almost routinely funded and built. Furthermore, the professional American astronomers were beginning to establish themselves as the preeminent authority on solar system research. The debate was not just about what telescopes could or could not do, but about who held sway (the amateur or the profession) with respect to promulgating astronomical and physical wisdom.

expressed in his *Telescopic Work for Starlight Evenings*. There he writes [1], "we must judge of large glasses by their revelations; their capacity must be estimated by results. The fruit of their employment is rarely prolific to the extent anticipated,[2] because the observers have been defeated in their efforts by impediments which inseparably attend the use of such huge constructions". These views are entirely consistent with Denning's belief that new discoveries, and insight must automatically follow the application of a sound observing program and the investment of observing time. Denning most often used a 10-inch reflecting telescope for his planetary work, and he believed that this was about the optimum size (Fig. 3.2). Going to much larger aperture sizes, he believed, would only be a hindrance. To the modern astronomer it seems strange to find Denning writing in 1891 that, "with my 10-inch in a sadly deteriorated state I have obtained views of the Moon, Venus and Jupiter that could hardly be surpassed. The moderate reflections from a tarnished mirror evidently improves the image of a bright object by eliminating the glare and allowing the fainter details to be seen" [2].[3] Denning was not alone in promoting such ideas concerning telescopic optics, but lacking any formal understanding of theoretical optics he let personal opinion (admittedly gained through long hours of observing experience) dictate his public stance. Indeed, he promoted a pseudo-scientific theory of optical glare, arguing (incorrectly) that the eye was overwhelmed by the brighter images produced by larger aperture telescopes, and that this glare resulted in a loss of image clarity. This particular issue of image glare and clarity became a particularly important topic in the debate concerning Martian markings [3], although this was a subject largely taken-up by other observers.

The dialog relating to the abilities of small versus large aperture telescopes was both longwinded, and largely the result of much personal bluster and misunderstandings. Many letters were written to journals and magazine expressing opinions and experiences, but very few observers actually bothered to experimentally test the issue at hand.[4] Indeed, the whole debate could have

---

[2] In the 8 March 1883 issue of *Nature*, Denning also expressed the concern that, "some of our best instruments are merely erected as play things serving to gratify popular curiosity".

[3] At other times, Denning bemoaned the poor state of his telescope. One entry in his Jupiter observing journal for 31 May, 1898 reads, "Cleared up in afternoon after rain. Air very transparent, sky deep blue - swept up Jupiter at about 6. 45 with power 32. Observed planet with 312 but though definition was good the image was too faint for details to be certainty seen - Had the mirror had a good surface it would have been different. Windy NW and very cold".

[4] This being said, Denning did exactly this. Writing in *Telescopic Work* (page 151), Denning explains that he compared the image of Venus with a "2-, 3-, and 4 ¼ - inch refractors and 4 -, and 10-inch reflectors", noting that while the smaller scopes suggested the appearance of markings, the greatly improved image quality of his 10-inch telescope revealed no markings at all. As a point of note, it is not clear what happened to Denning's With-Browning telescope. These are relatively rare instruments, but no trace of its subsequent history has been unearthed. The Royal Astronomical Society archive

Fig. 3.2 Denning's 10-inch With-Browning telescope (see also Fig. 1.3). The telescope mount is in an unusual altitude-azimuth configuration. Illustration from *Telescopic Work for Starlight Evenings*

been easily resolved by bringing a group of respected observers to the same observatory location, at the same time, tasking them to look through the same series of telescopes, with different apertures, small to big, at the same object. Rather than being resolved in a reasonable scientific fashion, however, the debate, with both sides digging in their heels, became a shambles, and then ultimately just faded away.

In the course of his early observing career, Denning made regular studies of Martian surface features, and he was particularly interested in determining the planets rotation period. Using observations collected between 1869 and 1884, he determined [4] a rotation period of 24 h 37.372 min (this is in good agreement with the accepted modern-day value of 24 h 37.44 min). Denning was not strongly drawn into the debate concerning the detection of

holds a 6-inch objective lens attributed to Denning (RAS reference number 163). This is a somewhat anomalous, however, since Denning only reported owning refracting telescopes with objectives of diameter 3-inchs and 4½-inches. The lens was donated to the RAS by his sister Marry Willets (née Denning) in 1942.

Martian canals, but he did report seeing them (Fig. 3.3), and he did occasionally note surface changes (Fig. 3.4). Denning's feelings as to the origin of the canals was that they were of natural, rather than artificial origin. As to the observation of planetary details, however, the essential point that Denning raised, along with other observers, was that those planetary markings, such as the Martian canals, which were prominent in small aperture telescopes were not apparent to observers using large apertures (to Denning this meant that apertures greater than about 10-inches). In the case of Mars this was clearly a crucial point, and observers such as Percival Lowell would insist, for example, that only small aperture telescopes should be used to study surface features. Lowell believed that large telescopes did not show intricate surface features because they suffered from a "fine imperceptible blurring" [3]. Likewise, Denning argued that the observer using a large aperture telescope would see images that were, "blazing disks, affected by incessant molding and flaring and nearly devoid of reliable markings". While Lowell was guilty of trying to force the observations to fit his theory of Martian canal origins, Denning simply reported what he saw. It is now recognized that what many early observers failed to fully appreciate (but see Denning's comments on Saturn below) was that at the threshold of telescope resolution physiological effects and observer-bias will determine what the eye perceives. In essence the telescope debate was not about the failure of large aperture telescopes being able to resolve planetary surface features seen in small telescope, but rather, it was that they did not resolve the surface features that specific (and prominent) astronomers wanted others to see.

In recognition of his long-running series of observations concerning Mars, a 165-km diameter Martian crater was named in Denning's honor (Fig. 3.5) by the International Astronomical Union in 1973. This particular crater most probably formed during the Late Heavy Bombardment—a period of high, inner solar system cratering that occurred some 4 billion years ago (or about 500 million years after the formation of the Earth). No doubt pushing visual imagery and metaphors too far, we note that just like Denning's largely lost and obscured biographical details, his Martian crater is as equally eroded and obscure, with more recent, smaller impacts and surface alteration deleting much of the original crater outline from our view.

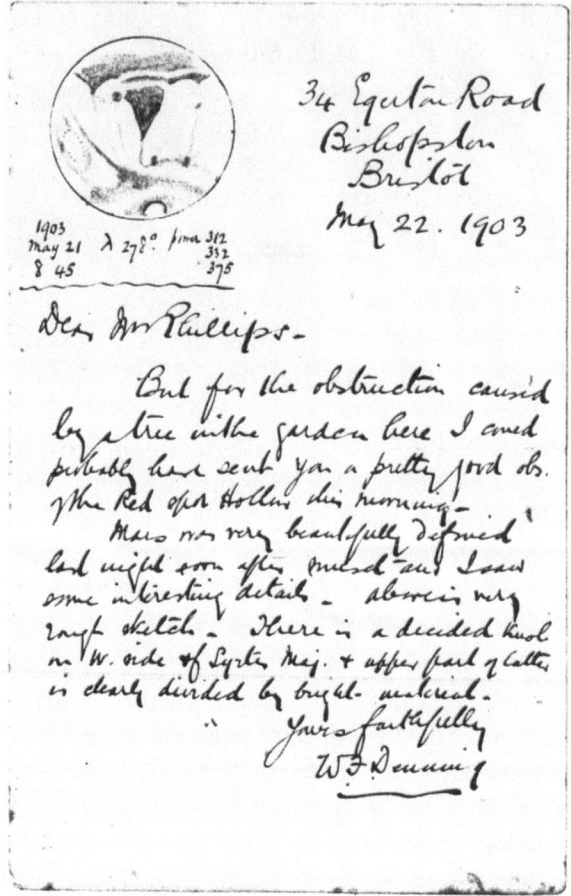

Fig. 3.3 Letter from Denning to T. E. R. Phillips showing a drawing of Mars (upper lefthand corner) made on 21 May 1903—running between darker albedo patches are canal-like, linear features. In the letter Denning complains about a tree that is blocking his view, but states that, "Mars was very beautifully defined"

3.2 Comets

Denning dedicated a considerable amount of his telescope time to the search for new comets, and he is credited with the initial discovery of four such bodies. These being 1881 V (comet 72P/Denning-Fujikawa = C/1881 T1), 1890 VI (now designated C/1890 O2), 1892 II (now designated C/1892 F1), and 1894 I (now designated as a lost comet: D/1894 F1). Additionally, he was pre-empted, by a matter of hours, by American astronomer Edward Barnard, in the discovery of comet 1891 I (now designated C/1891 F1). Of the comets discovered by Denning, only one, 72P/, is a short period comet,

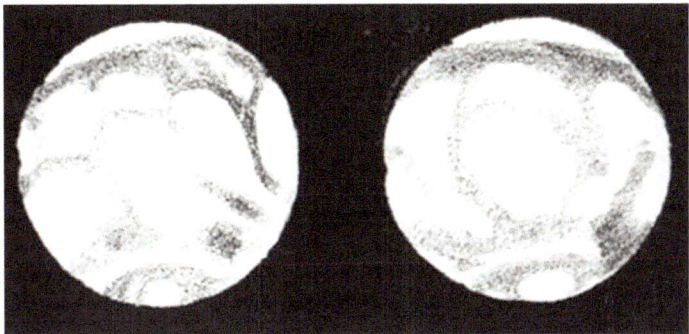

Fig. 3.4 Change in surface markings on Mars, at the same phase, as recorded by Denning on May 4 (left) and May 12 (right), 1888. Images from the *Journal of the Liverpool Astronomical Society* (6, plate XXI, 1887)

Fig. 3.5 The Martian crater Denning. The crater is located in the Terra Sabaea region of Mars's southern highlands. Since the main crater formed it has been subsequently modified by numerous smaller impacts, some of which indicate fluidized ejecta blankets. Image https://commons.wikimedia.org/wiki/File:DenningMartianCrater.jpg NASA, Public domain, via Wikimedia Commons

two have parabolic orbital elements indicating that they are visitors from the Oort cloud, and one that is now designated as lost (Fig. 3.6). Observations of the latter comet, D/1894 F1, discovered by Denning on 26 March, 1894, indicate an elliptical orbit, with an orbital period of about 7½ years. For all this, the comet has not been recovered since 5 June 1894, and it is presumably either destroyed (that is catastrophically disrupted) or entirely dormant. Comet 72P/ was discovered by Denning on 4 October, 1881, and has an orbital period of 9.02 years. Following its discovery, however, the comet was not seen again until 1978, when, after 10 missed perihelion returns, it was swept-up by Japanese amateur astronomer Shigehisa Fujikawa. The comet was again missed during its expected returns in 1987, 1996 and 2005 indicating that it is only occasional active at the time of perihelion passage. The comet, however, was successfully recovered during its 2014 perihelion return. Interestingly, comet 72P/ spends about one year per orbital revolution within the main belt asteroid region, and its activity at perihelion might be dependent upon collisions and surface-gardening suffered while passing through this zone. An investigation by the author concerning the possibility of 72P/ generating a meteor shower on Earth found that while the comet's orbit passes within about 0.1 AU of the Earth's orbit, strong gravitational perturbations due to Jupiter are such that no strong, or even weak, annual meteor shower is to be expected. The theoretical radiant for meteors derived from the orbit of comet 72P/ is located in Sagittarius, but any meteor activity would be expected to peak during daytime hours, making the shower a target for radar instrumentation rather than visual observers. Again, however, there is no indication of any specific activity within the radar data. Interestingly, the dust detector carried aboard the HELIOS I spacecraft recorded enhanced impact rates in 1978 and 1980 at times when the spacecraft passed close to the meteoroid stream location expected for comet 72P/debris. The comet's ascending node passes particularly close to the orbit of Venus, and it is possible that its associated meteoroid stream occasional produces a meteor display in the upper atmosphere of that planet.

In Chap. 1 it was noted that Denning was awarded Bronze Donohoe Comet Medals, by the Astronomical Society of the Pacific (ASP), for his discoveries of 1890, 1892, and 1894. His reaction to such awards, however, was not necessarily one of celebration. Indeed, Denning held strong and seemingly contradictory views about the value of such awards. Writing in the *Observatory* magazine for July 1882, a letter by Denning indicates that the newly announced Warner Prizes for the discovery of comets and meteorites was an issue of, "considerable interest", but further suggested that the conditions that applied to its award required "amendment". Financed

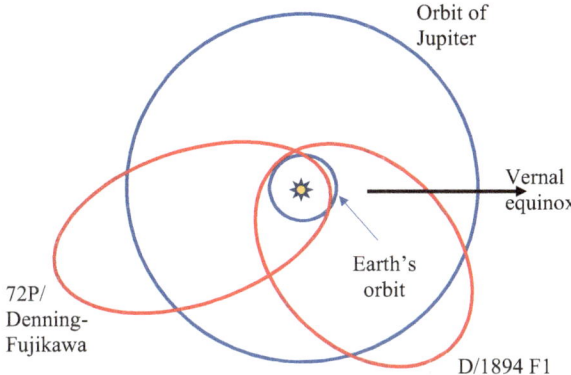

Fig. 3.6 The orbits of comets 72P/Denning-Fujikawa, and D/1894 F1

by businessman Hubert Warner, founder of the Warner Observatory in Rochester, New York, the award of $200 was to be given to the discoverer of any new comet (the award amount, due to many new discoveries being made, was later reduced to $100 in 1886). While Denning acknowledged that the, "sole purpose is doubtless to encourage original research and accelerate the progress of these [comet and meteorite] discoveries", he noted that the award was conditional on the Warner Observatory being given first and exclusive knowledge of any discovery. This, Denning, argued would delay information about any new comet discoveries being disseminated to other observers. On this point Denning writes, "it sounds like an attempt to buy up the first intimation of discovery, which by courtesy, and in common fairness, should come through … official astronomical representative[s]". Clearly, Denning was by inclination an astronomical socialist. For all this, Denning's main complaint seems to have been that British observers were disadvantaged by the award conditions. Seven years later, in the September 1889 issue of the *Observatory*, Denning noted in his *Comets and Comet-Seeking* column that the Warner prize had been suspended—indeed, Warner was forced into bankruptcy in 1893. It is in this same September column that Denning makes note of a new comet discovery prize (the Donohoe medal) to be offered by the ASP.

The 9 April, 1892 issue of the satirical magazine *Punch* carried an image (Fig. 3.7) entitled *A New Comet*. The image was accompanied by the text, "Mr. Denning, whose name is well known as a comet-finder, discovered a small Faint Comet on Friday March 18, at Bishopton [sic], Bristol". The image and caption were based upon an announcement made in the *Times of London* newspaper for 22 March. The *Times* article reads: "Discovery of a comet. A circular from the Royal Observatory, Edinburgh, says on March 18 at 11.8 pm Mr. W. F. Denning telegraphed from Bristol as follows. 'Discovered small faint comet, ascension 341 deg., declination 59 N., motion rather quick eastwards'". This new comet (now designated C/1892 F1) turned out to be a long period comet with a slightly hyperbolic orbit indicative of it being an Oort Cloud member likely on its first passage around the Sun.

The cartoon (Fig. 3.7) that accompanied the *Punch* article was drawn by Edward Tennyson Reed, who was a regular contributor and staff member to the magazine. The key to unraveling the cartoon is in the wording "small faint comet", which accounts for the humanized cometary coma: showing a fainting, even death-like facial expression. The observer, with a contented smile, is not a particularly good rendition of Denning, but Reed has very definitely caught the Great Coat that Denning characteristically wore when out observing (recall Fig. 1.1). *Punch*, being a satirical magazine was not interested in promoting the sciences, and the cartoon is simply humorous filler. However, it is revealing, that by 1892 Denning is described as being, "well-known as a comet-finder". Indeed, Denning's fame as an amateur astronomer, and recognition as a public figure, was built as much through his many contributions to newspapers and magazines, as it was through his discoveries and scientific publications.

Remarkably, Denning's comet discovery was the second such find of March 18th. Rudolf Spitaler at the Vienna Observatory picked-up another comet

Fig. 3.7 Mr. Denning discovers a new comet. Cartoon by Edward Tennyson Reed, from the 9 April, 1892 issue of *Punch* magazine

(in a different part of the sky) at almost the same instant Denning found C/ 1892 F1. The second comet, however, was in fact the recovery of 7P/Pons-Winnecke, a short period comet with an orbital period of 6.37 years. This particular comet was first discovered by Jean Louis Pons (in Marseilles) in 1819, and later re-discovered by Friedrich Winnecke (in Bonn) in 1858. Comet 7P/Pons-Winnecke is recognized as the parent comet to the June Boötid meteor shower—a shower whose radiant location was first identified and described by Denning in June of 1916. Indeed, this somewhat erratic shower put-on a spectacular display in 1916, and Denning wrote in the July issue of the *Monthly Notices* that, "large meteors came in quick succession … [they] were moderately slow, white with yellowish trains, and paths rather short … Several of the meteors burst or acquired a great intensification of light near the termination of their flight, and gave flashes like distant lightning". The discovery of the June Boötid shower could easily have been missed, since Denning indicated that the day (28 June) had been cloudy, and that he had only stepped out to observe after noticing a few breaks in the clouds. He further records that he began observing at 25 minutes past 10, and that it was immediately obvious that something interesting was going on, "a very special shower was in progress". So overtaken by what he was seeing, Denning notes that he called a friend (not named, but presumably a neighbor) to help with the meteor count, while he concentrated on plotting trails on his celestial globe (recall Fig. 2.3). The days and nights following the 28th were all cloudy, but on 30 June Denning was able to observed for 95 min, and he detected one bright meteor emanating from the region of Boötes. Fortunately, Fiammetta Wilson, in Totteridge, also witnessed this same meteor. The two-station reductions (see Chap. 5) for this meteor indicated a trail length of 44 miles (71-km), a velocity of 28 miles per second (45-km/s), and a radiant point situated close to the star β Boötes.

Having associated the June Boötids with comet 7P/Pons-Winnecke, Denning suggested that the shower might again show distinctive activity in 1921 and/or 1922—that is, strong shower activity was expected to occur at intervals corresponding to the orbital period of the parent comet. While the shower was active in 1921, as reported by observers in Japan, no enhanced, or distinctive outburst activity was seen. Indeed, the June Boötid shower has continued to prove highly erratic in its behavior, although outbursts were recorded in 1998 and 2004.

While someone has to be the first observer to catch a specific meteor shower in outburst, full credit for the 1916 June Boötids is given to Denning, and his never-waste-a-moment of observing time attitude. Even at 68 years of age, it was he, and of all the observers in the world, he alone, that noticed, realized the significance of, and analyzed the 28 June 1916 Boötid outburst. Certainly, others must have seen meteors on that night, but at the time of the outburst, the world was deeply entrenched in the second year of the Great War, with Europe steadily tearing itself apart. One report, a letter to the *Birmingham Daily Post* newspaper, from a correspondent identified simply as "observer", reported a rich shower of meteors on the night of the 28th between 11 p.m. and midnight. And, Denning, writing in the August 1916 issue of the *Observatory* magazine, noted that munition-workers (at an unidentified location) had reported seeing meteors on the night of the 28th, the effect appearing like a, "bombardment of the stars by the meteors which spread out into waves of light".

The Donohoe Medal was presented in accordance to a monetary gift to the ASP by well-to-do San Francisco banker Joseph Donohoe in 1890. The first medal recipient was William Brooks for his discovery of comet 16P/Brooks, with the last medal being given to Rudolph Minkowski for his discovery of comet C/1950 K1. Following Minkowski's award, the ASP council felt that so many new comets were being discovered (predominantly through the employment of photographic techniques) that it was no longer practical to physically acknowledge each new find. While it seems that Denning happily accepted the ASP medals for his discovery of comets in 1890 and 1892, he was less magnanimous concerning the award for his discovery made on 26 March, 1894. The circumstances surrounding this fall-out seem to have precipitated around the discovery announcement of a new comet, by Edwin Holmes, on 9 April 1894, and a letter, by a correspondent using the pseudonym of "Truth", published in the 20 April, 1894 issue of *The English Mechanic and World of Science*. The letter by "Truth" suggested that the comet discovered by Denning had attracted little attention from observers, while that by Holmes had garnered more interest. "Why should the chance doings of one man [Holmes] create such a stir, and another far more deserving gentleman [Denning] meet with hardly any acknowledgment", queried "Truth". The point of this question being that Holmes's discovery had turned out to be one of mistaken identity—his supposed new comet was in fact a faint gaseous nebula. Denning replied to "Truth" in the 27 April issue of *The Mechanic*, and didn't mince his words. Firstly, he argued that his comet discovery had generated much attention, and second, he defended Holmes in acknowledging that his sighting was a genuine mistake,[5] and that the nebula in question (N.G.C. 6503) did, in fact, have an elongated and comet-like appearance. Interestingly, Denning adds, however, that he had

---

[5] Holmes did have observing pedigree, and had earlier discovery a short-period comet, comet 17P/Holmes, on 6 November 1892.

been surprised to receive a telegram announcing the discovery by Holmes from the Astronomer Royal (then Sir William Christie) on April 10th, since just a few nights before he had scanned the region in question and seen nothing. Mistakes, in spite of skill and best intentions, are inevitable, seems to have been Denning's summary of the situation. Rather than leaving the topic there, however, Denning went on to write,

> I have written to Prof. Holden, of the Lick Observatory, declining the comet medal of the Astronomical Society of the Pacific. I quite fail to appreciate the utility of such awards. They afford no stimulus or encouragement to observations, and they are certainly no recompense for the labour involved in finding a comet. A man has his reward in the fact of success, and a bit of bronze is not likely to influence him in any way

These are strong words, and they reflect back on Denning's ideal image of the individual observer in which it is the persona that matters, and not the existence of any specific awards and/or prizes. The sentiments in Denning's letter of 27 April, reveal an individual somewhat frustrated, but one that strongly believes that it is the attitude and resolve of the individual that really matters, and that true success, and reward, will inevitably follow in the footsteps of hard work, irrespective of any honorary awards that might be on offer. These basic ideas were outlined by Denning as early as 1882, when he was to write [5], "success in this, as in other departments of research, depends, in a very large measure, upon the energy with which it is pursued. To an observer who devotes himself closely to it, and avails himself of every chance presented, there is an encouraging prospect of success". It is instructive to note, however, that a decade earlier, in 1872, Denning had written [6] that, "comets are not interesting objects in telescopes".[6] This comment is partly true, since comets, through a small telescope, are typically no more than faint, fuzzy patches of light. Perhaps, finally, Denning, in defending Holmes in 1894, was additionally celebrating the concept of the glorious failure, where it is reasoned that it is better to have tried and failed, than to have never tried at all.

Starting in 1891, Denning became the first Director of the British Astronomical Association's Comet Section, holding this office through to 1893. Even though his official involvement with the Comet Section was but brief, Denning continued to promote cometary sweeping well into the twentieth century. Indeed, he wrote a long-running column on cometary matters

---

[6] Denning soon changed his opinions on comets, and comet-seeking, however, and later remarked in the August 31st, 1895, issue of *Tit Bits*, that "comet-seeking is the most exciting work of any in which I have indulged".

for the *Observatory* magazine, and likewise for the popular journal *Knowledge*. Some measure of how much time Denning dedicated to cometary sweeping can be gained from a calculation that he presented [7] in 1894. At that time, he noted that in 596 h of comet-sweeping he had discovered five comets.[7] This naively averages to some 119 h of searching per comet. Although Denning continually tried to promote cometary studies among English amateur astronomers, he found that they did not easily turn to the subject. As early as 1881 Denning was encouraging amateurs to start looking for new comets. Writing in the November 1881 issue of the *Observatory*, Denning appealed to national pride, and suggested that, "English observers, who, like myself, have the necessary time and instruments … should make some attempt to enter into friendly competition with foreigners in the field of cometary discovery". To this he added, in the hope of covering more regions of the sky that, "we might act in unison, and arrange a division of labour". His appeal fell upon deaf ears, but in a letter to the March 1882 issue of the *Observatory*, Denning noted that, while he had received no response from English observers, "the Americans have, with characteristic promptitude, taken up the subject energetically, so that on March 1 last a concerted system of sweeping was commenced". Undaunted, however, Denning kept-up his appeals to British observers, noting that, "with Mr. Stanley Williams of Brighton, I propose starting, early in April, a combined system of sweeping similar to that adopted by the Americans, and I shall be glad to hear from [other observers and volunteers]". Denning's companion observer, Arthur Stanley Williams was a solicitor, amateur astronomer, and keen yachtsman. He is predominantly remembered for his later observational work (from the 1890s onward) concerning Jupiter's atmospheric belts. He was elected a Fellow of the Royal Astronomical Society in 1884, and elected a corresponding member of the Royal Astronomical Society of Canada in 1909. Williams was awarded the Jackson-Gwilt Medal of the RAS in 1923 for his work on planetary observations and variable stars. In addition to announcing the recruitment of Williams, Denning also indicated that he had been promised help and assistance from the Dun Echt Observatory in Aberdeenshire if any new comets were discovered. The Dun Echt observatory had been founded by James Ludovic (Lord) Lindsay (later Earl of Crawford and Balcarres) in 1872, and by 1882 the staff astronomer there was Ralph Copeland. While Denning had managed to attract at least one, then 21 years old, observer to his comet sweeping team, and the promise of assistance from a well-respected observatory, nothing positive or long-lasting emerged from

---

[7] Note, however, he was pre-empted by a matter of hours in the detection of one of these comets—comet C/1891 F1.

his efforts. Dun Echt observatory closed in 1892, and it appears that Williams took more to planetary observing than comet-sweeping.

Undaunted, as ever, Denning once more tried to establish a comet observing group in 1889. Writing in the July (1889) issue of his "notes on comets and comet-seeking" column in the *Observatory*, Denning observed that, "several observers have volunteered assistance in the proposed systematic search for comets". In total, including Denning, 6 observers had agreed to periodically and systematically examine set regions of the sky. The observers in this new comet-seeking group were:

- W. F. Denning, in Bristol
- Mr. Common & Mr. Taylor, in Ealing
- David Booth, in Leeds
- Rev. W. R. Waugh, in Portland
- Mr. R. I [sic]. Ryle, in Barnet.

While a small group of observers, the individuals were all men of note and indeed of some importance in amateur astronomy in the 1890s. David Booth was to become the first Director of the BAA's Meteor Section, from 1890 to 1892, and along with the Reverend Waugh was a founding member of the BAA's Comet Section, to which Denning was the first Director (1891–1893). Booth and Waugh were additionally founding committee members of the BAA. The Reverend William Robert (Maurice) Waugh was the first Director of the BAA's Jupiter Section. He was a long-standing Director of the Colored Star Section of the Liverpool Astronomical Society, which he joined in 1887, and was elected a Fellows of the RAS in 1888. The initials given for Mr. Ryle of Barnet includes a typo, and should read R. J., for Reginald John. Ryle was son of the John Charles Ryle, the first Bishop of Liverpool (installed in 1880), and his brother, Herbert Edward, was later to be appointed Bishop of Winchester (in 1903) and Dean of Westminster Abbey in 1910. Educated at Trinity College, Oxford, and Guy's Hospital Medical School in London, Ryle opened medical practices first in Barnet, in 1884, and later in Brighton, where he also served on the Town Council. It appears that Ryle published but one article on astronomy—a short note on the history of Jupiter's red spot, which appeared in the December 1890 issue of the *Observatory*. Mr. Common & Mr. Taylor in Ealing, is a reference to the observatory established by Andrew Ainslie Common in 1876 (Fig. 3.8). A sanitary engineer by profession, Common's observatory was initially equipped with an 18-inch aperture telescope built by George Calver. In 1879 the observatory was further equipped with a 36-inch aperture, George Calver reflecting telescope, and this was used chiefly for astrophotography. Common was a pioneer of astrophotography, and his images of the Orion Nebula earned him the Gold Medal of the RAS in 1884 (he had been elected a Fellow of the RAS in 1876, served as its treasurer from 1884 to 1895, and was its President in 1895 and 1896). In 1885 Common decided to commission the construction of a 60-inch aperture telescope, and after some delays and many setbacks this was eventually completed in 1890. After Common's death in 1903, the telescope mirror and parts were acquired by Harvard College Observatory. After refiguring, and the construction of an improved mount, the telescope was re-assembled (as the 60-inch Rockefeller Reflector) at the Boyden Observatory in South Africa in 1933.

Albert Taylor provides for an interesting study, and might, in some sense, be considered an unfortunate transitional figure. Born in 1865, we first hear of Taylor when he was elected a Fellow of the RAS on 14 December 1888. At the time of Taylor's election his address was given as Hurstside, West Molesey, Surrey. This address is significant in that it corresponds to that of Sir Henry Thompson, who had constructed an observatory at Hurstside equipped with a 12-inch aperture Cooke refractor. Thompson was a famed surgeon and polymath, and while not a grand amateur astronomer himself, he was both wealthy and prepared to spend money on state-of-the-art equipment. Work at the observatory began in May 1888, and Taylor was the staff astronomer.

Fig. 3.8 The garden observatory of Andrew Common in Ealing, London, established in the mid-1860's. The 18-inch aperture, George Calver reflector is seen in this image. Image https://commons.wikimedia.org/wiki/File:18_inch_newtonian_telesc ope_back_yard_Ealing_London_Andrew_Ainslie_Common.png. See page for author, Public domain, via Wikimedia Commons

Indeed, Taylor's first publication, which appeared in the January 1889 issue of *Monthly Notices*, was concerned with the spectroscopic study of various nebulae made with the Hurstside refractor. For reasons that have not been recorded, Thompson appears to have abandoned his observatory almost as soon as it was finished, and he eventually donated the 12-inch telescope to the Greenwich Observatory in 1890. With Hurstside closed-down, Taylor moved to Ealing, and became staff astronomer to Andrew Common. Taylor's second research paper of 1889 appeared in the June issue of the *Monthly Notices*, and was concerned with the very first spectral analysis of planet Uranus. In late 1889 Taylor left Liverpool aboard the mail steamer *Boony* for south-west Africa (Angola) in order to make spectroscopic observations of the Sun's corona during the 22 December eclipse. Later, in 1893, Taylor embarked upon a joint Royal Society and Royal Astronomical Society sponsored expedition to Brazil, in order to make spectroscopic and coronagraph images of the Sun at the time of the 16 April eclipse. Indeed, expenses to the order of £120 were made to Taylor to cover his costs. After the 1893 eclipse expedition, Taylor essentially disappears from view. We next hear of Taylor, in fact, in a small pamphlet, *The Story of the Telescope*, written by amateur astronomer, author and journalist Arthur Mee in 1909. Mee notes of Taylor that he is, "an astronomer of great ability, now one of His Majesty's Welsh Inspectors of Schools". While clearly talented and experienced, it appears that Taylor was unable to find full-time employment as an astronomer. Mee provides further information on Taylor in his *Who's Who in Wales*, published in 1921, where we learn that Taylor was an associate of the Normal School of Science, and the Royal School of Mines, in London. Indeed, Taylor had worked as a demonstrator in physics at the Royal College of Science (now Imperial College), and had been briefly an assistant to Sir Howard Grubb in Dublin. In the modern era, someone with Taylor's demonstrated expertise, and publishing record might well expect to find full-time employment as an astronomer or research scientist. This was not yet the case, however, at the beginning of the twentieth century. Even though Taylor was unable to find a professional position, it seems he preserved his interest in astronomy, and joined the Astronomical Society of Wales (ASW) in 1898. Formed largely through the efforts of Arthur Mee, the ASW held meetings in Cardiff from its inception, in 1894, to the beginning of the Great War, at which time it disbanded. Taylor delivered talks on spectroscopy at a number of annual meetings of the ASW, and he joined its council in 1910. Later he served as President to the Society, and his death is briefly recorded in the February 1931 issue of the *Monthly Notices*.

For all of the latent talent within his newly assembled team of comet-sweepers, Denning's announcement of 1889 paid no dividends, and we hear no more of its activities. As late as 1922, Denning can be found complaining in the journal *Nature* that, "it is remarkable that English astronomers appear hitherto to have taken little interest in cometary work, and that very few comets have been discovered from this country. … there are a great number of telescopic observers in the United Kingdom who have the means and the time at their disposal to accomplish valuable work in this department if they would only engage in it in an earnest manner". These are clearly the words of a frustrated organizer.

When Denning outlined the role of the Cometary Section in the June 1891 issue of the *Journal of the British Astronomical Association*, he argued that besides searching for comets its main aims were to discover new nebulae and record telescopic meteors. A knowledge of diffuse nebulae is of interest to the would-be comet-searcher since they can be confused with a new comet (as described earlier with respect to Edwin Holmes). Denning, echoing the historical complaint of Charles Messier, referred to nebulae [8] as, "the bane of the comet-seeker". For all this, however, Denning did discover and make notes upon the new nebulae that he discovered while sweeping for comets [9]. His first foray into this area consisted of a list of ten new nebulae, the details of which were published in the November 1890 issue of the *Monthly Notices*. In this article, Denning gives detailed positions and descriptions of each nebula's appearance. Interestingly he has enlisted the help of several professional astronomers to determine accurate coordinates, including Ralph Copeland, then Astronomer Royal for Scotland. Furthermore, and indicating an extensive network of international correspondence, some of the observational data in his nebula paper were made, upon request, by Auguste Charlois at the Nice Observatory, France.

The term nebula was not well defined in Denning's time, and objects such as galaxies, globular clusters, galactic star clusters, and diffuse interstellar clouds were included under the nebula umbrella. One contentious issue concerning nebula at that time was the issue of their apparent variability in brightness. Denning made a few comments on the supposed variability recorded in the nuclear region of the spiral galaxy in Andromeda [10]. While the variability that had been ascribed to the nebula (M31) was based upon photographic observations, Denning showed typical disregard for such hi-tech results, and commented that from his experiences, the supposed variability was probably due to "atmospheric disturbances". In this case Denning was correct, but instrumental techniques would soon outstrip the human observer in both sensitivity, and versatility.

3.3   The Planets

Before he abstained from further telescopic work, circa 1906, Denning devoted a large fraction of his observing time to the study of planetary albedo/atmosphere features. He directed most of his attention towards planet Jupiter, and indeed, his publications on this body run second only to his work on meteors. Denning's primary reason for studying distinctive planetary features was to determine rotation periods, and this requires that the observer makes repeated transit timings. Clearly, in order to perform such work, the planet under investigation must show either long-lived atmospheric features, or prominent surface (that is albedo) markings. For planets such as Jupiter, Saturn and Mars many such features are available; for Venus and Mercury, however, the situation is much more difficult since they tend to be uniformly bright.

Of the two inner planets, Mercury is the more difficult to observe since it is never far from the Sun when viewed from the Earth, and because it is a physically small planet. Denning noted [11] in 1900, however, that he had observed Mercury with his un-aided eye on no less than 102 occasions between February 1868 and December 1899. In addition to his naked-eye sightings of Mercury, Denning also reported seeing dark surface features of Mercury's disk [12]. Indeed, writing in *Telescopic Work*, Denning records [13] that he had, "occasionally seen Mercury, about two or three hours after rising, with outlines of extreme sharpness" (Fig. 3.9). After making a series of observations of Mercury in November of 1882, Denning forwarded his results to Giovanni Schiaparelli at the Brera Observatory in Milan. Schiaparelli agreed with Denning that albedo features were detectable, even with moderately large telescopes (10-inch apertures and larger), and that the continued observations of such features could lead to a better determination of the planet's rotation rate. Schiaparelli continued gathering data on Mercury from 1882 through to 1889 and concluded that Mercury was in synchronous rotation— that is, its rotation period was the same as its orbital period about the Sun (some 88 days). This was distinctly different to the 24 h and 4 min rotation period derived by German astronomer Johann Schröter in 1800. While in his *Telescopic Work* Denning encouraged amateur astronomers to view Mercury on all possible occasions, it was not a planet that he spent much observational time upon. Writing some 40 years after his 1882 observations, in *Hutchinson's Splendors of the Heavens* (edited by T. E. R. Phillips and W. H. Steavenson, published in 1923), the value of Mercury's rotation period was still far from clear, and, Denning made the almost standard plea that this was a topic calling-out to the amateur community, and that the field needed a "capable student" to move the study along. Denning indicated that he had observed motion of the markings on Mercury that were inconsistent with the long rotation period derived by Schiaparelli. Indeed, while acknowledging that the rotation period of Mercury was not known with any certainty, he suggested that, "it is probably about twenty-five hours". This guess at the rotation rate was not inconsistent with those offered by other observers at the time, but the definitive determination was not to be forthcoming until the mid-1960's, and the development of radar telescope technologies. As it turned out, the problem of determining Mercury's rotation period was something that lay well beyond the grasp of the amateur astronomer, but then, even professional observer's, with state of the art, large aperture optical telescopes also failed to resolve the problem. It is noteworthy that the two historically suggested rotation periods, either synchronous (at about 88 days) or about the same as that of Earth (24 h or so) were wrong. The radar telescope observations were able to show that the rotation rate was about 56.6 days. In an entirely unsuspected and unanticipated manner the rotation period was 2/3^{rds} that of the orbital period.

Fig. 3.9 Surface albedo features on planet Mercury, as recorded by Denning on (left) 5 November 1882, and (right) 6 November 1882. The observations were made with a 10-inch With-Browning telescope, using a power of 212. Illustration from *Telescopic Work for Starlight Evenings*

3.3.1 Jupiter and Its Red Spot

Of all the planets, Jupiter held special interest for Denning. He began observing this planet in the early 1880s, and wrote [14] of it in impassioned tones,

> Beyond the sphere of Mars, in distant skies,
>   Revolves the mighty magnitude of Jove,
>   With kingly state, the rival of the Sun.

The greater bulk of Denning's work on Jupiter was concerned with recording transit times of atmospheric features, and much of this data was published circa 1900—accounting for the distinctive peak in the publication versus year diagram shown in Fig. 2.2. Denning, along with many other members of the astronomical community, was drawn to the study of Jupiter in 1878, when what is now known as the red spot (Fig. 3.10) came into striking prominence. Not only did Denning record transit times for this feature over many decades, he also produced detailed sketches of the planet's appearance (Fig. 3.11). In addition to his transit studies, Denning also undertook an extensive literature survey in an attempt to trace historical reports of the red spot's appearance. He reasoned [15] on the basis of his survey work, published in 1899, that the "great red spot of recent years may be identical with the large spot discovered on Jupiter in 1664 by Robert Hooke". There is still some debate as to what exactly Hooke may or may not have actually seen, but it is generally agreed that the red spot, as now known, is a long-lived, multi-century long, phenomenon situated in Jupiter's upper cloud deck.

The point of measuring transit times, logged when a distinctive feature crosses the central meridian of a planet's disk, is to determine the rotation rate of the atmosphere (at the latitude of the monitored feature). Denning applied himself to these observations, and over many decades made repeated estimates of the Jupiter's rotation period. His first publication on the topic appeared in the February issue of the *Observatory* in 1881, where the transit times of the red spot and a bright equatorial spot were compared. The observations described in this first paper were collected from October 1880 through to mid-January 1881, and Denning deduced a period of 9 h 55 min and 34 s for the red spot, and 9 h 50 min and 25 s for the equatorial bright spot—a result that implies differential rotation. A few months later, in the April 1881 issue of the *Monthly Notices*, Denning collected his transit timings for the red spot recorded between September 1880 and March 1881. In this

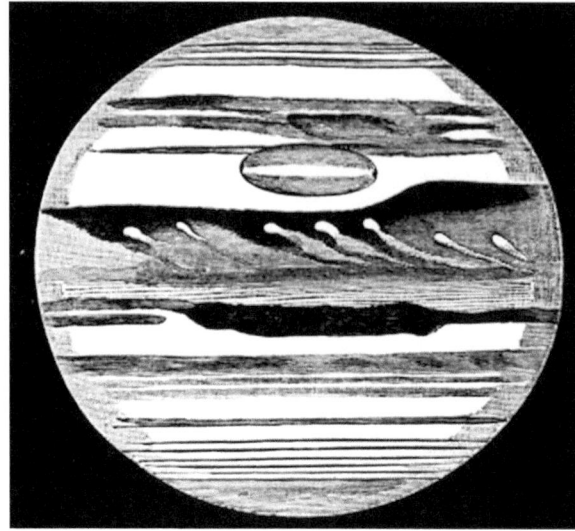

Fig. 3.10 Drawing of Jupiter's disk, including the red spot (oval feature), as recorded by Denning on the night of 9 April, 1886. The observation was made with his 10-inch aperture With-Browning reflector (Fig. 3.7) using a magnifying power of 252. Illustration from *Telescopic Work for Starlight Evenings*

Fig. 3.11 Drawing by Denning of the red spot (at center) and surrounding markings as observed on 30 October, 1882. Image from the *Journal of the Liverpool Astronomical Society* (5, plate XVI, 1887)

paper he commented upon his observing equipment being, "a Browning-With reflector of 10-inches aperture, power 300, and a Barlow lens[8] which increased it to 450". From these observations Denning determined that the

---

[8] Named after Peter Barlow, Mathematics Tutor at the Royal Military College, Woolwich, this lens acts to increase the effective focal length of the telescope and greatly magnifies the eyepiece image.

time for the red spot to cross Jupiter's meridian was 55.5 min. Returning to a discussion of transit times in his *Telescopic Work* in 1891, Denning, on reviewing 10 years' worth of observational data deduced a rotation period of 9 h 55 min and 39 s for the red spot. While the motion of the red spot is known to be variable, the standard (averaged) rotation rate is now taken as being 9 h 55 min and 42 s.

In late 1901, Denning noticed something strange was happening with respect to the motion of red spot. At this time, he was making numerous transit observations (recall Fig. 2.2) and, in the November, 1901 issue of *Popular Astronomy* he revealed that his transit timings indicated that the motion of the red spot was accelerating. A year later, writing in the December 18, 1902 issue of *Nature*, he commented that after some 23-years of uninterrupted retardation, the red spot's motion was now very definitely increasing. Denning suggested that the red spots acceleration might be due to the appearance of, "a large, irregular or multiple marking of a dusk hue, in the same latitude [as the red spot]". This new atmospheric feature, he conjectured, "may have forced the red spot along at a more rapid rate than that which it exhibited in previous years". French astronomer Michel Antoniadi noted the same phenomenon, and similarly suggested that the new disturbance had transmitted some of its *vis viva* to the red spot.

Only one notebook concerning Denning's Jupiter observations has survived, and this is held in the archive of the Royal Astronomical Society Library (RAS MSS ADD 174). This notebook concerns transit times and drawings of Jupiter made in 1898—at which time Denning would have been 50 years old. Collected over 121 nights (see Fig. 3.12) starting on 1 April running through to the end of July (with 2 additional observations being made in November), the notebook details 50 sets of transit observations. In total the minimum number of hours Denning committed to observing transits (taken from the time of the first observation on each night to the last) amounted to 76 h over the 121 nights, in which time he recorded 269 transit times. The results of these observations were published in the May 1899 issue of the *Monthly Notices*, where it is revealed that the notebook data indicated a rotation period of 9 h 55 min and 41.71 s for the red spot.

The data in the Jupiter notebook is well-ordered, and methodically arranged, and written in a clear hand. The sketches are crisp, and drawn with an expert eye for the necessary detail. Beyond the raw data, however, we find snippets of information about the local weather, and seeing conditions, with comments ranging from "nigh very clear", "very cold night", "windy, very cold", "splendid night", "definition all good", and "eclipse of the Moon—a beautiful night". We also learn of more mundane trials and tribulations, such

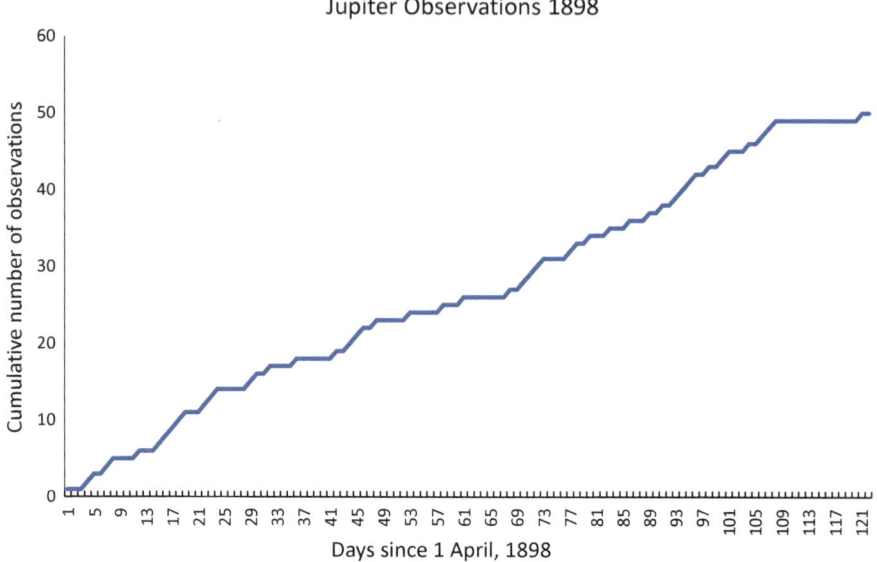

Fig. 3.12 Cumulative number of transit observations by Denning, versus day number since 1 April 1898. The data indicates that Denning must have been observing almost every night, with very few, presumably weather related, breaks

as, "shifted telescope in garden", as recorded on June 22—the reasons for this shift are unclear, but presumably relate to the viewing restrictions from neighboring houses and/or trees. Indeed, Denning's Jupiter notebook reveals some remarkable insight in to the activity of a dedicated amateur observer. Moreover, the notebook gives a clear indication as to Denning's commitment to his astronomical work. Not only was he observing Jovian transits over the April to July time interval, but he was also making observations of Mars and Saturn, and performing meteor survey counts as well.

In addition to providing a glimpse into Denning's observing practices, the Jupiter notebook also reveals Denning as a collector of information, and memorabilia. Pasted into the first 2 pages of the notebook are images of Jupiter (as drawn by Denning on 12 February, 1888) and Saturn (as drawn by Denning in 1880). Pasted onto pages 3, 4 and 5 are a set of images cut from various issues of *Popular Astronomy* (specifically volume 5, 1897, and volume 9, 1901) showing commemorative medals. Among the medals shown are several presented to Lewis Swift, including the Donohoe Medal of the astronomical Society of the Pacific, and the Jackson-Gwilt Gift and Bronze Medal of the RAS (awarded in 1897),[9] and two medals from the Imperial Académie of Sciences in Vienna. Several medals presented to Edward E. Barnard are also illustrated, including his RAS Gold Medal (presented in 1897), and the Prix Lalande of the Académie des Sciences of France (presented in 1892). Pasted on page 6 are a set of 4 drawings of Jupiter, made between April and June 1898, by the Reverend T. E. R. Phillips. These images being accompanied by a clipping concerning the various prizes being offered by the Paris Academy of Sciences in 1900. Pasted on page 7 is a full-disk pencil drawing of Jupiter made by T. E. R. Phillips on 16 January, 1899. A small undated (although one observation listed dates to 4 April 1899) notecard from Phillips is pasted on to page 9 of the notebook, and this discusses the variable rotation rate of the red sport.

In addition to the cuttings pasted into his Jupiter notebook, we again encounter the sad reminder that the vast majority of Denning's original notes and records are lost. Indeed, on page 8 of the notebook we find the written comment, "see other book for material and some results on the early history of the Red Spot", and the very last page of the Notebook contains the comment, "Observations of Jupiter 1899 are continued in the other book". There are also reminders of our incomplete knowledge about Denning's friendship with local enthusiasts. On 9 June, for example, Denning records that he had, "observed Jupiter using Mr. Fields telescope 6 3/8 [inch] refractor". Likewise, on the nights of 24 and 30 November he records that he was observing Jupiter, "in Webbs 4-in Cooke". No record or further information about Mr. Field has been found, but we suggest that Mr. Webb may have been J. Webb of Bristol, who is credited with the portrait picture of Denning published in the October 1897 issue of the *Observatory* (see also our Fig. A.2 in the Epilogue). Indeed, very little information exists about any of Denning's Bristolian friends—one fortuitous find, however, concerns a short article on "A Large Meteor" found in the November 1881 issue of the *Observatory*. This article is by William Barrett Roue, and begins, "as I was returning from observing Jupiter, in company with my friend Mr. W. F. Denning, F.R.A.S., of this town". William Roue (1850–1911) was a doctor and surgeon at the Bristol Royal Hospital for Sick Children and Women, and was elected a Fellow of the RAS in 1882. Various advertisements in *Hardwicke's Science-Gossip* for 1877 and 1878 indicate that Roue was an amateur naturalist, offering "British and Foreign Birds' Skins", birds' eggs and

---

[9] The page showing the medals presented to Lewis Swift must have been pasted into the notebook several years after the Jupiter transit observations of 1898 had been completed, since the pictures are from the November 1901 issue of *Popular Astronomy*. This same issue of *Popular Astronomy* contained an article by Denning on the accelerating motion of Jupiter's red spot.

"splendid slides of Algae", for exchange. How Roue met Denning is unclear—but it may have been through the Bristol Naturalists' Society (BNS). Founded in 1862, the first President of the BNS was corn merchant and amateur geologist William Sanders (1799–1875), and it is known that Barrett Roue read a paper "On comets and comet seeking" to the Society at its December 1883 meeting. Likewise, Denning had earlier published an article on "Shooting Stars" in the *Proceedings* of the BNS in 1879.

In happy contrast to his long-running complaint that British observers had not, in general, shown much interest in cometary searching, Denning was to praise them for their tenacity in observing Jupiter. Indeed, in 1920, with his health in serious decline, Denning was to write, "my abstention from planetary work has been practically enforced, but amid the regret caused thereby, I feel great satisfaction in the fact that others are pursuing it with much ability and energy" [16].

3.3.2   Saturn and Its White Spots

American astronomer Asaph Hall first recorded [17] the appearance of a large, and distinctive white spot in the atmosphere of Saturn on the night of December 7th, 1876. This was a completely new result, and at the time an unexpected mystery. More white spots, however, were soon to be observed in 1903. First detected by Edward E. Barnard on June 15 with the 40-inch Yerkes refractor in Chicago, this bright atmospheric feature was independently spotted by Denning in Bristol on July 1st (Fig. 3.13). In a letter published in the 9 July issue of *Nature*, Denning, ever eagle-eyed, announced his discovery, "On July 1, after observing Jupiter for some time, I directed my 10-inch reflector to Saturn, and found the details sharply defined. … I soon noticed a large bright spot on the north side of its [equatorial] belt". Although the origin of the white spots (or ovals) was unknown,[10] it was immediately realized that they allowed for the determination of Saturn's rotation period. Denning wrote on this specific topic in the December 1903 issues of *Popular Astronomy*, and the *Monthly Notices*, concluding in the latter that his transit timings at Bristol (made over 78 nights between 1 July and 11 December, 1903) indicated a rotation period for the white spots of 10 h 38.4 min.

Denning warned that great care should be exercised when studying Saturn, and especially so when reporting atmospheric markings. Indeed, he

---

[10] The exact mechanism behind the appearance of the white spots is still debated, but they are now known to be correlated with the times at which Saturn's northern hemisphere is pointed towards the Sun, and they are presumed derived from an atmospheric heating effect. The spots appear at intervals corresponding to Saturn's orbital period of 28.5 years, and were last observed in 2010.

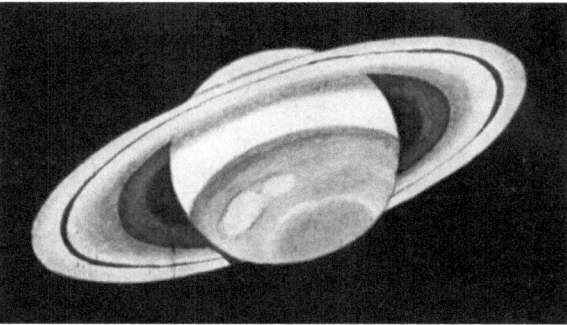

Fig. 3.13 Saturn showing its distinctive white spots (slightly left and down from center) drawn by Denning on 12 July 1903. Illustration from *Telescopic Work for Starlight Evenings*

commented in light of the 1876 white spot detection that, "perhaps there is no object upon which it is easier to exercise the imagination than upon Saturn. And there is probably no orb in reference to which more errors in detail have been made" [18]. Interestingly, Denning continued, "many of the abnormal results reported in recent years, and due to small instruments, may be safely dismissed, for they are not only doubtful but, when all the conditions are considered, ridiculous, and palpably the outcome of unconscious suggestions of the imagination". These comments show that Denning clearly appreciated that physiological effects could be very important when making planetary observations. They also reflect a change in Denning's attitude concerning the type of instrument that should be used in planetary studies. As discussed earlier (Sect. 3.1) Denning, like Lowell had argued that only small aperture telescopes should be used for planetary work. That Denning had changed his mind concerning the capacity of large telescopes is further exemplified by his comments[6] printed in the *Journal of the Royal Astronomical Society of Canada* in 1918. There he explained, "more than a generation ago there was an animated discussion as to the relative merits of large and small telescopes in dealing with detail on bright planets, and I argued that, judging from published observations and drawings, the great instruments previously in use could be regarded as possessing very little, if any, superiority.... But better instruments have undoubtedly been constructed in later years and there seems no reason to doubt that the great refractors of the present day are decidedly more effective in studies of planetary markings than smaller instruments". We see here a different Denning—a Denning prepared to capitulate on his ideas in light of new data. Throughout the stationary radiant debate (recall Chap. 2), for example, Denning had refused to review and/or refine his methods, steadfastly sticking to his beliefs—no matter how much at odds those beliefs might be with respect to accepted wisdom. When it concerned telescopes, however, we find Denning to be an observer who could appreciate advancements in optics, and construction methods. Once more, Denning's outlook was directed by a Baconian, conservative approach, the truth being revealed by time and experienced observation.

3.4   Novae and Nebulae

By their very nature the appearance of nova and supernova cannot be predicted, and their discovery must rely on serendipitous circumstances [19]. When Denning wrote his *Telescopic Work for Starlight Evenings* the mechanisms underlying nova and supernova outbursts were completely unknown, but he, like others, knew that these new and temporary stars required an exceptional explanation. Denning commented [20] in *Telescopic Work* that he had, "frequently, while watching for meteors, reviewed the different constellations in the hope of picking up a new object, but have never succeeded in doing so". Thirty years after writing those words he was to finally realize his wishes.

In a remarkable three-year period between 1918 and 1920, Denning was witness to the discovery of two nova. While his priority of discovery for the nova of June 1918 (nova V603 Aquilae) was not to be established, Denning was certainly one of the first independent observers to see it.[11] He wrote [21] of its discovery, "on commencing a watch for meteors on June 8th, I immediately observed a new star of considerable brilliancy had made its appearance in the western border of Aquila". That the nova was "immediately obvious" bears testament to Denning's intimate knowledge of the constellations. He observed the nova for three hours that June night, and made comparative estimates of its brightness.

Nova Cygni (V476 Cygni) was first observed by Denning on the night of August 20th, 1920. Again, he had set out to begin a meteor watch but had quickly noticed a new magnitude 3.5 star in the constellation of Cygnus [22]. Upon realizing that a new star had appeared, Denning wasted no time, and immediately sent a telegram to the Royal Observatory, at Greenwich. An extended visual and photographic study of the nova was soon initiated at that observatory, and a light curve was published [23, 24] by Willem J. Luyten in November of that year. Additional data on brightness was also gathered

---

[11] First sighting is generally given to Polish amateur Zygmunt Laskowski. Alice Grace Cook (see Chap. 5) was also one of the first observers to record this nova.

by the AAVSO[12] to the end of 1926 (Fig. 3.14). Denning received a great deal of correspondence concerning nova Cygni, and he wrote to his niece (Christine Gravely) on September 26th that, "the new star brought me about 100 letters extra, and the event seems to be regarded as a very important one in the astronomical world".[13] That Denning had received so many letters because of this one phenomenon underscores the fact that, even at age 72 years, he was still the focus of a great deal of correspondence.

Temporary stars (new stars = novae) have been observed throughout history, but astronomers only began to take an interest in them from the late sixteenth century onwards. The new star of 1572, which appeared in the constellation of Cassiopeia, caught the eye of Tycho Brahe, and his extensive investigations, and attempts at parallax measurements clearly revealed that the new object, whatever it was, was in the celestial realm, and not some atmospheric phenomenon. That novelty could occur in the heavens, literally that new stars might appear, was an important philosophical point in

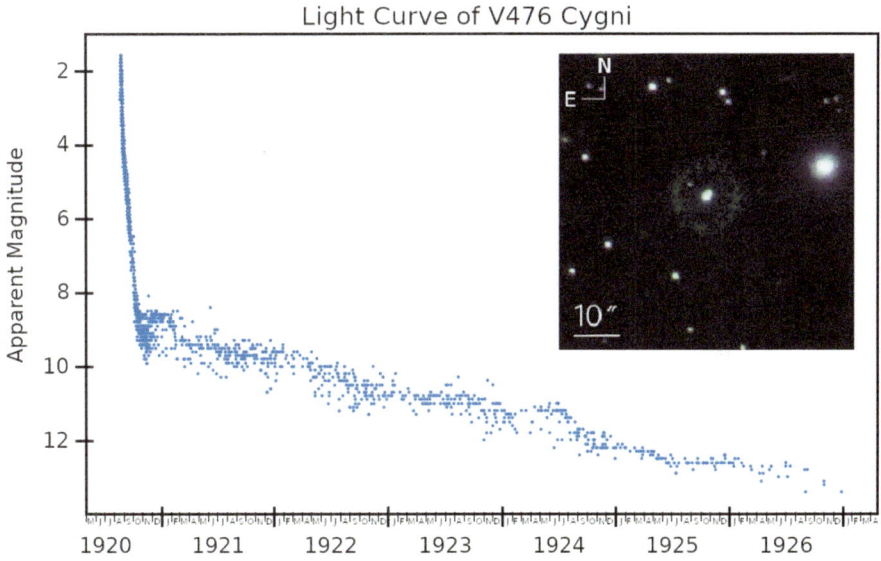

Fig. 3.14 Light curve of nova Cygni (V476 Cygni). Data and light curve image courtesy of the AAVSO. The inset shows nebula structure (faint ring around central star) in 2018, some 92 years after the nova's appearance. Inset image courtesy of E. Santamaria et al. 2020. Angular expansion of nova shells. *The Astrophysical Journal*, **892**, id 60

[12] The American Association of 'Variable Star Observers (AAVSO) had been founded by William Olcot in 1911, and continues to thrive to this vary day.

[13] Maurice Brain, personal communication, 1989. The letters from Denning to his niece date from November 25th, 1919, September 4th, 1923, and September 26th, 1923.

dislodging ancient dogma, and helped in part, to usher-in the new astronomy being promoted by Copernicus, Kepler and Galileo. The origin of new stars, however, remained entirely unknown, and it was not until the late nineteenth century that the first formation models were proposed. Indeed, the first integrated model of nova, or temporary star production was that articulated by Scottish mathematical physicist Peter Tait in the 16 December, 1869 issue of the journal *Nature*. Drawing upon the new spectroscopic results relating to nebulae (to be discussed shortly), Tate argued that many (but not all) nebulae were in reality glowing clouds of hot gas, and he suggested that temporary stars were either a cooled-off star, with a shell of glowing gas (or nebulosity) about it, or that they were a, "vast system of small cosmical masses in the act of grouping themselves by mutual gravitation, impact and friction into a new star". Novae accordingly represented either the reinvigoration of an old star, or the *bona fide* formation of a brand-new star. Key to both of these formation mechanisms was the idea that the space between the stars (the interstellar medium) was filled with meteoric streams. These streams were composed of myriad small cosmic masses, or meteorite-like bodies. Tait, and later Joseph Norman Locker, developed the idea that the heating and effusion of hot gases required to account for the spectroscopic observations of temporary stars, were the result of collisions between the meteoritic masses from within intersecting swarms and streams. Lockyer, in his *The Meteoritic Hypothesis* (published in 1890), for example, argued that the appearance of some temporary stars (he specifically had the nova of 1876, Q Cygni, in mind), might be explained by the intersection of numerous streams or sheets of meteoritic matter—that is, by "the clash of meteor swarms". Arguing that, if the cosmic streams could be compelled to orbit around a common center (a seed cosmic swarm of meteorites) then a planetary nebula like appearance would result (Fig. 3.15).

Drawing upon his intimate knowledge of the night sky, Denning made an interesting, and prescient observation in *Telescopic Work* concerning the distribution of novae. Specifically, he noted, "it is remarkable that nearly all the temporary stars have appeared in the region of the Milky Way". The data that Denning had to work with was limited—just 8 novae (Table 3.1), but it is the case that their sky distribution does approximately follow the plain of the galaxy (Fig. 3.16). This was fortuitous since the novae of 1860 and 1885 actually occurred in galaxies beyond our own—although the term galaxy, as representing a distinct system of stars, had not been recognized when Denning was writing (see below). Likewise, nor was it then known that there were different types of novae.

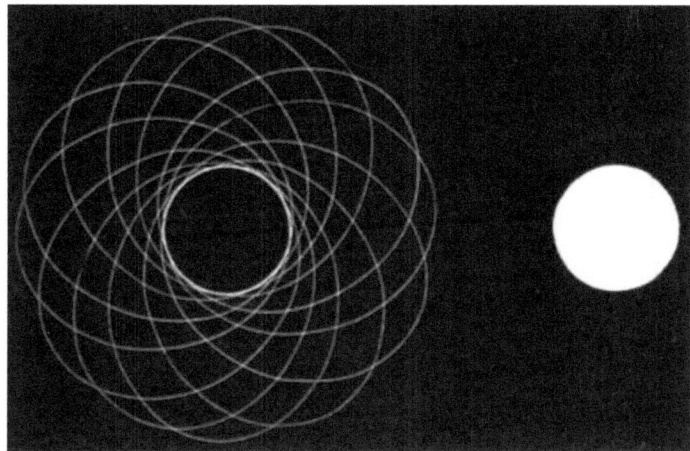

Fig. 3.15 Lockyer's scheme for the appearance of planetary nebula and temporary stars. "The luminosity is due to the collisions occurring along the sphere of intersection of the elliptical orbits of the meteorites. The left-hand diagram is a cross-section of the meteoric system, and the right-hand one shows the appearance of the collision shell as seen from a point outside". Illustration from Lockyer's *The Meteoritic Hypothesis*

Table 3.1 Galactic nova during the time interval 1572–1885

Year	Constellation	System	Type	d (°)	Comments
1572	Cassiopeia	–	SN	0	Tycho's supernova
1604	Ophiuchus	–	SN	20	Kepler's supernova
1670	Vulpecula	CK Vul	N	0	
1848	Ophiuchus	V841 Oph	N	20	
1860	Scorpius	T Sco	N	0	Nova in M80
1866	Corona Borealis	T CrB	RN	45	Last outburst in 1946
1876	Cygnus	Q Cyg	N	0	
1885	Andromeda	–	SN	15	Supernova in M31

Key: N = nova with a poorly known light curve and speed class, RN = recurrent nova, SN = supernova. Column 5 is the approximate angular distance, in degrees, of the nova away from the plane of the Milky Way on the sky

When Denning made his comments concerning the distribution of novae in 1890, the physical understanding of stars, star distributions, and nebula were still in their infancy. The study of nebulae effectively began in 1715, when Edmund Halley read a short paper concerning lucid spots to the assembled Fellows of the Royal Society in London. These lucid or glowing clouds, Halley argued, represented luminous entities that shone under their own light, and, as such, they were a complete mystery. Halley's list of six prominent lucid clouds contained, in modern terms, a star formation HII complex

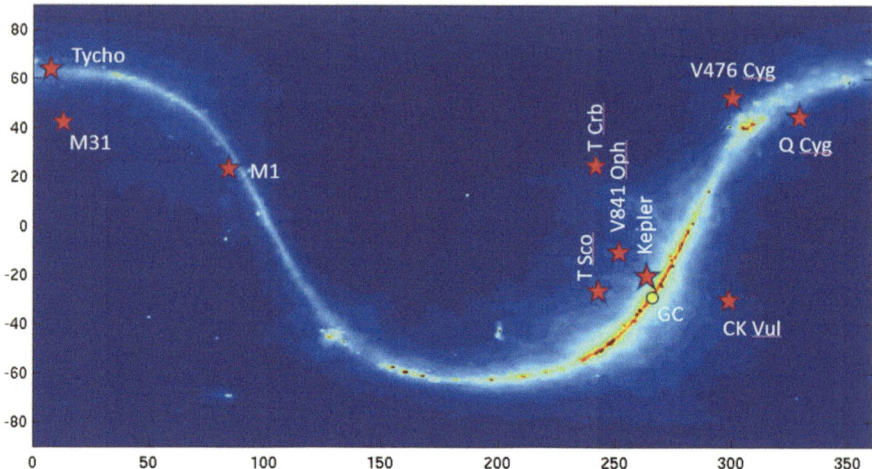

Fig. 3.16 Denning's list of novae plotted according to their right ascension and declination. The U-shaped arc corresponds to the plane of the Milky Way. The galactic center is marked as GC. The position of the Crab Nebula (M 1) and Nova Cygni (V476 Cyg) are also indicated in the diagram (see text for details)

(in fact, the Orion Nebula = M 42), a spiral galaxy (the Andromeda galaxy = M 31), three globular clusters (including ω Centauri = NGC 5139), and a galactic (or open) cluster (the Wild Duck cluster = NGC 6705 = M 11). All that Halley (and his contemporaries) could say about these nebulous objects was that they were novelties (that is, they did not have the appearance of stars, planets, asteroids or comets), importantly, however, the fact that they showed no parallactic motion indicated that they must be located within the starry realm. The first attempt to make sense of the distribution of stars on the sky, and specifically the appearance of the Milky Way was that by itinerant lecturer Thomas Wright of Durham. Writing in his *An Original Theory or New Hypothesis of the Universe* (published in 1750). Wright argued that the appearance of the Milky Way indicated that the Sun and planetary system was located within of a flattened disk of enormous size and composed of numerous stars. At play, Wright, suggested was an 'optical effect'. When looking into the depth of the disk an observer would see multitudes of stars; when looking above or below the disk, however, an observer would see relatively few stars. Wright ultimately changed his mind on the disk-like star distribution, but it was later revised and expanded upon by William Herschel. Building upon his numerous star gauges, Herschel, in 1785, introduced the idea that the stars were indeed distributed in a disk-like manner, with the added feature of a split, or cloven, section in the direction of the constellation of Sagittarius (Fig. 3.17). While Herschel was busy counting stars and cataloging new nebula, German philosopher and enlightenment thinker, Emmanuel Kant, had mis-interpreted Wright's ideas in his *An Original Theory*, and introduced the idea that some of the nebulae might be 'island universes' composed of myriad stars—objects that in the modern era we would call galaxies. Furthermore, Kant, in his *Universal Natural History and Theory of the Heavens* (published in 1755), speculated on the origin of stars and planetary systems, outlining the basic concept that became the nebula hypothesis.

In addition to studying the distribution of stars in space, astronomers in the late eighteenth century were continuing to find more and more nebulae. There was little consensus as to what these nebulous clouds were, but it was generally accepted that there were at least four basic types. These being spiral nebulae, planetary nebulae, dark nebulae, and the extended and irregular shaped nebulae. William Herschel introduced the term planetary nebula in 1782 after studying the so-called Saturn Nebula (NGC 7009), describing it as "a star surrounded by a cloud of true nebulosity", suggesting that it might be a star in the process of forming. This was a daring idea since, if true, it implied, for the first time, that the stars had individual times of birth, and possibly different ages and even lifespans (all these possibilities becoming certainties only later with the rise of astrophysics in the early twentieth century). The dark nebula, now recognized as dense clouds of interstellar dust, were dark maculations set against the backdrop of stars in the Milky Way—famous amongst these is the Coalsack Nebula in the southern constellation of Crux. By the mid- to late-eighteenth century, Charles Messier had begun the process of cataloging nebulae, and produced a series of catalogs, between 1774 and 1781, of these, so-called, nuisance objects—objects, that is, that should be avoided lest they be confused with the appearance of a new comet. The very

Fig. 3.17 The cloven-disk distribution of stars as envisioned by Richard Proctor in 1870. The Sun is located at the center of the disk, which in turn is cloven, or split, into distinct slices

first object in Messier's catalog, M 1, is located in the constellation of Taurus, and is now recognized as the remnant of an exploded star, a supernova, that occurred on 4 July 1054.

Among the 30 papers that William Herschel read at the Royal Society in 1780, one was concerned with his General Catalog—a compendium of some 500 nebulae. His great paper on the construction of the heavens, however, was read to the Royal Society in 1810. In this monumental work, consisting of some 67 pages in the *Philosophical Transactions*, Herschel set about describing the large-scale structure of the universe. Herschel, however, vacillated with respect to his interpretations of the nebula, but in grand Newtonian fashion, was to write that it was through the "action of gravitation" that nebulous matter was being brought together to form new stars and planets. Extending his father's work on star mapping and nebulae 'bagging', John Herschel brought-out his *General Catalog of Nebulae and Clusters of Stars* in 1864. This catalog contained information on some 5079 nebulae. John being more cautious than his father, however, made no attempt to speculate upon the various nebulae types nor how they might have formed.

While William and John Herschel directed their attention towards cataloging and mapping, other observers began to study, with increasingly large telescopes, individual nebula and star clusters. A pivotal player in this respect was William Parsons, the Third Earl of Rosse. Independently wealthy, university trained in mathematics, and looking to make astronomical observations, Parsons experimented with the casting of large speculum mirrors, and in the 1840's had built a series of large telescopes. Working from his estate of Birr Castle in Ireland, Parsons scanned the heavens with unprecedented telescopic power, and novelty soon appeared. It was in 1844 that Parsons, using a 24-inch aperture telescope, observed M 1, the first nebula in Messier's catalog, and from amid its complex glowing arcs and milky nebulosity saw hints of a crab-like body, coining the name Crab Nebula (NGC 1952), which has since become one of the most celebrated objects in the entire sky.[14] In the following year, Parsons oversaw the construction and commissioning of a giant 72-inch telescope—the Leviathan of Parsonstown. This was then the largest telescope in the world, and Parsons was determined to use it in the study of nebulae. Accordingly, in 1855 he turned the great telescope towards M 51, and found that he could resolve spiral features within the glowing nebulae. Parsons went-on to study other supposedly diffuse nebulae with the Leviathan, and was able to resolve many of them in to vast stellar systems. Parsons new observations resulted in the general suspicion that all nebulae

---

[14] It has been said that astronomy can be divided into two main branches: the study of the Crab Nebula, and the study everything else.

were, in fact, star clusters of one form or another. Their fuzzy, or nebulous appearance being simply a consequence of their great distances, and a lack of telescopic resolving power. The spectroscopic studies of William Huggins and William Miller in the mid-1860s (see Chap. 4), however, soon showed this idea to be wrong—some nebulae were genuine clouds of glowing gas, while others, and especially the spiral nebulae, were vast star systems.

Moving into the late 1860s, it became increasingly clear that not only were the spiral nebulae composed of stars, but that their distribution on the sky was different to that of other nebulae. While the glowing gas clouds were invariably set against the backdrop of the Milky Way (recall Fig. 3.19), the spiral nebulae, in contrast, were seemingly only to be found well above or well below it. A zone of avoidance was evident, and Richard Proctor in his *Other Worlds than Ours* (published in 1870) drew particular attention to this fact, writing, "if this peculiarity is accidental, the coincidence involved is a most remarkable one". Indeed, the question was a profound one, and even at the close of the nineteenth century, Charles Young in his widely read *A Text-Book of General Astronomy* (published in 1899) was to write, "why the [spiral] nebula avoid the region thickly starred is not yet clear". Ultimately, in the 1920s, Edward Barnard and Robert Trumpler were able to demonstrate that the zone of avoidance was cause by the absorption of starlight by interstellar dust in the plane of the Milky Way galaxy. Furthermore, moving deeper into the twentieth century, the understanding of the cosmos evolved dramatically. New methods of distance estimation emerged, and it became clear that the universe was filled with remote and isolated galaxies, the Milky Way being just one of many billions.

When Denning was writing his *Telescopic Work* (in 1890) very little was known with any certainty about galactic structure. His observation that novae are always found close to, or within the Milky Way (recall Fig. 3.19) is therefore an understated, but revolutionary claim. It is the case that the circumstances leading to the production of novae were not known when Denning was writing, but he was drawing attention to something new within the zone of avoidance that others had missed. The zone of avoidance contained novae, but not spiral nebulae. Denning's idea about the restricted sky distribution of novae would remain true until the early 1920s. Indeed, starting in 1917, Heber Curtis, in America, realized that the available photographic survey data indicated that novae in the Andromeda nebula (M 31) were systematically fainter than their counterparts seen against the backdrop of the Milky Way. This, he argued, implied that M 31 was a distinct assembly of stars (a galaxy) independent of the Milky Way. These observations became part of the (so-called) Great Debate between Curtis and Harlow Shapley in

1920, although it was only in 1925 that Edwin Hubble was able to show, through the study of Cephid variable stars, that both M 31 and M 33 were located at vast distances from the Sun, and were accordingly individual star systems in their own right.

3.5   Wasps, Birds, and Natural History

Denning's inquiring gaze was not always directed skyward and it would seem that he was a keen natural historian. As indicated in Chap. 1, Denning was interest in natural history from an early age, and it is reasonably clear that he continued such interests throughout his life. Most of what we know of Denning's studies in these other areas is contained in the few surviving letters to his niece, and in a handful of articles published in *Nature* and the *Observatory* magazine.

Virtually nothing is recorded with respect to Denning's early interests in natural history, and it is not until circa 1912 (when Denning would have been 64 years old) that we find clear evidence of any dedicated work in this area. The details of his interests first became evident in 1916, at which time Denning was drawn into a debate concerning annual wasp populations. The question of wasp scarcity had been raised by Davis [25] in the 12 October issue of *Nature*, and Denning responded in the 14 October issue [26], writing,

> I may say that in this district ordinary wasps have been decidedly scarce this year... I make a point of cultivating these insects, as they are extremely interesting to watch, and destroy myriads of flies every summer. There were six embryo nests in my garden in May last, but only one (Vespa vulgaris) managed to withstand the vicissitudes of the inclement weather. This nest was a weak one, for when I dug it out on September 20 it consisted of four layers of cells, the top one alone being for small working wasps (1000 cells), while the others were exclusively for queens and drones (1250 cells). This proportion is quite exceptional according to my own observation, for I have commonly found the smaller cells greatly in excess of the others.

Denning continued in his letter that wasps were not as aggressive as commonly supposed, and that they "display remarkable industry and activity". To this he added that, "on a bright summer day in 1913 I carefully watched the entrance of a wasp's nest in my garden, and concluded that the insects brought home at least 2000 flies". Clearly Denning had made some careful observations of hive activity.[15] He wrote to *Nature* again [27] on his observations in 1920, and in this communique, we learn that Denning had been, "observing wasps during the past eight years" i.e., since 1912. In this second letter Denning presented a table of his observations of hive activity for the summers of 1915 and 1918.

It is clear from his writings that Denning was disheartened by the common practice of destroying wasp nests, and he wrote, "Man often misapprehends the benefits from certain forms of animate nature. Birds are destroyed and noxious insects enabled to multiply. Efforts are ever being made to exterminate the wasp, and hordes of pestiferous flies naturally become the bane of our summers" [27]. Such sentiments would suggest that Denning was concerned for the conservation of nature, and that he appreciated the dangers of blindly meddling with ecological systems.

Denning's ornithological interests are evident from a letter he wrote [28] to the *Observatory* magazine in May of 1915. Writing on *Birds that Pass in the Night*, Denning suggested that an appropriately experienced astronomer might study the passage of birds, and thereby gain some useful knowledge on nocturnal bird behavior, and migration. From his own experience Denning explained, "the nightjar at certain times of the year is often in evidence; but the bird which, more prominently than any other, makes its presence known is the redwing, for every spring and autumn, during many weeks, droves of these fugitive nocturnal itinerants may be heard passing above almost incessantly".

During the closing years of his life Denning only rarely left the confines of his home, and this isolation, although self-imposed, did cause him some regrets. He still made observations, however, and in a letter to his niece Christine Gravely dated November 25th, 1919 (the day of his 71st birthday) he commented, "I have to observe from my garden but one needs to go farther afield, although it is quite astonishing what one can see of bird life even in a limited place like the surroundings by houses". Just over one year later he was again to write to his niece that, "I hear few birds in this locality just at present except the Robin and we have them singing every day—I like their song better than that of any other British bird". Interestingly, in this same letter Denning comments that he has been thinking that he would, "like to cultivate bees" and that he intends to enquire about the cost of hives. It is not known if he followed through on this desire.

---

[15] If Denning was ever to have a coat of arms, surely a bee should be at its center. The parallels between hive industry and activity and Denning's approach to observing seem remarkably congruent.

3.6   Meteorology

Sunshine is delicious, rain is refreshing, wind braces us up,
  Snow is exhilarating: there is really no such thing as bad weather,
  Only different kinds of good weather.
John Ruskin

The documentation and discussion of weather conditions, be it rainfall amounts or hours of sunshine, is an obsessive and near limitless human pastime. In line with this, and parallel to Francis Galton's passion for mensuration and statistics, extensive datasets of meteorological observations were collected during the nineteenth century. Increasingly, however, once the measurements were made, the overriding question became what can all this rain-gauge data tell us? That is, what does all the data actually say about weather patterns, climate variations, and the Earth's atmosphere? From such questions, coupled with the Victorian zeal to quantify, order and organize, meteorology as a branch of science, rather than a field of folk-lore, began to evolve. As one might well expect, Denning became fully caught-up in this new enthusiasm for the scientific investigation of wind, rain, sunshine, and snow.

At the very first 1931 gathering of the British Association for the Advancement of Science (BAAS), the weather was on the agenda. Indeed, the dismal state of meteorology, as a scientific subject, was decried by Scottish physicists and glaciologist James Forbes. What was needed, Forbes argued, was a dramatic transformation—meteorological observations needed to be standardized, more reports were needed from geographically diverse locations, and the data so collected needed to be carefully analyzed in order to reveal theories and laws. Meteorology, Forbes enthused, was the ideal subject for amateur and professional collaboration, and it held the potential for great and new scientific discovery—indeed, it was a topic crying out for support and encouragement. Rather than build observatories, however, the BAAS initially set about funding experiments on standardization, self-logging data machines, and equipping observers with the apparatus needed to supply data. Perhaps the first major success to follow from BAAS funding was that supplied to Sir John Herschel, and his long-time mathematical assistant William Radcliffe Birt, in 1839. Specifically, the BAAS funds had enabled Herschel and Birt to study barometric observations gathered from across Europe between 1835 and 1838. The analysis was complicated, tiresome, and time-consuming, but the end result (presented to the BAAS in 1843) supported the idea that Earth's atmosphere could support the propagation of extensive pressure waves, and waves, of course, were something that mathematical techniques could be set against. The science of meteorology was beginning to emerge, but the enthusiasm to conduct the tiresome analysis waned, with both Herschel and Birt turning the attention to other topics after 1850. Indeed, Birt became a renowned selenologist, writing extensively on the topic in the *Astronomical Register* (see Chap. 6), and he founded the (short-lived) Selenographical Society in 1878. Birt also produced a number of reports for the BAAS, from 1865 to 1869, on mapping the Moon, this project being carried out with the assistance of astronomer Edmund Neison (see Chap. 7).

In 1842 the BAAS took over the running of Kew Observatory in London (with Birt initially being involved in the recording of both meteorological and magnetic measurements), and this office they continued to fund until 1872, after which time the Royal Society assumed responsibility for operations. In the between time, the British Meteorological Society (BMS) was founded in 1850. Meeting in the plush library setting of John Lee's Hartwell House, in Buckinghamshire, brewer Samuel Whitbread became the first President, but it was James Glaisher who became its public figure-head and driving force. The aims of the BMS were straightforward and practical, looking to promote, "the advancement and extension of meteorological science by determining the laws of climate and of meteorological phenomena in general". Glaisher was superintendent of the Department of Meteorology and Magnetism at the Royal Observatory in Greenwich, and a founding member of the Aeronautical Society of Great Britain, in 1866. Indeed, Glaisher rose to great public notice through his balloon ascents (and adventures) conducted with Henry Coxwell, between 1862 and 1866. Funded under the auspices of the BAAS, some 28 balloon flights were made in order to measure variations of temperature, barometric pressure, and humidity with height in the Earth's atmosphere. Glaisher and Coxwell nearly died during one early ascent, to a then record-breaking height of 8.8-km, on 5 September 1862. With the jamming of the balloon's descent valve, Glaisher fell into unconsciousness, and Coxwell, had to climb from the balloon's basket in order to release the stuck valve chord with his teeth—his hands being frozen and numb (Fig. 3.18). The BMS underwent a number of name changes, becoming the Meteorological Society in 1866, and the Royal Meteorological Society (RMS) in 1883. Denning joined the Meteorological Society in 1872—the same year that he applied, albeit unsuccessfully, to become a Fellow of the Royal Astronomical Society.

While the aims of the BMS (later RMS) were broad-based, the more data-orientated British Rainfall Organization (BRO) was formed by the young,

Fig. 3.18 The perils of early scientific ballooning. During a record-breaking ascent made on 5 September 1862, Henry Coxwell had to climb from the balloon's gondola in order to release a stuck release valve. James Glaisher (right) had, by this time, lapsed into unconsciousness

20 years old, George James Symons in 1859. The aims of the BRO were simple, at 9 a.m. each morning observers (scattered across the British Isles) would record the amount of rain that had fallen during the past 24 h, and then send their measurements to Symons.[16] The data so gathered would then be analyzed by Symons in a standardized fashion. Starting in 1862 Symons produced the first volume of *British Rainfall*—a series of annual reviews than ran until 1968. In the same year, Symons approached the BAAS for funding

---

[16] Symons became a prominent contributor to the British Association, providing various reports on lightning conductor research, and contributions relating to the great eruption of Krakatoa in 1883. He was elected a Fellow of the Royal Society in 1878.

in order to off-set the costs of producing standardized rain gauges, observatory inspection visits, and publications. This they dually consented to do, recognizing that the BRO provided an important source of meteorological data. Indeed, in 1865 the BAAS established, under the chairmanship of James Glaisher, a Rainfall Committee to oversee its spending in the area, and to report on new advances. The membership of the BRO went from strength to strength, with the initial 168 observers in 1860, becoming some 500 by 1861; rising to some 2000 by 1876, and nearly 3500 by 1900, and 5000 by 1915. Such growth is remarkable and highlights the great public interest in amateur science at the close of the nineteenth century. Furthermore, the enthusiasm of so many observers, engaged in a national group-effort, could only act to encourage the growth of national amateur societies in general, whether they be meteorological, geological, or astronomical. BAAS funding of the BRO continued until 1875, at which time it was deemed desirable that the British Government should take-over the responsibility of maintaining costs—this eventually happened in 1919, when the BRO became a branch of the Meteorological Office.

Denning submitted many notes on rainfall and local weather conditions to various meteorological journals, including *Symons Meteorological Magazine* (first published in 1866) and the *Quarterly Journal* of the RMS (first published in 1871). Indicating that he was dedicated to the spirit of the BRO, Denning published in the *Quarterly Journal* of the RMS (in 1877) a paper concerning a sixteen-month review of the rainfall data he had gathered at Bristol between July 1874 and November 1875. The mean daily rainfall during the sixteen-month interval was 0.122 inches, and the total rainfall in the time interval concerned exceed the 20-year average by 18.946 inches. The 20-year average was derived from the data collected by George and William Burder, at Clifton between 1853 and 1872.

William Corbett Burder (1822–1865) trained as an architect in Bristol and apprenticed as an engraver in London under the tutelage of J. H. Le Keux. Following a bout of ill health in 1852 Burder left the engraving profession, and turned his attention to landscape painting. An acute observer of nature he additionally turned his attention to meteorology. Starting in 1853 he kept a daily record of rainfall, temperatures and barometric variations. His first monograph, *The Meteorology of Clifton* was published in 1864, and this summarized ten years' worth of continuous daily meteorological records.[17] Burder was elected a Fellow of the RAS in 1852, and was known as a keen instrument maker, and an inventor of mechanical aids for

---

[17] The author chanced to purchase of copy of Burder's book in 2022, with the unexpected, but pleasing find, that it contained Denning's signature on the inside front cover.

the infirmed. The anonymous compiler of Burder's obituary [29], concludes his account by writing: "sociable without frivolity, and well informed without a trace of pedantry, of a cheerful disposition, a ready humor and transparent honesty of purpose". George Forster Burder (1824–1892) was brother to William, and trained as a medical doctor. With a practice in the Clifton area, he continued William's daily meteorological observations, and published, in 1872, the results of 20 years' worth of data. In his obituary [30], George Burder is described as, "a thoroughly clear, logical thinker, and endowed with good common sense, Dr. Burder possessed as a speaker the rare power of expressing his thoughts in forcible, well-chosen language".

While Denning's rainfall accounts are all matter-of-fact, he did occasionally observer more unusual and interesting meteorological phenomena. In the July 1915 issue of *Symons Meteorological Magazine*, for example, Denning reviewed the circumstances of a tremendous thunderstorm that had occurred in the Bristol area, producing much damage, on July 4. He specifically noted the occurrence of large hail stones, some being more than 1-inch across. Indeed, he had collected a sample of these hail stones, and recorded that many were pyramidal in shape, rather than being spherical or oval. He additionally included a diagram of several hail stones in which he revealed their markings, shape and structure. In addition to daytime storms, Denning also observed many remarkable nighttime events while out recording meteors, including the sighting of lunar rainbows, and parselene [31]. Perhaps one of the most spectacular sights that he reported on was that of the twilight glows associated with the Tunguska event [32] of 30 June 1908. This remarkable catastrophe remains largely unsolved to this day. It is clear, however, that on that June day an explosion of some 10–15 megatons of TNT equivalent energy occurred at an altitude of about 10-km above the Tunguska region of Siberia. It is still debated as to whether the impactor was a small (some 30–50 m across) cometary nucleus, or a similar sized stony asteroid. Certainly, some 2150 km^2 of forest were decimated by the explosion, but no impact crater, or impactor material, has ever been unambiguously identified.[18] For all this, however, following a notice in the May 1930 issue of the *Observatory*, concerning a talk by F. J. W. Whipple on the seismic and atmospheric waves associated with the Tunguska event, Denning recounted his experiences, noting that he was observing meteors on the evening in question, and that the, "illumination of the atmosphere was remarkable" [33]. Indeed, he continues, "at midnight a game of cricket was played at Durdham Down,

---

[18] It has been suggested that the impactor may have been a fragment, dislodged from comet 2P/Encke, that became embedded within the beta Taurid meteoroid stream. The latter daytime meteor shower reaches its maximum activity on June 30th each year.

Clifton, and various other avocations, only possible in the half-light of an ordinary night, were freely indulged in". Concerning the evening of 1 July 1908, he recollected, "during more than 60 years of night observations I cannot recollect seeing the firmament so light".

During his many meteor watches, Denning often recorded that he had seen auroral activity. Indeed, in 1885, he commented that, "scarcely a very clear night passes but that there may be traced, with a critical eye, some feeble traces of aurora". This is no-doubt an exaggeration, with strong or noticeable auroral activity being relatively rare at the latitude of Bristol [34]. None the less, the aurora was a sky phenomenon that was only poorly understood, in terms of physical theory, well into the twentieth century. Not only was the physical origin of the aurora unclear in the nineteenth century, so too was it unclear if auroral activity might have some influence on lower atmosphere weather phenomena. In this respect, it is not unreasonable to find Denning, as well as many other observers, speculating on possible correlations. Writing in the 6 September 1872, issue of the *English Mechanic and World of Science*, for example, Denning, building upon a comment that he had read in the *Western Mercury* newspaper, suggested that strong winds and blustery weather often followed 2–3 days after an auroral event. He draws upon his own observations, made in August 1871, and suggested that other observers might have additional information to support, or discredit, the correlation. In similar vein, writing a few months later in the 7 February, 1873 issue of the *English Mechanic*, W. R. Birt was to suggest that a correlation might exist between rain storm strength and the presence, or not, of auroral activity. Even though Birt ended his short note with the comment that, "perhaps some bold theorist may find a fitting place for these odd facts". No theory was forthcoming—indeed, it is now clear that no such correlation should be forthcoming. Likewise, some observers have reported an apparent correlation between auroral brilliance and the passage of a meteor. Indeed, Irish astronomer William H. S. Monck, commented on such a possibility in the February, 1887 issue of the *Journal of the Liverpool Astronomical Society* (JLAS). Responding to a letter by Major A. Veeder (Lyons, New York) published in the 9 December 1886 issue of *Nature*, indicating that he had witnessed meteors effecting aurora brightness, Monck asked JLAS readers if they too had any anecdotes to pass on. Indeed, Monck noted that, "if meteors consisted of some magnetic substance which was pulverized or dissipated in the upper regions of the atmosphere, such a causal connection might well occur. Denning's response in the March, 1887 issue of the JLAS was pithy— "I believe there is no good reason for considering that any physical correlation exists between these phenomena".[19] Indeed, Denning is correct [35], but the exchange of letters indicates the generally poor understanding of meteoroid structure and the ablation process in the late 1880s (see Chap. 5).

While the state of meteorological knowledge in the late nineteenth century did allow for much physical understanding and progress to be made, one important correlation between meteors and the atmosphere was derived. Specifically, it was realized that meteors could be used to probe the density variation of the Earth's upper atmosphere—we shall pick up on this topic more fully in the next chapter. Additionally, it was realized that upper atmospheric winds could be studied by monitoring the deformation of meteor trains (see Chap. 5). In exceptional circumstances, bright fireballs can leave glowing trains along their atmospheric flight path, with the trains being visible for many minutes to in extreme cases hours on end. Such trains will often become warped and distorted according to the different wind speeds operating at different atmospheric heights. One exceptional case of a long-duration meteor train was studied by Denning in 1909. In this case a very bright (brighter than the full-moon) fireball was observed off the coast of France, at 7:33 in the evening, on February 22nd. Associated with this fireball was a glowing train that lasted for some 3-h. Using some 122 observational reports [36], Denning deduced that the initial fireball trail was some 155-miles long, traversing a region of the atmosphere between 56 and 41 miles in height, remarking that, "the phenomenon may be aptly described as the meteoric spectacle of a generation". Using the reports from observes situated in France, the Channel Islands, Devon, Wales and across England, Denning was able to map-out the trail's configuration as it drifted northward (Fig. 3.19). As the train moved northward, at a speed of some 80–100 miles per hour, its upper and lower portions became noticeable curved. While it might seem reasonable to attribute the curved ends of the train to differential wind speeds in the beginning and end height regions, Denning seemingly missed this point, and with respect to its western most bend (corresponding to lower atmospheric heights) he questioned if the ablating meteoroid had exploded, or had been deflected from its original course by, "some potent influence" [36]. That meteoroids might be deflected as they travelled through the atmosphere was not a new idea, and there have been many historic reports of

---

[19] One situation where meteors may be related to a meteorological phenomenon is that concerning sprites, jets and elves. These elusive phenomena are associated with strong thunderstorm systems, and are seen as upward propagating, lightning-like, discharges [35].

curved or spiraling meteor trails.[20] In this latter situation it is not a long-lived train that is undergoing distortion, but it is the flight path of the meteoroid, as it ablates in the atmosphere, that deviates from a straight-line path. Denning, by invoking such an in-flight deflection for the 1909 fireball, was to a certain extent reversing the opinions that he had previously expressed on "erratic meteors". Writing in the January 1887 issue of the *Monthly Notices*, Denning had argued that apparently, "crooked paths are nothing more than mere impressions". This being said, he did report seeing what are now called non-linear trails, noting specifically that in 1885 out of 1334 meteors observed, just 4 (that is 0.3%) appeared as being, "conspicuously curved".[21] Of course, there is nothing wrong with changing one's opinion, but it remains the case that no-good understanding of the curved or sinuous meteor trail phenomenon exists even to this very day [37].

One question that has repeatedly occurred to meteor astronomers (indeed, right up to the modern era) is that which asks if annual meteor showers can produce noticeable effects on local weather. It has been suggested, for example, that the small dust grains left in the trail of an ablated meteoroid might act as condensation nuclei and thereby enable extended periods of rain. In principle this idea can be tested by examining the weather conditions surrounding the times of peak annual meteor shower activity, but no fully convincing evidence for any rain-meteor-shower correlation has even been presented. It has also been suggested that meteor shower activity might initiate distinctive cold spells [38]. Ernst Chladni, Alexander von Humbolt, and Deniel Kirkwood, for example, considered the possibility of so called "dark days" being caused when a meteoroid stream chanced to transit the Sun's disk. In this sense, it was suggested, the stream of meteoroids would partially block some of the Sun's light from reaching the Earth and thereby cause a short-lived cold snap. Interested in this possibility, in 1915, Denning attempted [39] a statistical study of past periodic cold spells. He drew specific reference to the work by German meteorologist George Adolf Erman, who had argued, in 1839, that periodical cold spells occurred in May and February each year because of the passage of, "falling stars between us and the sun". Denning concluded that there was some statistical evidence to support Erman's claim for a periodic cold spell in February. He also suggested that

---

[20] In his excellent *Catalog of Astronomical Anomalies* (The Sourcebook Project, Glen Arm, 1986), William R. Corliss, lists two apparent occasions where two meteors appeared to collide. Both accounts are from observes located in India, with one 'collision' being seen in 1832, and the other being witnessed in 1959.

[21] In a survey conducted by the author [37] it was found that approximately 1 in every 200 meteors reported is described as having a curved or spiraling path (that is about 0.5%). A suggestion was made that the curving trails might be the result of the Magnus Effect, or supersonic yawing, similar to that demonstrated by bullets and missiles. The spiraling effect might further be related to some torque-free precession or a jetting effect.

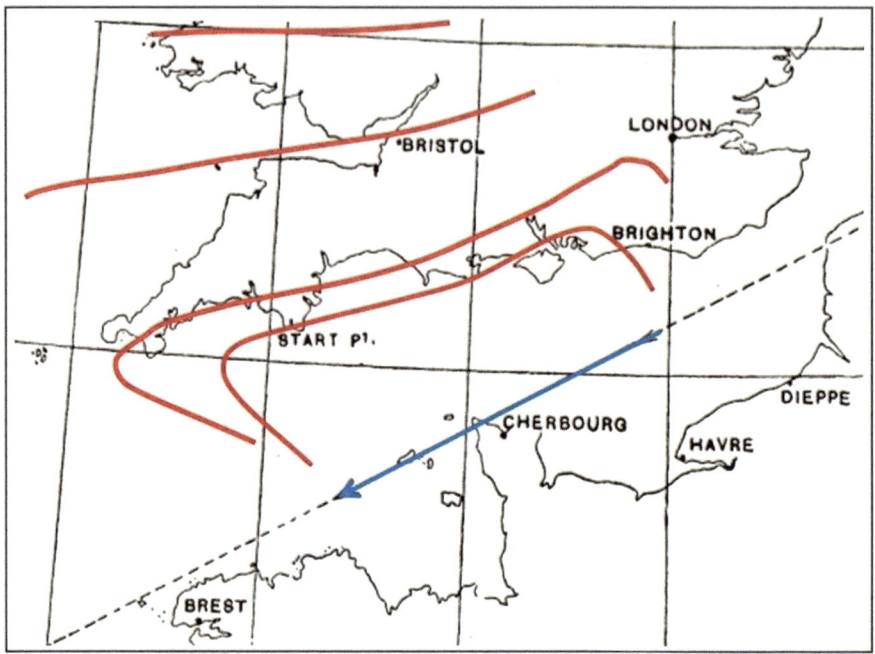

Fig. 3.19 Diagram illustrating the shifting location and shape (red lines) of the meteor train associated with the 22 February, 1909 fireball. The fireball (blue line) was observed at 7:33 in the evening off the coast of northern France, and the resultant train drifted northward over the next several hours. The drawing was prepared by Denning, and used in *Hutchinson's Splendours of the Heavens* (1923); here, we have picked-out the original fireball path along with the location and train as deduced at 8:15, 8:25, 8:55 and 9:25 as it drifted over southern England, Devon and Cornwall and eventually southern and central Wales

there was some evidence for the occurrence of several other periodic cold spells. In light of this apparent discovery Denning questioned, "is there a meteoric swarm with a periodic time of about 30.5 days, and sufficiently distended to occupy about six days in passing the sun, revolving around that luminary at little inclination, but with necessary density to moderate the solar rays to an appreciable degree" [39].

It is now clear that no meteoroid stream is anywhere near dense enough to produce the "dark days" effect that Denning and others were suggesting. It is also clear, in the light of more recent analyses, that periodic cold spells (as described by Erman) do not, in fact, occur. That Denning could entertain the idea that a meteoroid stream might have a period of 30.5 days is also interesting. Such a meteoroid stream is quite impossible, and this perhaps underscores Denning's complete disregard for theoretical constraints. As with stationary radiants, Denning was only concerned with what he believed he saw not so much with what was required to explain the data actually acquired. As far as Denning was concerned the observations implied a 30.5-day periodicity, and that, in spite of any theoretical counter argument, was the period that the postulated meteoroid stream must have.

Bibliography

1 Denning, W. F. (1891). *Telescopic work for starlight evenings* (p. 29). Taylor and Francis.
2 Denning, W. F. (1891). *Telescopic work for starlight evenings* (p. 60). Taylor and Francis.
3 Sheehan, W. (1988). *Planets and perception: Telescopic views and interpretations, 1609–1909.* The University of Arizona Press.
4 Denning, W. F. (1891). *Telescopic work for starlight evenings* (p. 162). Taylor and Francis.
5 Denning, W. F. (1882). Comet-Seeking. *Observatory, 5*, 285–289.
6 Denning, W. F. *Astronomical phenomena in 1872*. Wyman and Sons.
7 Denning, W. F. (1894). The discovery of comets. *Monthly Notices of the Royal Astronomical Society, 54*, 544–546.
8 Denning, W. F. (1922). Observation of comets. *Nature, 109*, 613.
9 Denning, W. F. (1891). *Telescopic work for starlight evenings* (p. 341). Taylor and Francis.
10 Denning, W. F. Variations in nebulae. *Observatory, 14*, 196–197.
11 Denning, W. F. (1900). Mercury as a naked eye object. *Nature, 61*, 430.
12 Denning, W. F. (1883). Note on observations of mercury. *Monthly Notices of the Royal Astronomical Society, 43*, 300–301.
13 Denning, W. F. (1891). *Telescopic work for starlight evenings* (p. 141). Taylor and Francis.
14 Denning, W. F. (1891). *Telescopic work for starlight evenings* (p. 170). Taylor and Francis.
15 Denning, W. F. (1899). Early history of the great red spot on Jupiter. *Monthly Notices of the Royal Astronomical Society, 54*, 574–584.
16 Denning, W. F. (1920). The great red spot on Jupiter. *Nature, 105*, 423–424.
17 Sanchez-Lavega, A. (1989). Saturn's great white spots. *Sky and Telescope Magazine, 78*, 141–142.
18 Denning, W. F. (1876). Notes on Saturn and his markings. *Nature, 62*, 237–238.
19 Stephenson, R. R., & Clark D. H. (1978). *Applications of early astronomical records* (Chap. 3). Oxford University Press.
20 Denning, W. F. (1891). *Telescopic work for starlight evenings* (p. 351). Taylor and Francis.

21 Denning, W. F. (1918). Observations of Nova Aquilae. *Monthly Notices of the Royal Astronomical Society, 78*, 570.

22 Anonymous (1920). Our astronomy column. *Nature, 105*, 838.

23 Luyten, W. J. (1920). Visual and photographic observations of Nova Cygni-3, made at the royal observatory, Greenwich. *Monthly Notices of the Royal Astronomical Society, 81*, 61–65.

24 Beech, M. (1993). Denning on Novae. *Journal of the British Astronomical Association, 103*, 130.

25 Davis, H. V. (1916). Scarcity of Wasps. *Nature, 98*, 109.

26 Denning, W. F. (1916). Letter to the editor. *Nature, 98*, 149.

27 Denning, W. F. (1920). Wasps. *Nature, 105*, 328.

28 Denning, W. F. (1915). Birds that pass in the night. *Observatory, 38*, 220–221.

29 Anonymous (1866). *Proceedings of the British Meteorological Society, 3*, 225–227.

30 Denning, W. F. (1892). *Proceedings of the Bristol Naturalists' Society, 7*, 61–63.

31 Denning, W. F. (1914). Lunar rainbows. *Meteorological Magazine, 49*, 147–147.

32 Jenniskens, P., et al. (2019). Tunguska eyewitness accounts, injuries and casualties. *Icarus, 327*, 4–18.

33 Denning, W. F. (1885). Letter to the editor. *Nature, 33*, 152.

34 Livesey, R. (1989). The visibility of auroral light in Southern England. *Quarterly Journal of the Royal Astronomical Society, 30*, 489–491.

35 Symbalisty, E., et al. (2000). Meteor trails and columniform Sprites. *Icarus, 148*, 65–79.

36 Denning, W. F. (1909). *Monthly Notices of the Royal Astronomical Society, 69*, 539–542.

37 Beech, M. (1988). Non-linear meteor trails. *Earth, Moon, and Planets, 42*, 185–199.

38 Dean Fyfe, J. D., & Hawkes, R. L. (1986). *Planetary and Space Science, 34*, 1201–1212.

39 Denning, W. F. (1915). The seasons—Recurring cold periods. *Meteorological Magazine, 50*, 44–45.

4

The Amateur Astronomer

An important factor in distinguishing the amateur astronomer from the mere dilettante is that the amateur sets out to follow some form of research program. This was certainly Denning's viewpoint. For all this, however, Denning stressed the importance of setting time aside for teaching, and the bringing of new observers into astronomy. Writing [1] in *Telescopic Work for Starlight Evenings*, Denning explained, "it is the duty of all of us to encourage a laudable interest in the science". These sentiments were qualified, however, with the warning that, "the utility of an observer constituting himself a showman, and sacrificing many valuable hours which might be spent in useful observations, may be seriously questioned". Although he proffered this warning, Denning clearly saw educational activities as an important part of an astronomical society's mandate. In this respect, Denning was fully supportive of the suggestion [2] made by the Liverpool Astronomical Society that they, "institute periodical examinations in astronomy, and that the Council should be empowered to grant certificates of competency". As early as 1883, Denning had written [3] to the journal *Nature* on this very topic. There he wrote, "It seems a thing to be deplored that in this country there is no establishment where astronomy is made a special subject for teaching, and where those who early evince a taste in this direction may be educated in conformity with inclination". Indeed, Denning was effectively 100 years ahead of his time in advocating the open access to astronomy education—such general astronomy courses now being routinely offered, via the internet, by numerous universities and colleges from around the world.

Denning was a dedicated observer and he continually encouraged the enrollment of new members to the amateur astronomy fold. He firmly

© The Author(s), under exclusive license to Springer Nature Switzerland AG 2023
M. Beech, *William Frederick Denning*, Springer Biographies,
https://doi.org/10.1007/978-3-031-44443-2_4

believed that the scientific study of the heavens was not just a noble pastime, but also an essential human activity. To this end he was often critical of the casual observer. He was to write [4], for example, in tones similar to expressed by Conan Doyle's Sherlock Holmes that, seeing an object is not the same as observing it. The mere seeing, as such, counts for nothing from a scientific stand-point, though it may doubtless afford some sense of personal satisfaction and wonder. The practical astronomer, with, the interest of science at heart, will "require something more". Denning clearly wanted to see purpose, order, and resolve in amateur astronomy. Indeed, in 1883, Denning was to write [3], "the fault with amateurs seems to be that they are devoid of organization, and generally of proper education to the work in hand. ... it seems desirable to make some attempt to organize the labors of amateurs in directions suitable to their means and inclinations, and to utilize such results for the benefit of astronomy". These comments capture the mood of the times, and were written some ten years after the demise (circa 1872) of the Observing Astronomical Society (see Chap. 7), but seven years before the formation (in 1890) of the British Astronomical Association.

In 1897 Denning was again to write to *Nature* on the subject of "Organized or Sectional Work in Astronomy" [5]. This time he responded to an article written by S. C. Chandler on the successes that organized observation work had brought to variable star research. Denning reasoned, "in some other departments of observation, there does not appear to exist the same necessity for organized effort. In fact, I think that it can be shown from results—the best of all tests—that it has been a comparative failure as far as affects the progress of astronomy". To this he added (in the male-dominated tones of the time), "a little reflection will prove that all the best work has been accomplished by individual and independent effort. A good man will persevere in his labors, just the same, whether he belongs to any combination or not; and it is really much better for such a person to be isolated, so that he may perform the work of his choice in his own way". Denning did temper these arguments slightly by suggesting that, "beginners sectional work is often most beneficial, as it affords them a useful preliminary training". In these writings we find the essential Denning. To Denning it was the act of observing that was all important, indeed, it is "the best test of all". It was to these ends of 'observing with purpose' that Denning encouraged the development of amateur astronomy. Indeed, Denning felt that the amateur astronomer could both assist and, in some areas, best the professional observers at their own game. These latter observers, fussing, as many of them (appropriately) did, over precision instrument measurements and complex correction terms, had little time for anything other than producing (important as they were) ever more refined star charts. The amateur, in contrast, could adapt to many different areas of study, and with relatively inexpensive equipment make new discoveries (such as new comets and nebulae), and, in addition, provide long-term monitoring of astronomical bodies such as the planets, mapping their surface albedo markings and atmospheric shadings, and in variable star monitoring.

There is little doubt that Denning was a dedicated and life-long promoter of amateur astronomy, and, indeed, amateur astronomy blossomed over the course of his lifetime. He was not the only player in the game, of course, but his voice, as witnessed through provincial and national newspapers, magazines and journals, carried weight, and acted to inspire amateur participation in many areas of astronomy. Denning's ambition was to encourage well organized observational work, and to foster a sound, and wide-ranging interest in astronomy. He was driven in his outlook by the indefatigable Victorian belief that method, discipline, and application would bring a continued stream of new discoveries—if one just diligently explored, then the heavens would provide for novelty and gain. In his 70th year (1918) Denning still held true to such ideals, and writing [6] in the *Journal of the Royal Astronomical Society of Canada* he argued, "it may be taken as certain that the amateur will never be out of the running. So long as there remains a new comet to be found, a temporary [nova] or variable star to be discovered, a new or old planetary feature to be investigated or other important work to be accomplished, he may be expected to take his share. ... [the amateur] is a sort of free-lance in astronomy, and his more or less vigilant examination and watching of the heavens must necessarily result in discovery". Denning's sentiments are as true today as they were in his time, and indeed, the dedicated amateur can still hope to make important contributions to astronomy, and even discover new comets that will carry their name forward in perpetuity. Citizen science is not something new to the internet generation—all that has changed in the modern era is the role of the observer. The eye of the modern citizen scientist, rather than being glued to the eyepiece of a telescope, is more often glued to a computer screen displaying data obtained by other researchers.

Addressing the issue of topics suitable for the amateur astronomer to pursue, Denning made an interesting list of activities in his *Telescopic Work for Starlight Evenings*. Writing in parallel columns, Denning indicated various observing projects that the amateur astronomer might fruitfully explore, and contrasted them against projects that the amateur should carefully avoid. The list runs as given in Table 4.1.

Table 4.1 Denning's list of observing projects that the amateur astronomer should avoid (column 1) contrasted against those that might yield favorable results (column 2)

Projects to avoid	Projects to work on
Satellites of Venus	Satellites of Uranus and Neptune
Vulcan	Ultra-Neptunian planet
Active volcanoes on the Moon	Changes on the Moon
Detached cusps of Venus and Mercury indicating high mountains	Rotation of Mercury, Venus, Uranus, and Neptune
Rings of Uranus and Neptune	Minor planets [asteroids], and comets
Multiple companions to Polars and Vega	Nebulae and double stars

The left hand side of Table 4.1 indicates the usual list of suspects, and it is a combination of topics that had long distracted astronomers, professional and amateur alike. With respect to searching for new planetary moons, that associated with the hypothetical satellite of Venus had essential run its course by the time that Denning produced his list in 1891. The idea that Venus might have attendant moons dates back to an observation by Italian astronomer Francesco Fontana in 1645. Similar such sightings of a supposed Cytherean companion were reported throughout the 18th century by, amongst others, Giovanni Cassini, James Short, Joseph Lagrange and Christian Horrebrow. American astronomer Edward E. Barnard even reported seeing such a moon in 1892. These apparent detections, no doubt, relate to either spurious reflections within a telescope's optics, and/or the misidentification of a background star. For all this, William Herschel conducted a survey of Venus, in 1768, in the hopes of finding a new companion moon—with nothing, of course, being found. For more than a century after Herschel's survey, and Horrebrow's last supposed sighting (also in 1768), the subject lay essentially dormant. In 1884, however, Belgian-born astronomer Jean-Charles Houzeau suggested in the journal *Ciel et Terre* (to which he was one of the editors) that rather than seeing a Cytherean moon, the supposed sightings were actually related to a planet that orbited the Sun between Venus and the Earth with a period of 283 days (orbital radius of 0.844 AU). Indeed, Houzeau named the new planet Neith, after the Egyptian goddess of the cosmos, fate, childbirth and wisdom. While Houzeau's new planet idea never gained strong traction amongst astronomers, the initial article was reported upon in the 6 June, 1884 issue of the *English Mechanic and World of Science*. It is from this latter article that Denning possibly drew his information, and having spent (by 1891) many years fruitlessly searching for planet Vulcan (see below and Chap. 7), supposedly located interior to the orbit of Mercury, he probably felt that to recommend searching for yet another hypothetical planet, that may or may not exist, would likely result in observer disappointment, and act to discourage future survey work. Such a project is also somewhat ill founded, in that it is not clear how to resolve the continued lack of any detections into an actual conclusion that there is no moon. This being said, the circumstances of Asaph Hall's discovery of the moons of Mars on August 12th, 1877, could have acted to inspire the amateur in the search for Cytherean moons. Indeed, Hall was to write that he would have abandoned his search for Martian moons except for the continued encouragement of his wife. What Hall, and the many other astronomers who had looked for Martian moons, had not initially taken into account was how small (and therefore inconspicuous) the moons might be. Spoilt by the a-typical example of Earth's Moon, it was generally though that the moons of other planets should likewise be large and bright. Indeed, once you know what you are looking for, then detection is more straightforward, and, while Hall was using the United States Naval Observatory 26-inch aperture telescope when he discovered the moons of Mars, amateur astronomer and landscape artist John Brett was to write in the 11 October 1877 issue of *Nature* that he had readily observed them with a 9-inch aperture telescope—an instrument well within the reach of many amateurs. Once you know what you are looking for, it is much easier to see.

Rather than looking for a companion to Venus, Denning reasoned that the amateur astronomer might have more luck in the search for moons in orbit about Uranus and Neptune. In 1891, when Denning compiled his table, Uranus was known to have 4 moons: 2 had been discovered by William Herschel in 1787 and 2 had been discovered by William Lassell (Fig. 4.1) in 1851. Likewise, Neptune was known to have at least one moon, Triton—discovered by Lassell in 1846 (the moon being found just 17 days after the planet itself was first swept-up by Johann Galle in Berlin). Lassell had also discovered a new Saturnian moon, Hyperion, in 1848. It is now known that all the Jovian planets support abundant moon systems, in the case of Uranus and Neptune, however, the 5th and 2nd moons were only discovered, by Gerard Kuiper, in 1948 and 1949 respectively. Denning certainly had the right idea in suggesting that the search for new moons in orbit around the Jovian planets might be fruitful, but the reality was that the detection of any such moons would require equipment well beyond the reach of even the best-equipped amateur astronomer of the late 19th and even early 20th century.

The storied life of William Herschel is well known, and need not be repeated here. That of William Lassell, however, is more pertinent to the times in which Denning was writing. Lassell, like Denning was an amateur astronomer. He had no formal training in astronomy, and he held no

Fig. 4.1 Grand amateur astronomer William Lassell (1799–1880)

teaching office with respect to any university. What distinguished Lassell from Denning was financial means. Owner of a successful brewing company, Lassell was able to indulge his astronomical passions without the worry of making a living. Additionally, Lassell had the financial means to push the technology of observing to its limits. Denning had to strain his finances to afford a 10-inch With-Browning telescope (and after its purchase, in 1871, he used it exclusively, never buying another, larger instrument). In contrast, Lassell, in 1845, was able to fund an entire observatory, at his home of *Starfield* just outside of Liverpool, housing a custom built 24-inch aperture telescope. Furthermore, unlike Denning, who initially struggled to gain membership, Lassell was feted by Royal Astronomical Society, being elected a Fellow in 1839. Lassell was additionally elected a Fellow of the Royal Society in 1849, respectively receiving from these august bodies their Gold Medal in 1849 and Royal Medal in 1858. Lassell also served as President to the RAS from 1870 to 1872.

In 1855, as a follow-on from his 'two-foot telescope', Lassell had constructed an even larger instrument with an aperture of 48-inches, and this instrument he had installed at a custom-built observatory on the island

Fig. 4.2 William Lassell's, self-financed, 48-inch telescope, built in 1855 and installed at an observing site in Malta

of Malta (Fig. 4.2). Remarkably, this massive telescope was only used for a handful of years, and yielded no notable discoveries—it was eventually dismantled and sold for scrap. Lassell wrote of its demise in the *Observatory* magazine for September 1877: "I ultimately consigned the cast and wrought iron to the furnace and tilt hammer of the engineer, and the specula to the crucible of the bell founder". The story behind Lassell's 48-inch telescope only acts to reinforce Denning's point (recall Chap. 3) that the best telescope is the one that someone actually uses, and that this is not necessarily the largest one. Lassell's 24-inch telescope fared a little better than its larger, short-lived, cousin, and was eventually donated to the Royal Greenwich Observatory in 1883 (it was later scrapped, however, in 1895). While both amateur astronomers from Britain, Denning and Lassell were in many ways from totally different worlds—worlds separated purely by finance, rather than innate ability or through any difference in their enthusiasm for astronomy.

The idea that large mountain ranges might be the reason for the occasional observation of detached cusps associated with crescent phase observations of Mercury and Venus, was short-lived, and soon dismissed as a feature of poor telescope resolution, and atmospheric-seeing—although some observers held strongly to the apparent observation of displaced cusps. Denning was correct to position astronomers away from reporting such phenomena, and he was certainly correct in suggesting that more useful work could be done in observing planetary discs with the aim of determining rotation periods. For all this, the definitive values for the rotation periods for Mercury and Venus were not to be derived until the 1960s, and this result required the advent of radio astronomy (and specifically telescopes working at radar-wavelengths). Likewise, the rotation periods for Uranus and Neptune were not to be accurately derived until the magnetic field survey measurements conducted by the Voyager 2 spacecraft in the last decade of the 20th century. It is additionally the case that Denning was correct to steer amateurs away from the search for Uranian and Neptunian ring systems. This being said, British amateur astronomer the Reverend William Dawes had co-discovered (along with William Bond in America) Saturn's Crepe-ring in 1850. In contrast, William Lassell had claimed the detection of a Neptunian ring system in 1846, but this was an unconfirmed observation, and ultimately dismissed as the combined result of poor resolution (even though he was using a 24-inch aperture telescope), observer zeal, and atmospheric-seeing[1]. The Uranian and Neptune ring systems were first imaged in the mid to late 1980s during flybys of the Voyager 2 spacecraft.

The search for a planet located interior to the orbit of Mercury preoccupied many astronomers (both amateur and professional) from circa 1860 to at least the first decade of the 20th century, and Denning played an important role in both encouraging and organizing British amateur astronomers in the hope of recording it in transit across the Sun's disk. While this subject, and Denning's role in organizing observational searches, will be discussed more thoroughly in Chap. 7, that Vulcan should be listed as an object for amateurs to avoid points more to frustration than strong belief. That the search for Vulcan was set in contrast with the suggestion to look for planets beyond Neptune, however, was an idea many decades ahead of its time. Indeed, it was not to be until 1930 that Clyde Tombaugh discovered (134340) Pluto, and it was to be a further 62 years before Jane Luu and Dave Jewitt discovered 1992 QB1 (now designated 15760 Albion)—the first of the trans-Neptunian, Kuiper Belt Objects[2]. These latter discoveries, however, were made with equipment well beyond that accessible to any amateur (or professional) astronomer in the

---

[1] Although Neptune does have a ring system (discovered via stellar occultation observations in 1984), it would have been quite impossible for Lasell to have resolved them with the instruments at his disposal.

[2] Although Pluto is now designated as a dwarf planet, its official catalog number is that of a small solar system body—technically, it was the first Kuiper Belt Object to be detected.

late 1800s. Indeed, these discoveries first required the development of photographic techniques (which first appeared circa 1900) and then, in the 1970s, the development of the charge coupled device (CCD electronic cameras).

Observing, mapping and drawing the Moon's surface features was a major preoccupation of amateur astronomers in the late 19th century—as it is to this very day. Whether any of its surface features changed over time was, however, a topic of great debate. Both arguments for change and for stasis were presented, but it is the case that the Moon's surface is capable of change through new surface impacts. Such impacts, and their resultant new craters, are now routinely discovered through artificial satellite imagery, and both amateur and professional observers have recorded the transient flashes associated with meteoroid impacts into the lunar regolith. William Herschel had championed the idea that lunar volcanoes existed in the late 18th century, describing observations of such objects in eruption in May of 1783 and April of 1787. Indeed, it was generally thought at the time Denning compiled his list of projects that all craters were likely produced through volcanic activity, and it seems a little odd, therefore, that Denning would suggest that the amateur should avoid the topic. Indeed, the detection of volcanic activity would have been of great interest with respect to lunar geology. Once, again, Denning was probably trying to steer amateur observers away from potentially controversial topics—indeed, the reporting of transient lunar phenomena (TLPs) is still problematic to this very day. In the late 20th century, one lunar crater in particular, crater Linné, drew much attention and controversy. The controversy began when Johann Schmidt suggested, in 1866, that the appearance of this small, 2.4-km diameter crater (Fig. 4.3), had changed—morphing from a distinct crater to that of a bright spot that varied in diameter. Much journal space was taken-up with the discussion of possible changes in the shape and appearance of Linné, by both professional and amateur observers, and it was the subject of study by many members of the Observing Astronomical Society (see Chap. 7).

Denning was cutting a fine line when he recommended that amateur observers should search for changes on the Moon's surface, rather than monitoring for lunar volcanos. Certainly, very few observers suggested that they had witnessed lunar volcanoes in eruption, but in Denning's-day the possibility of physically witnessing such events had not been physically ruled out. Indeed, in 1890 when Denning was making his list, the existence, or not, of a lunar atmosphere had not been settled. Lunar observer's, in fact, reported the apparent detection of lunar meteors (that is meteors thought to be ablating in a supposed lunar atmosphere) as late as the 1940s.

Fig. 4.3 Lunar crater Linné as observed during the Apollo 15 mission. Image courtesy of NASA

One area of discovery where amateur astronomers have historically excelled is that of comet sweeping. The art of such sweeping is not technically challenging (once one is practiced in it), but success typically requires many hundreds of hours of tiresome and non-rewarding observation. The rewards of discovery, however, if *Fortuna* is smiling, are great. Indeed, new detections will be named after the observer. When Denning compiled his table in 1890, the number of known periodic comets was just 16 bodies. Today the number of known periodic comets is in excess of 400, and while most detections, since at least the onset of the 21st century, have been made by dedicated survey telescopes, operated by professional astronomers, the amateur observer can still make new discoveries. Most significantly, in 2019, the first *bona fide* interstellar comet 2I/Borisov was detected by Gennadiy Borisov, an amateur astronomer working in the Crimean[3].

Denning was never greatly concerned with making observations relating to the stars, either as single objects or in multiple systems. Indeed, Denning was to write dismissively in his *Telescopic Work* that, stars, "exhibit a sameness and lack of detail that is not satisfying to the tastes of every observer". For all this, the identification of stars displaying distinct colors was a topic of great activity

---

[3] The physical status of the first interstellar object 1I/Oumaumua, detected in 2018 with the PAN-STARRS telescope, remains unclear. It showed no coma or tail when passing perihelion, and is technically, therefore, not classified as a comet, but as an asteroid.

in the late 19th century. At that time, however, it was not realized that the color of a star is related to its temperature. Likewise, in the late 19th century, double star observations were routinely published by amateur observers, with such observations indicating changes in star separations and positions. Such positional data, when carried out over many decades, can reveal both proper motion effects, and enable the determination of visual and astrometric binary star orbits. This field has been dominated by professional observers, but was in principle open to the amateur. Well known examples of visual binaries, in which the two stars can be observed in orbit about a common center, include 70 Ophiuchi (with the two stars having an orbital period of 87.7 years) and Krueger 60 (with the two stars having an orbital period of 44.6 years). In terms of astrometric binaries, the famous 19th century example is that provided by the star Sirius. In this case, only one star is seen (the bright star Sirius A), but an unseen companion (Sirius B) is inferred by the non-linear proper motion track that Sirius A makes upon the sky. This feature of the proper motion of Sirius A (along with that deduced for the star Procyon) was reported by German astronomer Friedrich Bessel in 1844.

When Denning suggested in his projects table (our Table 4.1) that amateur astronomers should avoid looking for companions to Polaris and Vega, he was actually referring to the simple counting of faint background stars. Indeed, counting the number of faint companions to these two stars, was a popular means of gauging the light-gathering power of a telescope. Such counting, purely for the sake of it, is rather pointless, and provides no useful astronomical information.

Conspicuous by its very absence, the most remarkable oversight in Denning's table of projects is that of variable star research. Indeed, Denning completely missed what has become one of the most important branches of present-day amateur astronomy. Certainly, many variable stars had been discovered by the time that Denning was writing *Telescopic Work*, but he gives them short shrift. This is all the more remarkably given the important work that was in the process of being initiated in this area by British Astronomical Association (BAA) members. Indeed, between 1900 and 1910 the members of the BAA variable star section gathered more than 26,000 brightness esti-mates. Likewise, following its formation in 1911, the American Association of Variable Star Observers (AAVSO) can boast that its members have gathered many tens of millions of brightness estimates of variable stars—a remarkable result and undertaking.

The discovery of asteroids has never been a field in which the amateur observer has excelled. Indeed, almost exclusively the discovery of such bodies has been the domain of professional astronomers working from well-equipped national observatories. At the time of Denning's writing, in 1890, 302 asteroids were known, but within just a few years this number was to increase dramatically. Indeed, in 1891 the first asteroid to be discovered via astrophotography—asteroid 323 Brucia—was reported by German astronomer Max Wolf, and by the close of 1895 the number of known asteroids had grown to 409. The number of known asteroids was to reach 604, by February 1906. While in principle the amateur wasn't excluded from the possibility of asteroid discovery post 1891, it is the case that none have actually been discovered by any amateur astronomer working with a backyard telescope. Even as he wrote *Telescopic Work*, Denning was reversing his earlier opinions on asteroids. Writing in the book, with the lengthy title of, *Astronomy for Amateurs—a practical manual of telescopic research, in all latitudes, adapted to the powers of moderate instruments*, edited by John Westwood (Longmans, Green, and Co. London, 1888), Denning wrote of asteroids that they, "can hardly be considered to afford much practical interest to amateurs … searching for them involves so much labour and systematic application that it justly falls on professional shoulders". One might question, given these words, where Denning's general appeal to systematic and diligent observing had gone to. Such views concerning asteroids were not uncommon, however, and well-known amateur observer, and regular contributor to the *English Mechanic and World of Science*, George F. Chambers was to write in his popular book, *The Story of the Solar System Simply Told for General Readers* (George Newnes Ltd., London, 1895), that asteroids, "are of no interest to the casual amateur who dabbles in astronomy, and indeed, they are of very little interest to anybody". These very same words were echoed some 28 years later by Hector Macpherson in his book *The Romance of the Heavens* (Seeley, Service & Co. London, 1923). These are harsh words indeed, and especially so in the modern era where the study of asteroid origins and orbital dynamics is seen as being fundamental to our knowledge of solar system formation and planetary orbit evolution. None the less, suggesting that the search for asteroids was something that the amateur astronomer might pursue did see Denning at odds with most other practitioners of the time.

4.1   The Rise of the Professionals

It is with the asteroids that we begin to see a clear divide opening between the research expectations of amateur and professional astronomers. In other areas of astronomy, planetary observations, lunar topology studies, and comet discovery, the amateur and the professional worked on an essentially level playing field. Increasingly, however, as the 19th century moved towards its close, professional astronomers started to follow research programs in areas that required both a deep theoretical knowledge, and the construction and operation of specialist equipment—equipment typically well beyond the technical and financial reach of even the most well-to-do amateur observer. This division between the professional and amateur astronomer was only to grow throughout the 20th century. Indeed, the role of the scientist, and who it was that controlled the growth of knowledge, underwent fundamental change towards the close of the 19th century. Much of this change was brought about by the increasingly important position of scientists within society, universities, and the civil service. Furthermore, as the 19th century advanced, so scientists became increasingly recognized as public figures, and were increasingly being rewarded, through knighthoods and peerage, for their works. George Biddell Airy, Astronomer Royal from 1835 to 1881, through his many contributions to British society, for example, became a well-known and readily recognizable civil servant (Fig. 4.4). Other scientists became champions of industry through their work in engineering (e.g., Charles Babbage[4]), telegraphy (e.g., Charles Wheatstone), and numerous other areas that were both newsworthy and perceptible to the public's gaze. Other scientists, as a result of their public works and ground-breaking research, were honored as lords of the realm (e.g., William Thomson, who adopted the title Lord Kelvin, and John Strutt, who adopted the title Lord Rayleigh).

Some measure of what it meant to be a scientist prior to the last quarter of the 19th century is provided by the ever-numerating Francis Galton. Indeed, Galton presented an evening talk at the Royal Institution in London on Friday, 27 February 1874, with the title, "On men of science, their nature and their nurture". The purpose of the talk, "was to gauge the chief qualities by which English men [and he meant only men] of science are characterized". To achieve this goal Galton surveyed 180 members of the Royal Society, sending them a questionnaire concerning their upbringing, schooling and attitudes. Of the 180 surveyed, 115 replied, with some 90 of these providing, "minute detail". Interestingly, while Galton specifically sought-out the opinions of Fellows of the Royal Society, he estimated that there were about 300 individuals in the United Kingdom who might be described as leading scientists. This, he further estimated, implied that about 1 man in 10,000 qualified as

---

[4] Babbage is perhaps not the best example, given his now infamous clashes with the British Government and industry over the funding and manufacture of his difference and analytic engines. His most influential work relating to industry, *On the economy of Machinery and Manufacturing*, appeared in 1832.

Fig. 4.4 Caricature of George B. Airy, Astronomer Royal (1835–1881), as seen by readers of the November 1875 issue of *Vanity Fair*. https://commons.wik imedia.org/wiki/File:%27Astronomy%27_(Airy)_(caricature)_RMG_PT4010.jpg Brooks, Vincent, Day and Son; Carlo Pellegrini, Public domain, via Wikimedia Commons

being a scientist within the population of the United Kingdom at that time. Interestingly, from our historical perspective, Galton noted that many people might be surprised that the number of scientists in the country was as high as 300. Having estimated their number, Galton then set about analyzing his survey returns, finding that the main characteristics of scientists were:

- Energetic in body and mind
- Worked at night after being, "engaged all day in anxious business"
- Holders of good business habits, keeping-up with accounts and correspondence
- Of a smallness of head [which is an interesting reversal of the modern idea of scientists as being 'big-heads']—here we additionally learn from Galton that the average diameter of a British gentleman's head is between 22 ¼ and 22 ½ inches.
- Education: of those surveyed,

- One-third were from Oxford and Cambridge universities
- One-third from "other" universities
- One-third had no university training

- Generally, in good health since the onset of manhood
- Independent of character, with a strong taste for enquiry and the sciences
- Interested in mathematics and mechanics
- "Strongly anti-feminine"

Galton's list is revealing, although perhaps predictable, and (by modern standards) decidedly misogynistic. However, for all this, the information provided by Galton reveals that in the mid-1870s there was very little difference between the grand amateur and those practitioners that might be called leading scientists, accepting that is, upon the time and freedom they had to conduct scientific research. The scientist, as such, was typically university educated, but the research that they performed was not necessarily conducted at a university. For all this, however, Galton noted that, "those who select some branch of sciences as a professor must do so in spite of the fact that it is more un-remunerative than any other profession". It was the positive changes in the circumstances surrounding this latter condition, that is being able to earn an actual living, that ultimately enabled the professionalization of the scientist. While the position of Astronomer Royal, since its initiation in 1675, was a paid governmental office, it was only in 1881, with the employment of Joseph Normal Lockyer (Fig. 4.5) as a civil servant (with the specific title of being an astronomer), that the job of being a scientist, as a means of making a living, became possible. Not only did the title of scientist begin to mean something altogether more formal than that of a person interested in the sciences, it implied a person in the employ of some institute of higher learning. The scientist now actively sought out government funding in order to perform research, and additionally, the teaching and mentoring of graduate students was seen as a fundamental part of a scientist's identity and a means by which the interests of science could be made to grow. With the close of the 19th century, astronomical research, along with the promotion of astronomical knowledge was increasingly placed under the control of the professional scientists and their allied professional societies. While the Royal Astronomical Society was never explicitly an amateur society, the increasing professionalization of its constituent members, and the relative demise of the amateur, is writ large in the award of its most prestigious prize, the Gold Medal. Indeed, Denning's award of the Medal, in 1898, was the very last time that it was awarded to an amateur astronomer. Since Denning, only professional astronomers, directly allied to some university or national observatory, have received the Gold Medal.

Not only were scientist becoming increasingly allied to universities and professional bodies, but the word scientist, introduced by William Whewell at the 1833 meeting of the British Association, as a term intended to imply unification, began to split into rival domains. The astronomer now fought for funding and influence against the physicist, the chemist, the geologist, the botanist, and the engineer. Competition and research excellence became the name of the game, with an ever-increasing number of professional, career-orientated, scientist, chasing never enough government money to go around. Additionally, it also became the case that professional astronomers began to work together, in specialist research-groups, on narrowly-focused research topics.

David W. Hughes (University of Sheffield) and Richard de Grijs have estimated that circa 1900 there were some 2000 active (professional and/ or semi-professional) astronomers worldwide, working in about 100 observatories [7]. At the same time, the membership of the British Astronomical Association (having formed in 1890) had risen to a total of 1150 members. These numbers were to change dramatically over the ensuring 100 hundred years. The BAA membership in the year 2000 was around 3000 members,

Fig. 4.5 Joseph Norman Lockyer (1836–1920). Pioneer spectroscopist, and first person to be employed with the title of Professor of Astronomical Physics (at the Royal College of Science in 1885—now Imperial College, London). Lockyer was the founder (in 1869), and first editor of the journal *Nature*. Image courtesy of the Royal Astronomical Society

while the number of professional astronomers (as indicated by membership to the International Astronomical Union) was some 12,000, of which about 1000 are located in the United Kingdom. While these numbers are not definitive, they do act to encapsulate the dramatic change that took place in professionalized astronomy during the 20th century. While the realm and reach of institutionalized astronomy changed significantly during the 20th century, the role and practices of the amateur changed but little.

4.2 The Professional Amateur

When Denning was writing his list of topics, to which the amateur astronomer might apply themselves (recall Table 4.1), he clearly demonstrates a disconnect with respect to what might reasonably constitute the direction of future studies. There is nothing wrong *per se* with the topics that he is actually suggesting, but problematically, they are largely topics that fall squarely within the domain of professional research, and/or require significant funding, and specialist equipment that placed them well beyond the means of all but the very wealthiest of amateurs. This being said, not all pioneering work in the late 19th century was conducted through the professional office. The case of William Huggins (Fig. 4.6), for example, indicates the fundamental role that the amateur could still play in the advancement of astronomical knowledge. It was Huggins, along with his wife (Margaret Lindsay Murray), and chemist William Miller, who pioneered photographic and spectroscopic techniques suited to astronomy. These spectroscopic studies revealing the gaseous nature of planetary nebulae, the stellar nature of spiral galaxies, the compositions of the Sun, and the line-of-sight velocity of the stars. Feted by the Royal Society, the Royal Astronomical Society, and awarded a knighthood in 1897, William Huggins (perhaps) represents the acme of amateur contributions to the new and developing astronomy—the new astronomy, that is, that emerged as the Victorian era was beginning to slide into its demise. Huggins's research represented the future, in the same way that Denning's represented the past. To Denning the only instrument an amateur astronomer needed was a telescope, and then a telescope no more than 10–12 inches in aperture. The new-fangled, technical-astronomy that saw development in the closing years of the 19th century, was something that Denning could not hope to aspire to—and nor, for that matter, did he seem to openly trust. Denning, in contrast to the growing trend of forming collaborations, and for using highly specialized equipment, was the quintessential loner and all-round observer.

Fig. 4.6 William Huggins (1824–1910). https://commons.wikimedia.org/wiki/File:Wil liam_Huggins,_Vanity_Fair,_1903-04-09.jpg Leslie Ward, Public domain, via Wikimedia Commons

Denning returned to the amateur in astronomy topic in an article published in *Knowledge* magazine for January 1915. Writing some 24 years on from his musing in *Telescopic Work*, Denning still felt that the amateur had much to offer astronomy, "there is no doubt whatever", he argued, "that amateur work will always be pretty much to the fore, and possibly rank in value with that of the best professional talent". Such sentiments were more bravado than truism, but he was right in the sense that the amateur had not been fully written out of the picture—although, in terms of cutting-edge research, the professional astronomers were definitely playing a higher intellectual game. Once again, Denning argues that the amateur will continue to find new comets, nebulae, and novae, and be the key observers of meteors and aurora. He again misses variable star observations as a viable amateur contribution to astronomy, and he reflects more on sentimentality than practicality—"there is a sublimity and infinity about astronomy which attract the intelligent mind and induce a feeling of reverence and awe", he wrote. True, the sense of awe at the universe should never be forgotten, but sentimentality is not what astronomy research is about. One new point that Denning did make in his 1915 summary, however, was that concerning the contributions of amateur astronomers to the popularization and promotion of astronomy. Here he makes mention to a new breed of writer—the astronomy popularizer—naming specifically Agnes Mary Clerke, Nicolas Camille Flammarion, Richard Proctor, William Smyth, and Thomas Webb. Books by these authors are still very readable, and useful to the amateur observer to this very day (albeit dealing with older technical ideas). In particular, the following short selection of texts became popular upon their release, remain relevant, and are still available in print having stood the test of time:

- Agnes Mary Clerke (1842–1907)

  – *A Popular History of Astronomy During the Nineteenth Century* (first published 1885).
  – *Problems in Astrophysics* (first published in 1903)

- Nicolas Camille Flammarion (1842–1925)

  – *Real and Imaginary Words* (first published in English in 1865)
  – *Popular Astronomy* (first published in English in 1894)

- Richard Anthony Proctor (1837—1888)

  – *Half-Hours with the Telescope* (first published in 1868)
  – *Other Worlds than Ours* (first published in 1870)
  – *Myths and Marvels of Astronomy* (first published in 1877)

- William Henry Smyth (1788—1865)

  – *A Cycle of Celestial Objects for the use of Naval, Military, and Private Astronomers* (in two volumes, first published in 1844)
  – *Sidereal Chromatics* (first published in 1864)

- Thomas William Webb (1807—1885)

  – *Celestial Objects for Common Telescopes* (first published in 1859)
  – *The Sun, a familiar description of his phenomena* (first published in 1885)

The books included in the list above cater to the amateur astronomer in both the practical, how to, sense of observing, the finding of specific astronomical objects, and in providing introductions to new and developing fields of study. Clerke, Smyth, and Webb give us books describing the Sun as a star, and they outline the late 19th century developments concerning the stars as physical objects—objects with surface characteristics to measure (size, mass, luminosity, and temperature), and physical properties to determine (distances, and distributions). Flammarion and Proctor provided the reader with more speculative astronomy books relating to other worlds, while Clerke provided an informative history of astronomical discoveries and developments. All these topics, the stars, observing techniques, the Sun, the planets, other worlds, and the history of astronomy, in one form or another, are the bedrock of popular astronomy publications to this very day. All that has changed in the modern era is the physical detail. The basic motivations and desires of the majority of amateur astronomers is probably no more different now than it was in the mid-19th century.

4.3   In Decline No More

For all the strength and confidence that distinguished the sciences at the close of the 19th century, this was not how, at least in some quarters, the century began. Indeed, the cantankerous genius Charles Babbage[5], decried the sorry state of the sciences, in a rather rambling book with the provocative title *Reflections on the Decline of Science in England* (first published by Benjamin Fellows, London, in 1830). Babbage, in fact, makes little effort to justify the title of his tome, and much of its contents is related to the public airing of his personal spleen. For all this, he drew attention to the poor state of financial support for the sciences in England, commenting that, "the pursuit of science does not in England constitute a distinct profession". Singled out for much of Babbage's vitriol is the Royal Society of London. This august body he castigates for its lack of leadership in promoting the sciences, encouraging scientists, and in its management/executive structure (or "misrule" as he called it). The days in which the Royal Society could have any influence on the sciences, he argued are "long past". Furthermore, Babbage argues, there has been a gradual decline in mathematics and physical sciences in England since the days of Isaac Newton. Such claims could certainly be disputed, but Babbage was on a mission to embarrass others. He singles out the school system for its lack of scientific content, and notes that, "scientific knowledge scarcely exists amongst the higher classes of society"—a fact that he attempts to support in terms of his own dealings with the British Government.

In Chap. 3 of his *Reflections*, Babbage looks at the general state of "learned societies in England", identifying some 13 such bodies. This list includes the

---

[5] In his semi-biographical account, *Passages from the Life of a Philosopher* (published in 1864), Babbage explains that, as a young lad, he tried to raise the Devil, positioned from within a ring drawn of his own blood, as he, "desired an interview with the gentleman in black to convince my senses of his existence". Babbage claimed that no interview was afforded, but those who later experienced his wrath would probably have disagreed.

Royal Society of London (as well as those of Edinburgh and Dublin), the Antiquarian, the Linnean, the Geological, the Astronomical, the Zoological, and the Horticultural society. Indeed, as Babbage notes, this is not a specific indication of a declining interest in science, but sarcastically he adds, "an ambitious scientist could render their name a kind of comet, carrying with it a total of upwards of forty letters, at the average cost of 10 pounds, 9 Shillings and 9 pence per letter". Of the Societies that one could try to become a fellow, the Royal Society is singled out as being one of the most expensive and one of the most exclusive. Indeed, it was the elitism and leadership of the Royal Society that Babbage took umbrage with. While Babbage insists that the sciences are in decline (in the run-up to the 1830s), there is little evidence that his warnings were taken that seriously. Although, in sympathy with some of Babbage's suggestions, it was in 1831 that the British Association for the Advancement of Science was formed, and the formation of this association certainly heralded the onset of change (see Chap. 5.7).

Babbage made no great effort to determine the number of members or Fellows belonging to the 13 learned societies that he listed, only noting that the Royal Society of London had 685 Fellows, to which he noted, "in the United Kingdom every 32,000 inhabitants produces a fellow of the Royal Society"[6]. Of course, the number of Fellows per number of individuals in the population is not a particularly meaningful measure of anything. A more detailed analysis of society membership, and a measure of growth in membership over time, however, was provided by Leone Levi at the 38th meeting of the British Association for the Advancement of Science, in Norwich, in 1868. Levi was Professor of Commercial Law, and a keen statistician, at Kings College London. In his extensive review Levi considered the growth in learned societies, in the UK, from circa 1831 to 1867. Fig. 4.7 is a summary plot of his findings. Indeed, growth in the number of learned societies was dramatic, more than doubling in the 30 years from 1836. Of the three Royal Societies considered by Levi, the London and the Edinburgh Societies both saw a decrease in their membership between 1831 and 1867, falling by 18 and 2 percent respectively over the time interval considered. The Royal Irish Academy saw a slight increase of 3 percent in membership (indicating the addition of 13 more members) in the years from 1838 to 1867. While no new Royal scientific societies were established in the time interval considered by Levi, it was the case that other learned societies flourished. The greatest amount of growth was in the area of those societies catering to geologists and

---

[6] This is based upon a total population of 22,299,000. Currently there are approximately 1700 Fellows of the Royal Society of London, and with a UK population of 67.33 million, this implies that there is one Fellow per 40,000 individuals.

archaeologists, closely followed by the applied sciences (including such areas as manufacturing, civil engineering, pharmaceutical and medical). Societies catering to biological and natural history interests also showed considerable growth, along with those relating to mathematical and physical interests, the latter including the newly formed Statistical Society (1834), the London Mathematical Society (1865) the Chemical Society (1841), and the British Meteorological Society (1850). Usefully, Levi looked at society membership numbers, and estimated that the "total number of men directly contributing, by their learning or their wealth, to the promotion of science constitutes about 15 in every 10,000 of the population". On this basis, by 1868 approximately 1 in every 666 members of British society was associated with one learned society or another. Indeed, Levi notes with respect to Babbage's 1830 *Reflections*, that in the ensuing 38 years, "there has been a large increase in the number and membership of learned societies in the United Kingdom, a fact indicative of a decided advancement of science". Continuing his review with yet more statistics, Levi estimated that there were of order 120 learned societies in the UK in 1867, with a total membership amounting to some 45,000 individuals[7]. Furthermore, Levi estimated that the collective income of these societies (from membership fees and donations) ran to some £120,000 per year. In terms of societies with the greatest number of members, in 1868 the top 5 were, the Royal Agricultural Society, the British Association for the Advancement of Science, the Royal Horticultural Society, the Royal United Services Institution (basically a defense and security *Think Tank* founded in 1831), and the Society for the Encouragement of Arts, Manufactures and Commerce, with 5525, 3629, 3595, 3283, 3278 members respectively. The Royal Astronomical Society, at this same time, had a total of 528 members, being about half that of the Geological Society of London, some five times larger than that of the London Mathematical Society, but comparable in size to the membership of the Institute for Mechanical Engineers, and the Royal Institute of British Architects.

Clearly, important changes were taking place, with respect to the advancement of science in England during the middle years of the 19th century. Not only were practitioners looking to ally themselves with the ever-increasing number of learned societies and professional bodies, they were also prepared to spend good money to do so. Furthermore, there was also an increasing sense of needing to belonging to an organized society, composed of like-minded individuals, with these societies acting as specific repositories of

---

[7] The total number of members in all of the societies listed by Levi ran to some 60,000 subscriptions, but, of course, some individuals belonged to more than one society.

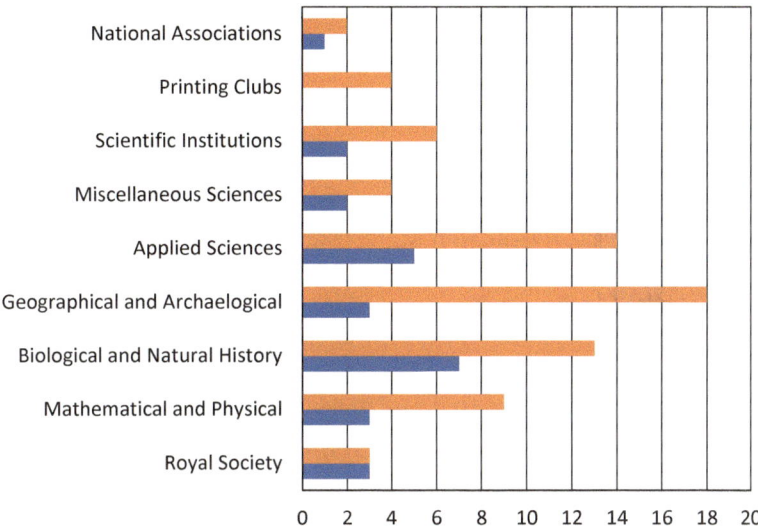

Fig. 4.7 The growth of learned societies in the United Kingdom between 1831 (blue bars) and 1867 (orange bars) as deduced by Leone Levi. See text for details

knowledge and expertise. At the time that Denning was beginning his astronomical studies, in the mid-1800s, the UK provided for a vibrant spectrum of scientific interests and societies for both the amateur and the professional practitioner.

Not only was the profession of astronomy undergoing dramatic change as the 19th century drew to its close, but so too was the scope and range of astronomy itself. Along with the development of the professional/amateur divide, so astronomy became increasing allied to the disciplinary practices of mathematics, physics, and chemistry. No longer was it the case that astronomy was solely concerned with positional work—through the measurement of angles on the sky and the determination of times—rather, it became increasingly anchored to applied physical theory. Stars were no longer simple pinpoints of light, to which the astronomer determined ever more accurate positions on the celestial sphere, but they became objects that must obey the increasingly precise laws of physics and chemistry that were being investigated in terrestrial laboratories. The stars must have a mass, a physical size, a surface temperature, a central temperature, a composition, and an energy source. And, the interiors of the stars must obey the laws of thermodynamics, and the conservation of energy, and their atmosphere's must obey the same laws of spectroscopy that applied to the study of tube-trapped gases in the laboratory. The rise of astrophysics opened-up an ever more subtle and ever more complex universe for astronomers to contemplate. Not only was the deep-structure of the heavens under increased scrutiny, at the turn of the 19th century, however, but so too was the deeper and fundamental structure of matter, and harping back to medieval times, what happened below (that is, in the laboratory, and in the equations of physics), must also take place in the heavens above.

There was no such sub-field of astrophysics in the mid-19th century. What theoretical work there was at that time was entirely mathematical in form, and then directly related to the study of Newtonian dynamics. The motion of the planets, the Moon, satellites, asteroids, and comets were entirely subjugated to the domain of gravitational interactions, and the solution of those equations needed to explain observed motions and perturbations. Indeed, the nature of celestial dynamics had been deftly captured in the form of differential calculus by Pierre Simon Laplace in his landmark *Traité de Méchanique Celesté*, published in 5 volumes between 1799 and 1825. Built upon the pioneering works of Isaac Newton, Joseph Lagrange, Leonhard Euler, and Laplace, classical mechanics was, by the mid-19th century, a finely developed branch of mathematics, and astronomical observations merely (but importantly) provided the boundary conditions necessary to fully-solve the equations of motion. How meteor astronomy was brought into the realm of physical theory and incorporated within the fold of classical mechanics will be discussed in Chap. 5.

The Victorian scientist loved nothing better than to classify, unify, and coral a field of study—with everything in its place, and there being a place for everything. Such enthusiasm was evident in the doctrine of positivism introduced by August Comte, one of the first philosophers to develop a theory of the sciences, and one of the pioneers of social studies. In his *Positive Philosophy*, a series of texts published between 1830 and 1842, Comte outlined a systematic and hierarchical classification of the sciences according to the degree by which their associated phenomena could be exactly determined. In this manner the various sciences were effectively ordered according to how well their subject matter could be described mathematically. In terms of decreasing generality, therefore, Comte argued, that the sciences should be arranged according to the hierarchy of mathematics, astronomy, physics, chemistry, biology and then sociology. Astronomy, throughout the majority of the 19th century, could reasonably be placed above physics, since it effectively had no astrophysical (or theoretical) component—today, astronomy is very much a sub-branch of physics, but theoretical physics is still essentially a sub-branch of mathematics[8]. Moving towards the close of the 19th

---

[8] Arthur Eddington was to write of theoretical physics in his *The Philosophy of Physical Science* (published in 1939) that it is a highly mathematical subject, but that, "the mathematics is not there

century, however, astrophysics as a *bona fide* field of study began to emerge, and in step with this, Comte's positivist outlook, along with that of the deterministic approach of classical mechanics were called into question. While the emergence of spectroscopy, as a fundamental tool of the astronomer, in the later part of the 19th century, had negated Comte's positivist argument that humanity would never know the composition of the stars, other distinguished practitioners began to signal a belief in the end of scientific discovery itself. Lord Kelvin reputedly argued at the British Association meeting in Bradford, in 1900, that there was nothing new to be discovered in physics, and that future generations of physicists could look forward to nothing more than the refinement of fundamental measurements. How wrong Lord Kelvin was. For indeed, as the 20th century emerged Max Planck introduced the concept of quantized energy. Initially formulated as a mathematical-trick to solve the problem of blackbody radiators, Planck's work ultimately resulted in the death of classical determinism, and ushered in the new science of quantum mechanics. Furthermore, in 1905 and 1915 Albert Einstein introduced his ideas on special and general relativity respectively. Einstein's theories not only introduced new ways of thinking about space and time (now amalgamated into spacetime), they revealed the limitations of Newtonian theory, and they set limits on speed and the possible structure of the universe. Ultimately, from Einstein's theories, and the new experimental discoveries relating to the structure of the atom, and the development of quantum theory, the astrophysicist has given us, in the more modern era, an understanding (well, a partial understanding) of such bizarre objects as white dwarfs, neutron stars and black holes. All of this, of course, takes us well away from our specific time of interest, but it illustrates the direction in which astronomy, and astrophysics in particular, was heading at the turn of the 19th century. Strange new physics leading to strange new worlds. Denning played no part in this new astronomy, however, and indeed, as the 20th century evolved so meteor astronomy, both theoretically and instrumentally, began to move in directions that he could not, or, perhaps, would not, follow.

For all of the fundamental change initiated towards the close of the 19th century, the amateur astronomer, indeed, the amateur scientist, was not written out of all contention. J. B. S. Haldane enthusiastically praised

---

till we put it there". That is, it is the mathematical description that follows from the discovered of some new physical phenomenon. In the modern era, the reverse situation has come to prominence, and it is generally asserted that if the equations describing some idea can be written down in a consistent and logical manner, then the equivalent physical object must exist somewhere in the universe. In other words, that mathematics is there from the start. More recently, physicist Steven Weinberg (1933–2021) championed the notion that if it can be imagined in the human mind, then it must exist in reality (somewhere, somehow, and somewhen).

the amateur scientist in his *Possible Worlds and Other Essays* (first published in 1927). Recognizing that while many theoretical and intensive laboratory topics, such as fundamental physics and chemistry, were closed to the amateur, Haldane noted that in areas such as biology, geology, animal behavior, speleology, meteorology, and astronomy, the amateur could still make important discoveries and contributions. Furthermore, Haldane used Denning as a specific example whereby the amateur still held numerous advantages over his professional colleagues. Where unaided observing was concerned, Haldane writes,

"Some day, no doubt, an instrument superior to the human eye will be invented for the observation of meteors; but until this invention is made, the equipment of an observatory will be worse than useless for this purpose. Hence the world's greatest meteor observer, and probably its greatest amateur scientist, Mr. W. F. Denning, owns no allegiance to any observatory. The observer of meteors requires a clear sky, a thick coat, a notebook, knowledge of the constellations, infinite patience, and a tendency to insomnia"

Haldane was entirely correct in his assessment, and it was only towards the close of the 20th century that low-light level photographic and videographic techniques began to replace (or more correctly, complement) the human observer with respect to meteor astronomy. Haldane's praise for Denning is also notable, and illustrative of how wide-spread his name, in both public and scientific society, had become. Interestingly, Haldane saw it as an advantage that Denning had no allegiance to an observatory, and this highlights a distinction, then prevalent, between British and American attitudes towards scientific authority. While Denning was indeed an independent, non-aligned amateur astronomer, his rival in America (recall Chap. 2), Charles P. Olivier, used his professional status, university and scientific society connections, as well as his allegiance to various observatories, as important factors in leveraging funding for his meteor work.

4.4   On Work and Longevity

In parallel to his strong views on what amateur astronomers should be doing, Denning also expected a great deal from the individual observer. He noted [8], "virtually the observer himself constitutes the most important part of his telescope: it is useless having a glass of great capacity at one end of the tube, and a man of small capacity at the other". To this end, Denning argued that serious amateur astronomers should continually strive to improve their observing skills. He also suggested that, "every amateur should practice drawing", and, habitually exercise their sight. On this later issue he remarked [9], "I invariable use the right eye on the markings of planets and the left on the minute stars and satellites. Practice has given each eye a superiority over the other in the special work to which it has been devoted". There is unfortunately no evidence to support Denning's claim that the eye can be trained in the manner he described. It does seem, however, that Denning had very good eyesight. That this was so is evidenced by one of Denning's early contributions [10] published in the *Monthly Notices of the Royal astronomical Society*, where he recorded the fact that he had been able to visually resolve without a telescope the Jovian satellites Ganymede and Callisto. Denning also commented in the same short note that he could distinguish 13 stars in the Pleiades, and had on occasion seen Jupiter in full sunshine. Such impressive accounts of eye sensitivity are rare but not unheard of.

Never afraid of measuring, collecting, and collating data, Francis Galton set out to test the efficacy of prayer in a research paper published in the 1 August issue of the *Fortnightly Review* for 1872. This was a revolutionary (if not controversial) study. Using numerous obituary accounts, Galton assembled data on the lifespan of individuals associated with various trades, traditions, and societal backgrounds. Accordingly, he found that members of the clergy lived to an average age of 69.49 years, while doctors, lawyers, gentry, and members of royalty had average lifespans of 67.31, 68.14, 70.22, and 64.04 years respectively. The average lifespan of those distinguished in the literary and scientific arts was a healthy 67.55 years. Form this, Galton concluded that prayers concerning the granting of longevity were ineffective—specifically he noted that if they were effective, then members of Royalty should have the longest lifespans—the members of Royalty being routinely prayed for by church and the populace. Galton's conclusions are not obvious, and are certainly open to question, and re-interpretation, but the point is that statistical analysis was being applied to extensive data sets (i.e., people), and it was being used to find answers to seemingly unanswerable questions. Indeed, Galton's study was an entirely new way of looking at how human society operates, and it demonstrated how data, and data analysis can be applied in the analysis of complex social problems.

Echoing Galton, Denning claimed in his *Telescopic Work for Starlight Evenings* [9] that, "a distinguishing trait among astronomers has been their keenness of vision, which in many cases they have retained to an advanced age"[9]. Such sweeping and unsubstantiated claims reflect an interesting side of Denning's character. While he was typically quite reserved in what he would claim from a set of observations, he would occasionally make generalizations on the basis of personal beliefs, and/or dubious statistics. An example of the latter case is exemplified in his comment, "night air is generally thought to be pernicious to health; but the longevity of astronomers is certainly opposed to this idea" [11][10]. In an attempt to substantiate this claim Denning published two articles [12, 13], in the style of Galton, one in 1897, and the other in 1917, which presented lists of astronomers that had lived to the age of 80, and beyond. From these he concluded, "astronomical pursuits were conducive to longevity". The promotion of such ideas, on the basis of poor (actually, none) statistical testing, highlights Denning's disregard for some aspects of scientific methodology (i.e., no non-astronomer age comparisons were made, and his data selection process was poorly defined). He did not even calculate an average lifespan. For all this, the topic of longevity has its interesting points, and seems worthy of a little more exploration.

Denning was not the first author to consider the apparent longevity of astronomers. Indeed, Abert Lancaster (Directory of the Royal Observatory of Belgium) wrote on the topic as early as 1884, but made a clear attempt at comparisons. Lancaster's opening sentence informs his readers that the average human lifespan "in civilized countries" is about 33 years [14]. Using bibliographic data on 1,741 astronomers (from ancient to modern times), however, Lancaster determined that the average lifespan of an astronomer was some 64.4 years—a remarkably better average lifespan than that attributed to the general populace. Lancaster further noted that when a similar analysis was performed with respect to famous mathematicians, the average lifespan increased to 69.5 years. In concluding his article, Lancaster writes, "become an astronomer, if you wish to live long" [14]—one presumably takes from this conclusion that becoming a mathematician was not preferable, in spite of the few extra years that such a profession seemingly afforded. In more recent times, historian of astronomy, D. B. Herrmann, has also written on the life expectancy of astronomers [15]. Using biographical data of astronomers born

[9] Denning does not clearly articulate what he means by the label 'astronomer', and certainly many of the historical figures that he includes in his tables would not have considered themselves as being astronomers *per se*.

[10] This is a somewhat contradictory claim, since Denning suffered many bouts of ill health throughout his life. He did comment, however, in an interview for *Tit Bits* magazine (August 31st, 1895, 386) that, "When I commenced habitual night-work I was very sensitive to cold, and the winter usually found me with a troublesome cough". He continued, however, "on the whole I think that 'star-gazing' may be beneficial to health, as well as intellectually profitable".

between 1715 and 1825, Herrmann finds an average lifespan of 71.6 years[11]. Denning is accordingly found to be correct in his hunch (if not his statistics) about the longevity of astronomers. Assuming that the average lifespan of astronomers has not changed dramatically into the modern era, a career change is not so advantageous now, since the average life expectancy (at least in wealthier nations) has risen to 73.4 years [16], although this does vary significant from one country to the next.

In spite of the fact that Denning received no formal scientific training, he was an enthusiastic promoter of scientific study. He saw the advancement of knowledge as a necessary and noble occupation. Indeed, on this subject he was to write in prosaic tones, "the progress of science may be compared to the ceaseless running of a stream. It continues its course uninterruptedly through the years. ... occasionally there is a slackening of velocity and a lowering of the waters but it ever goes on to its goal. So, the advance of scientific discovery is maintained through the ages. There may occur comparative lulls and temporary abatements but there is never absolute inactivity"[12]. Here, if ever there was one, is a genuine belief in scientific *plus ultra*. Denning's outlook is certainly consistent with the new, and much more expansive scientific outlook that began to develop towards the close of the 19th century. Science was finding answers to both old and new problems, and science, through engineering, technology, and chemistry was improving the human lot, as well as lining the pockets of industry.

Bibliography

1 Denning, W. F. (1891). *Telescopic work for starlight evenings* (p. 74). Taylor and Francis.

2 *Observatory, 11*, 181–182, 1888.

3 Denning, W. F. (1883). Amateurs and astronomical work. *Nature, 27*, 434–436.

4 Denning, W. F. (1891). *Telescopic work for starlight evenings* (p. 78). Taylor and Francis.

5 Denning, W. F. (1897). Organized or Sectional work in astronomy. *Nature, 56*, 9–10.

6 Denning, W. F. (1918). A few notes on amateur observers and observations. *Journal of the Royal Astronomical Society of Canada, 12*, 157–159.

---

[11] One notes that Denning lived to his 83rd year.
[12] This quotation is from a hand-written article entitled Recent Astronomical Discoveries. The article was destined for the journal *Knowledge*, and was written in 1928.

7 Hughes, D. W., & de Grijs, R. Top ten astronomical breakthroughs of the 20th century. See: https://www.capjournal.org/issues/01/11_17.pdf.

8 Denning, W. F. (1891). *Telescopic work for starlight evenings* (p. 33). Taylor and Francis.

9 Denning, W. F. (1891). *Telescopic work for starlight evenings* (p. 71). Taylor and Francis.

10 Denning, W. F. (1874). Naked-eye observations of Jupiter's satellites. *Monthly Notices of the Royal Astronomical Society, 34*, 309–310.

11 Denning, W. F. (1891). *Telescopic work for starlight evenings* (p. 75). Taylor and Francis.

12 Denning, W. F. (1897). Longevity of astronomers. *Observatory, 20*, 206.

13 Denning, W. F. (1917). Longevity of astronomers. *Observatory, 40*, 132–133.

14 Lancaster, A. B. M. (1884). Longevity of astronomers. *Popular Science Monthly, 25*, 60–62.

15 Herrmann, D. B. (March 1992). *The messenger.*

16 From https://www.worldometers.info/demographics/life-expectancy/

5

The Rise of Meteor Astronomy (1830–1930)

The aim of this chapter is to follow the early development of meteor astronomy. The timeframe considered, from 1830 to 1930, encapsulates, in broad brush form, the span over which meteor astronomy evolved from that of a minor science, even a trivial preoccupation, to one of the upmost scientific importance. Prior to 1830 the study of meteors was essentially incidental, and primarily driven by short-lived investigations of specific singular events. Our chosen timespan effectively covers the classical period of visual meteor astronomy; the time when all data was gathered by the human eye, and then mostly gathered by amateur astronomers. It was also in the time interval between 1830 and 1930 that meteor astronomy was tamed according to the theories of celestial mechanics. Not only were the orbital characteristics of meteoroid streams determined for the first time, but the perturbative effects of gravitational interactions with the planets were introduced to the study of stream evolution. It was post circa 1930 that the origins of our modern-day physical theory of meteoroid ablation were first established, and indeed, used to determine the physical nature of the Earth's upper atmosphere. The time interval from 1830 to 1930 also covers the working lifespan of William Frederick Denning.

In the seventeenth century, John Donne asserted that, "No man is an island entire of itself", and this certainly applies to the natural historian and scientist. Indeed, science is all about communication between knowledgeable practitioners. Denning, although a reclusive person, was no exception to Donne's devotional prose, and he worked with, and sometimes against, an ever-growing international community of both amateur and professional astronomers. The intent of this chapter is to take account of the greater

© The Author(s), under exclusive license to Springer Nature
Switzerland AG 2023
M. Beech, *William Frederick Denning*, Springer Biographies,
https://doi.org/10.1007/978-3-031-44443-2_5

community, the larger continent, if one will, to which the would-be island of Denning was closely anchored. In this respect our reach will often pass beyond the extremities of our set timeframe, but largely so as to determine precedencies, set the historical scene, and/or to provide some closing clarity.

5.1   A Very Brief History of Meteors

Many scientific fields of investigation can trace their origins back to Aristotle, and meteor astronomy, in spite of being one of the more recent branches of astronomy to develop, is no exception. For all this, Aristotle largely presents a dismissive story. Writing circa 350 B.C. Aristotle produced the remarkable text *Meteorologia*. The term remarkable is used since, as with most expositions written by Aristotle, it was a grand synthesis of meteorological phenomena, based upon detailed observations. And, as with Aristotle's other expositions in natural philosophy, it is wrong about pretty much everything. This, of course, is not for want of imagination or effort, or indeed, ability. Rather it reflects a largely non-experimental approach to interpreting natural phenomenon. Indeed, in his *Meteorologia* Aristotle asserts that meteors are a terrestrial/atmospheric phenomenon [1]. In terms of explanation, Aristotle was entirely correct, meteors, as a phenomenon of light emission, are a phenomenon of the upper atmosphere. That the substantive cause of the meteor is a meteoroid entering the Earth's upper atmosphere, at high speed, from outer space, however, is something that neither Aristotle nor any of his contemporaries could have countenanced. Indeed, to go from the observed meteor to a causal meteoroid is something that is a far from trivial step.

Aristotle explained meteors as a process involving the burning of combustible matter, or vapour, that had assembled in a line and then ignited in the upper regions of the terrestrial sphere (that is the region just below the first celestial sphere containing the Moon). According to the amount of combustible material, and how rapidly it burned, one would see either a lightning stroke, a meteor, a comet, or an auroral arc. Other meteorological phenomena, wind, rain, snow, thunderstorms and heat waves were all explained in Aristotle's program, and then explained in such a convincing, common-sense way, that his ideas remained accepted and in place for the best part of 2000 years.

It was Edmund Halley who first openly questioned the meteorological ideas presented by Aristotle [2]. Investigating several extremely bright fireballs that had raced through the skies of England on 6 March, 1716, and on 18 March 1718, Halley assembled enough eye-witness accounts to deduce approximate paths, heights and speeds. He was staggered by the results. Indeed, the heights and speeds were such that Halley questioned the ability of vapour to rise to such altitudes in the atmosphere, and the speeds seemed far too high for combustion along a train of matter. Halley made no fixed decision, alternately suggesting that the fireballs were composed of some solid matter that had entered the Earth's atmosphere from without, and then suggesting that perhaps they were long trains of sulphurous matter. This latter idea was more in line, in fact, with Isaac Newton's views, as explained in his *Opticks* (published in 1704).

Moving into the later half of the eighteenth century, practitioners of the new science of electricity soon offered an explanation for meteors. Indeed, John Perkins, from Boston, wrote to his long-time friend Benjamin Franklin in 1753, suggesting that meteors were nothing more that electrical fires. Along similar lines, Jean-Baptiste Le Roy, in France, suggested that the extraordinary meteor of 17 July, 1771 was some form of electrical discharge. Likewise, Charles Blagden, in England, argued that the bright fireball procession of 18 August, 1783 was caused by the movement of an electrical fluid. While the electrical discharge explanation for producing meteor phenomena had popularity on its side, this is not to say that the Aristotelian explanation had been abandoned, and indeed, neither had the idea that they might be the result of some extraterrestrial effect. Thomas Clap (President of Yale College: 1740–1766) suggested, for example, that perhaps fireballs were Earth-orbiting cometary bodies which occasionally dipped into the atmosphere, and then moved out again.

In terms of affording an explanation to the appearance of meteors, towards the close of the eighteenth century, the field essentially suffered from too much theoretical speculation, and not enough systematic observation. All this was to change, however, as the 1790's unfolded. Firstly, German physicist and musician Ernst Chladni (Fig. 5.1) published a number of highly influential works on meteors and meteorites. Second the first systematic study of meteor heights took place, and thirdly a dramatic fall of hundreds of meteorites occurred in L'Aigle, France, in 1803.

Often called the father of acoustic theory, Ernst Florence Friedrich Chladni was essentially a roving researcher [3]. He traveled throughout Europe, visiting the libraries in university towns, and communicating with many of the most respected natural philosophers of his time, examining specific topics as he chose and pleased. Following conversations with Georg Lichtenberg in Göttingen, however, in the early 1790s Chladni was inspired to embark on a study of the fall circumstances of meteorites. His conclusions, published in 1794, asserted that meteorites are extraterrestrial in origin, and that their

Fig. 5.1 Ernst Florens Friedrich Chladni (1756–1827). https://commons.wikimedia. org/wiki/File:Echladni.jpg Staatsbibliothek zu Berlin, Public domain, via Wikimedia Commons

arrival of Earth is presaged by the passage of a bright fireball through the sky. Chladni dismissed the rising vapors, and electrical discharge models for producing the light phenomenon, arguing instead that it was the result of a frictional heating effect produced through the rapid deceleration of a solid body entering the Earth's atmosphere from planetary space. Chladni's thesis was roundly ridiculed upon publication. Respected commentators such as Johann Goethe and Alexander von Humbolt soundly dismissed it, and even Lichtenberg was sceptical of its conclusions. Fortunately, within a year of its appearance, a bright daytime fireball was seen traveling across northern England, and an associated stony-meteorite fell at Wold Cottage in York-shire. Likewise, in 1803 another great fall of meteorites took place, this time in L'Aigle in France, and on this occasion the French Minister of the Interior sent astronomer Jean Baptiste Biot to investigate. Based upon numerous eye-witness accounts, Biot concluded that, indeed, the stones had fallen from the sky, with great speed, and that there had been an associated fireball. While Biot did not specifically endorse the ideas outlined by Chladni, his detailed account of the fall in France brought the topic of meteorites and fireballs to the consciousness of a large audience of natural philosophers.

One of the important experimental outcomes that resulted from the conversations between Lichtenberg and Chladni, was that Lichtenberg convinced two students, Heinrich Brandes and Johann Benzenberg, to undertake a systematic study of meteor heights. The hope was that two-station observations might reveal at what height in the atmosphere meteors were occurring, and whether they were generally traveling upwards or downward. A predominantly downward direction would imply an extraterrestrial origin, while a combination of directions would suggest a terrestrial origin.

5.2   Two-Station Observations

The study by Brandes and Benzenberg was the first proactive, rather than reactive investigation into meteor phenomena. It was literally the birth of practical meteor astronomy [4]. The method to be exploited was that of parallax, and to achieve this the two observers took-up observing positions some 8-km apart, and simultaneously viewed the same region of the sky—waiting for any meteors to appear.

To turn visual meteor observations into actual atmospheric flight data, a meteor trail must be recorded from at least two locations. Such simultaneous observations require each observer to record the position of the meteor trail, as seen on the sky, with respect to the background stars. Figure 5.2 indicates the essential two-dimension geometry of interest (in actuality, of course, 3-dimensional, spherical geometry should be applied—but such refinements are not needed here). The meteor ablation trail[1] (shown in red in Fig. 5.2) is recorded by two observers A and B a known distance d_{AB} apart. The beginning and end points on the celestial sphere are recorded for each trail Meteor A and Meteor B. The recorded beginning and end points of each meteor trail can be in either altitude and azimuth, or right ascension and declination—the two being readily inter-related from the recorded time of the specific event. The radiant point of the meteor can now be calculated directly, or, graphically, as Denning preferred to do, by drawing the two trails (Meteor A and Meteor B) on to the surface of a celestial globe, and projecting their paths backward (along great circle arcs) to find the location at which they cross—this intersection point being the sky location of the radiant. To determine the

---

[1] Here we are using modern terminology, and theory, which associates the ablation of a meteoroid as it descends through the atmosphere with the process responsible for the generation of light (the meteor).

actual atmospheric heigh H_B, at which vigorous ablation begins, the elevation angles α_A and α_B need to be measured (see Fig. 5.2). With the distance between the observers known, then straightforward trigonometry gives

$$H_B = d_{AB} \frac{\sin \alpha_A \sin \alpha_B}{\sin(\alpha_A + \alpha_B)}$$

A similar calculation (with appropriate angles being inserted) will determine the end height at which ablation stops. Additional calculations (not shown here) can further reveal the length of the ablation track between the beginning and end heights. The calculations required to determine the physical path of a meteor through the atmosphere are essentially straightforward to make, but being able to gather accurate estimates of a meteor's beginning and end points on the sky requires a great deal of patience and many hours of practice. As seen in Chap. 3, Denning collaborated with numerous observers, especially A. S. Herschel, for many decades, in order to gather data on meteor ablation paths (see Sect. 5.10).

The first two-station observations by Brandes and Benzenberg were conducted on six nights between 11 September and 4 November, 1798. Their original supposition was that if meteors were atmospheric in origin, then the expected height range of interest would have been about 10-km altitude in altitude [5]. They soon realized, however, that meteors were located

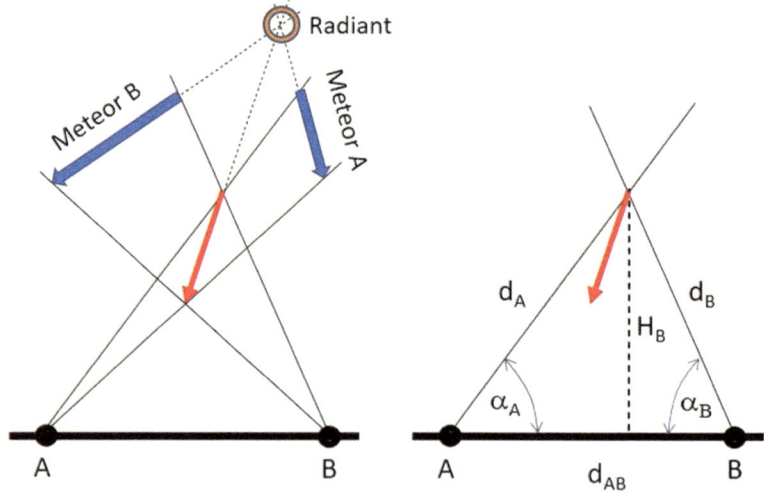

Fig. 5.2 A two-dimensional (simplified) geometrical representation for the determination of a meteor's radiant point, and its beginning height H_B. The ablation trail (corresponding to the actual meteor) is shown as the red arrow; the two observers are located a known distance d_{AB} apart at positions A and B

in much high regions of the atmosphere, and accordingly they increased the baseline between their observing locations to some 15.5-km. During the first observing session 204 meteors were observed of which 21 were deemed to be simultaneously recorded. From these they determined an average beginning height of 50 ± 30 miles, or about 91 km. Only 4 of the 21 simultaneously observed meteors were sufficiently well observed for both beginning and end height estimates to be made. Of these it was a 50/50 split between the end height being higher than the beginning height, and the end height being lower than the beginning height. From these early beginnings it was not clear if meteors moved upwards or downwards through the atmosphere. Clearly more observational data were needed. Chladni took the observational results at face-value, and in 1817 amended his thesis to include the possibility of two types of luminous meteor: one type entering Earth's atmosphere from outer space, and the other type being formed within the atmosphere.

Brandes returned to the study of meteors in 1823, this time with his own students in tow, and data was obtained on some 63 meteor trails. Of these trails 36 had both their beginning and end heights determined, with 26 apparently moving downwards in the atmosphere, 1 appearing to move horizontally, and 9 appearing to move upwards in the atmosphere. Chladni's thesis that meteors entered Earth's atmosphere from outer space was looking more likely, but the parallax method being employed by the observers, while mathematically sound, could not be applied with sufficient accuracy (by naked-eye observers) to provide a definitive answer to the origin of meteors. This very conclusion was, in fact, confirmed by mathematician Friedrich Bessel in 1839. Indeed, Bessel performed a detailed error analysis of the two-station observations, and concluded that the observers were using far-too short a baseline between their observing stations for accurate results to be obtained. This being said, the studies had clearly shown that meteors were a phenomenon of the upper atmosphere, and that it was likely that they moved downwards, towards the ground, in a manner consistent with entry from outer space. More observations were to follow, and eventually the originally sceptical Alexander Von Humboldt, in his many volumed *Cosmos*, triumphantly argued, in 1849, that the issue of apparently ascending meteors had been fully resolved, and that they were an artifact of the observational uncertainties. Meteors were a phenomenon associated with the entry of small objects (meteoroids) into the Earth's upper atmosphere from interplanetary (possibly interstellar) space.

Following the notable outburst of the Leonid meteor shower in 1833 (see the following section), Elias Loomis, tutor at Yale College, pondered in the pages of the July 1835 issue of the *American Journal of Science*, "Every person of a reflecting mind must have often asked himself the question, what are shooting stars". Building upon such thoughts, Loomis accordingly entered into a detailed discussion of the two-station reduction methodology for determination of meteor heights. To this theoretical discussion, Loomis appended a table of the meteor beginning and end heights derived by Brandes and his students in 1832. Likewise, Herbert A. Newton writing in the July 1864 issue of the *American Journal of Science*, reproduced the 1823 results from the Brandes team of observers. Clearly, while it was realized that systematic observations of meteor heights, speeds and directions were highly important, very few practitioners, after Brandes, were able to establish, and then staff new observational campaigns. This was not from a lack of interested parties, since many individuals were, by the mid-1800s, regularly observing and recording meteors. The problem seems to have been one entirely of organization. Eduard Heis in Göttingen (see Sect. 5.5) navigated his way around this organizational issue, however, by recruiting successive generations of university students to make the observations, although his primary interest in collecting meteor trail data was to determine rates and radiants. Some organization success was evident in America and England, however, during the early 1860s, with observational campaigns being established by Hubert A. Newton in the US, and by the Luminous Meteors Committee (LMC) of the British Association for the Advancement of Science (BAAS—see Sect. 5.6). The later group set out to systematically observe the August Perseid meteors in 1863. In England the campaign was led by A. S. Herschel and observations were made at the Royal Greenwich Observatory (RGO) in London, Cambridge University, and Hawkhurst in Kent. A total of 20 meteors were deemed to have been observed simultaneously and observed well enough to determine beginning (H_B) and end (H_E) heights. The results were presented in the 1863 Report of the LMC to the BAAS, with the averages being $<H_B>$ = 81.6 miles, $<H_E>$ = 57.7 miles, with the average estimated velocity being 39 miles per second. These averages, for visual observations, are in fact remarkably good when compared against instrumentally derived, present-day, data. This being said, large uncertainties are invariably associated with visual observations. A further August observing campaign was conducted in 1864, but only 1 meteor was doubly observed from Hawkhurst and the RGO. The next observational campaign organized by the LMC took place during the 1868 Perseid shower. This time observations were gathered from 11 locations distributed across England and Scotland. In all some 10 Perseids were simultaneously observed, and A. S. Herschel determined heights and velocities. The reductions from this campaign indicating that $<H_B>$ = 75.1 ± 17.6 miles, $<H_E>$ = 47.5 ± 6.1 miles. A final observational campaign was organized by the LMC in 1870, with a further 26 meteors being simultaneously observed. In spite of the valiant efforts of the LMC little headway was made in recruiting amateur observers to make regular double station observations. Indeed, this is perhaps not surprising since it required the development of very specific observing skills, and is a mostly unrewarding enterprise. Denning was eventually recruited by Herschel to conduct two station observations (recall Chap. 3), and several of the prominent observers of the Meteor Section of the British Astronomical Association (BAA) conducted regular two station campaigns through to the 1930s. While much can be achieved by skilled observers, the determination of definitive meteor heights and velocities is something that really requires the application of instrumental techniques. Such photographic methods were first developed in the last decade of the nineteenth century (see Sect. 5.10.1). For all this, however, precision observations were not generally available until the deployment of the Baker super-Schmidt widefield meteor cameras by Harvard University and the Smithsonian Observatory in the 1950s.

5.3   The 1833 Leonids and After

The popular origins of modern meteor astronomy can be readily dated to the night of 13 November 1833 [3, 4, 6]. On this night, across much of North America, meteors fell, with no prewarning, in great abundance. Caught off-guard both the public and the practitioners of natural science wanted answers: what was this phenomenon, how did it fit-in with respect to perceived astronomical wisdom, and why did it occur when it did? In terms of physical phenomenon this was easily answered, it was a display of meteors. The problem, however, was that it was not then clear (or rather, agreed upon) what a meteor actually was. Furthermore, it was not at all clear what role meteors played in the dynamics of the solar system, and it was not clear why, of all the nights in the year, November 13th should provide for such a remarkable display. Inspired by events, however, the answers to some of the most basic questions were soon forthcoming. Denison Olmsted and Alexander Twining, both associated with Yale College, along with other researchers, soon realized that the various accounts of the Leonid display carried an important piece of structural information [7, 8]. Indeed, many observers reported that it appeared as if the meteors, if their paths were traced backwards, radiated from a small localised region on the sky. This meteor radiant point (Fig. 5.3) was located in the constellation of Leo, and it was this very observation that resulted in the meteor display being called the Leonids. Indeed, meteor showers are now typically named after the constellation in which their radiant resides on the night of shower maximum.

Having established the observation of a radiant point, it was next necessary to understand why there was radiant in the first place. Writing in the

Fig. 5.3 The 1833 Leonid meteor storm as seen above the raging waters of Niagara Falls. The dramatic composition attempts to illustrate both the great number of meteors seen, as well as the presence of a radiant point on the sky (top center) to which, by tracing their paths backward, all the meteor trails appeared to diverge. The zig-zag lines correspond to long-lived meteor trains. Image from *Smith's Illustrated Astronomy* (1855) by Asa Smith

American Journal of Science, Olmstead and Twining had gathered together numerous eye-witness accounts, and realized that the radiant was a manifestation of a perspective effect, requiring that the bodies (meteoroids) responsible for producing the meteors must enter the Earth's atmosphere along parallel paths (Fig. 5.4). Not only were the Leonid meteors derived from material entering from outside of the Earth's atmosphere, they were also moving through the solar system in an organized manner—that is they were spread along and within a meteoroid stream.

Inspired by the 1833 Leonid outburst, researchers soon began to scour the historical literature for other accounts of meteor storms. Accordingly, it was soon revealed that there had been a Leonid meteor storm in 1799, as witnessed by explorer Alexander Humboldt when stationed in Cumana, on the coast of Venezuela. Furthermore, it was additionally found in subsequent years, that some activity from the Leonid shower could be observed each and ever year, indicating that it reappeared annually. In 1837, Heinrich Olbers reviewed the situation, and concluded that not only did the Leonids peak in November each year, but that spectacular storms and outbursts tended to occur at intervals of 33 years. Herbert Newton dove even deeper into the history of the Leonids, and by 1864 had established that they had been unwittingly observed for at least a thousand years. Indeed, the earliest known account of the Leonids dates as far back as 902 A.D., when from across North Africa it was recorded that 'stars scattered themselves like rain'. Other meteor storms were reported in the tenth century, 1582, 1602, and 1698. From

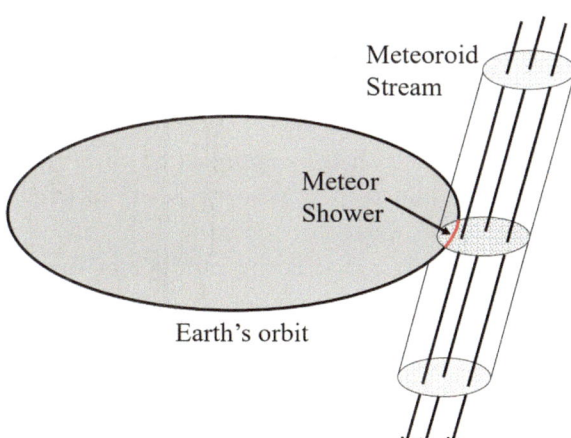

Fig. 5.4 To produce a radiant on the celestial sphere, the meteoroids must encounter the Earth along parallel paths. In this manner, a meteor shower will occur as a result of the Earth cutting through some portion of a meteoroid stream (along the short red arc)

these historical results, and Olbers suggestion of a 33-year outbursts interval, Newton confidently predicted that another Leonid storm should occur in 1866—unlike the situation in 1833, this time, astronomers and the public would be ready.

In the mean time, on the evening of 9 August 1837, American Edward Herrick chanced to notice the passage of a good number of meteors across the sky. Inspired by the story of the November Leonids, Herrick set about looking for historical accounts of abundant meteors being seen in August [4]. It was soon clear that August too yielded an annual meteor shower. Working upon similar (independent) lines of enquiry Belgian astronomer/mathematician Adolph Quetelet also discovered numerous accounts of meteors being seen each year in August. More directly, however, on the night of 9 August 1834 a Cincinnati schoolmaster, John Locke, chanced to notice that a distinct meteor display was in progress, and by careful study deduced a radiant in the constellation of Perseus. A result confirmed by Herrick in 1839. The Leonids of November, were now joined by the Perseids of August. Herrick continued his historical digging, and found additional evidence for another annual meteor shower in early December—this shower turned out to have a radiant in the constellation of Andromeda.

As the predicted time for the 1866 Leonid storm approached, astronomers could look back with some considerable degree of satisfaction concerning the surprise events of 1833. It was generally clear that the Earth experienced multiple annual meteor showers (the Leonids, the Perseids, and the Andromedids), and it was clear that meteors must be extraterrestrial in origin, and that at least in the case of the annual showers, the component meteoroids must enter the atmosphere travelling along parallel paths. How such streams formed, and what they might relate to in the solar system, was still, however, a mystery.

As the night of November 16th 1866 approached observatories from across Europe, where the maximum display was predicted to occur, marshalled a small army of volunteers to watch for meteors. The aim being to gather as much information about rates, that is the number of meteors observed in a given time interval (typically an hour), and to determine as accurately as possible the location of the radiant point on the celestial sphere. The public prepared themselves for a cosmic spectacle. Neither the public (Fig. 5.5) nor the astronomers (Fig. 5.6) were to be disappointed. It is estimated that at the time of peak activity the equivalent of some 2000–5000 meteors per hour could be seen. Likewise precise observations of the beginning and end points of meteor trails enabled good determinations to the radiant location to be made. There is no surviving record of what William Denning saw on the night of the 1866 Leonids, but, as is evident on numerous occasions there-after, it was a spectacle that he saw and never forgot, and indeed, a spectacle that set his ultimate career path as a meteor observer.

Not only did astronomers in 1866 see the prediction of a Leonid storm come true, they also saw a new and fundamental relationship established between the orbital characteristics of meteoroid streams and those of periodic comets. For all this, however, the physical nature, and the origins of comets were, at that time, entirely unknown. The idea that there could be a connection between meteoroid streams and comets was not new in 1866, indeed suggestions of the kind followed in the wake of the 1833 Leonid storm. Denison Olmstead suggested in 1836 that meteors might be derived from the material seen in cometary tails, and in 1865 Herbert Newton concluded that meteoroids probably moved along highly elliptical orbits similar to those orbits deduced for comets. Perhaps the first clear articulation of a possible association between meteors and comets was that made by American astronomer Daniel Kirkwood—Professor of Mathematics at Indiana University (Fig. 5.7). Writing in the spring issue of the *Danville Quarterly Review* for 1861, Kirkwood proposed that when a comet is observed, what is really seen is just the largest component within a much more extensive stream of

Fig. 5.5 The 1866 Leonid display over Greenwich Park, London. Image from *Sun. Moon, and Stars—a book for beginners*, by Agnes Giberne. (Seeley and Co., London, 1891)

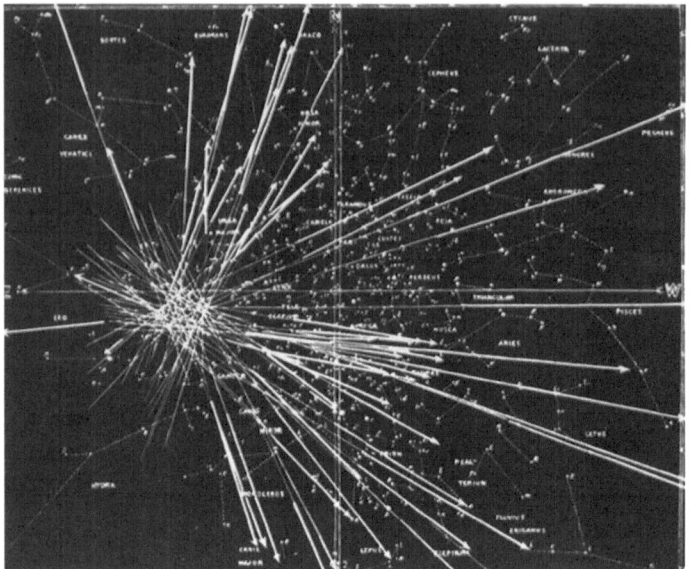

Fig. 5.6 Leonid meteor trails, and the Leonid radiant location, as recorded and deduced by observers at the Royal Greenwich Observatory in 1866. Inage from J. N. Lockyer *The Meteoritic Hypothesis* (1890)

smaller bodies. Basing his arguments on the fact that cometary nuclei are occasionally observed to split, such as observed with comet 3D/Biela in 1852, Kirkwood reasoned that through a process of gradual fragmentation an extensive trail of debris would disperse around the comet's orbit. It was the small debris component of the stream that produced meteors when they chanced to encounter the Earth's atmosphere. Kirkwood strengthened his ideas in a later, 1867, publication entitled, *Meteoric Astronomy—A treatise on shooting-stars, fireballs and aerolites*. This small monograph brought together a great deal of observational data, and presented a liberal amount of theoretical speculation concerning the differences and similarities between comets, meteors and meteorites.

It was not the 1866 Leonid storm, but rather observations of the more sedate 1866 August Perseid shower that enable the definitive link between comets and meteor showers to be made. The key player in this part of the story was Italian astronomer Giovanni Schiaparelli (Fig. 5.8). A professional astronomer, at home both behind the eyepiece of a telescope, and at the desk dealing with the equations of celestial mechanics, Schiaparelli was interested in the calculation of cometary orbits. To this end, he had determined the orbital parameters of a newly detected comet, comet 109P/Swift-Tuttle, observed in 1862. Furthermore, in calculating the orbital parameters for

Fig. 5.7 Daniel Kirkwood (1814–1895). Portrait sketch (1888). Courtesy of Indiana University

the Perseid meteoroids (from their deduced radiant location) he realized in 1867 that the two entities (the comet and the meteoroid stream) had almost identical orbits. Schiaparelli also determined the orbit associated with the 1866 Leonids, and along with German astronomer Carl F. W. Peters at the Hamburg Observatory, realized that the Leonid stream had orbital elements almost identical to those deduced for comet 55P/Temple-Tuttle. Independently of Schiaparelli and Peters, Austrian astronomer Edmund Wiess, at the Vienna Observatory, was also calculating orbits, and in 1868 found that the Andromedid meteoroid stream had similar orbital characteristics to those of comet 3D/Biela. Furthermore, Weiss determined that meteoroids from the April Lyrid meteor shower had orbital elements nearly identical to those of comet C/1861 G1 (Thatcher). Within just a few years nearly all of the then known periodic meteor showers had been linked to some specific parent comet. This association was further cemented when, in 1872, the Andromedid stream underwent a spectacular outburst—an outburst predicted by Weiss in 1868 on the basis of his study of comet 3D/Biela (which recall had been observed to split into two nuclei in 1852). Specifically, Weiss realized that a very close approach would take place between the remnants of 3D/Biela, and the Earth in 1872, and accordingly, if meteoroids were formed through the process of cometary decay, so a great meteor shower should be seen at that time. This did indeed come about, with large numbers of meteors being observed across Europe on the night of November 27th. Under less-than-optimal viewing conditions, Denning counted 79

IL GRANDE ASTRONOMO SCHIAPARELLI NELL' OSSERVATORIO DI BRERA.
(Disegno di A. Beltrame, dal vero)

Fig. 5.8 Giovanni Schiaparelli (1835–1910). Here depicted at work in the Brea Observatory in Milan, Italy. The image is of a lithograph by Achille Beltrame, produced for the magazine *La Domenica del Corriere*

Andromedid meteors over a 40-min time interval starting at 5:50 p.m. Situated further north, William Andrews in Coventry reported in the December issue of the *Astronomical Register* that he had counted some 749 Andromedid meteors in a one-hour interval starting at 6 p.m. Writing in the January 1873 issue of the *American Journal of Science*, Edward Lowe estimated that on the night of the 27th some 58,000 Andromedid meteors fell between the hours of 6 and 10:30 p.m.

5.3.1 A Humbling Prediction

Nature has a way of humbling scientists, and the seemingly safe prediction that the Leonids should produce, in step with its 33-year outburst cycle, a distinct storm on November 14th, 1899, turned out to be a bust. The public, rather than being taken by surprise, as in 1833, were now left in a state of annoyance. The astronomers were likewise dumfounded, and while Leonid meteors were certainly seen, nothing approaching the 1833, or 1866 levels of activity were reported anywhere on Earth. Writing later in his book *Meteors* (published in 1925), Charles Olivier (recall Chap. 2) commented that the failure of the 1899 prediction was the, "worst blow ever suffered by astronomy in the eyes of the public". While this seems to be a reasonable summary of the public response, the failure was really one of theoretical astronomy. It was patently clear from the historical record that the Leonids were only likely to undergo an outburst at intervals corresponding to the 33-year orbital period of its parent comet—it was not a certainty that storms must occur every 33 years. The failure was one in the ability to predict in detail the orbital change of Leonid meteoroid's due to planetary perturbations. This is not to say that the equations to describe such evolution were not then known, but rather that it was prohibitively time consuming to evaluate the numerical results.

That the 1899 Leonids might not live up to the hype of prediction was hinted-at during the 2 March 1899 meeting of the Royal Astronomical Society. At that time (retired) physicist George Johnson Stoney, and mathematician Arthur M. W. Downing (one of the founders of the BAA) presented the results from a new set of calculations concerning planetary perturbations on the Leonid stream. Their results indicated that parts of the stream had undergone close encounters with Saturn in 1870 and Jupiter in 1898, and accordingly the prediction of a strong 1899 display was placed in some doubt. Indicative of the complicated task in hand, the actual calculations reviewed by Johnson Stoney and Downey were carried out at the Nautical Almanac Office in London, by three computers (these being actual people: F.B. Cooper, J. H. Bell, and W. H. Walmsley) funded by the Royal Society. Even in the modern era, with the application of fast and efficient computer algorithms, it remains very difficult to predict meteor shower activity from one year to the next. Indeed, the task was only partially successful during the epoch of the most recent 1999 Leonid storm.

5.4 Rates and Variations

In a letter to Angelo Secchi, Director of the Collegio Romano Observatory, dated 10 August 1863, Schiaparelli distinguished between two categories of visual meteors: *systematic* and *occasional*. In modern terminology these would be shower and sporadic meteors. The systematic (annual shower) meteors could be linked to some specific radiant on the sky, while the occasional (sporadic) meteors were fewer in number and seemingly obeyed no specific rule with respect to where they might appear on the sky. Furthermore, leaning heavily of the data collected by Remi Armand Coulvier-Gravier in Paris, Schiaparelli reasoned that visual meteor counts could be understood in terms of statistical laws.

In general, visual meteor observers tend to concentrate on one of two activities. Either meteor paths are recorded, with the observer noting the beginning and end points of the meteor trail on the sky, or the number of meteors seen in a given time interval are counted. These two activates are not necessarily mutually exclusive, but to determine meteor rates an observer has to keep their eyes fixed to the sky at all times. In contrast, when plotting meteor paths on a star map a good fraction of an observer's time is spent looking away from the sky. It was only after the 1833 Leonid storm that astronomers set-about the systematic determination of meteor rates. Edward Herrick, for example, organized observing sessions with 4 and more observers, in the late 1830s, in an attempt to determine all-sky meteor counts. Other observers from across Europe began similar such counting programs, with, for example, Remi Armand Coulvier-Gravier in France, Cuno Hoffmeister in Germany, and Johann Schmidt at the National Observatory in Athens, eventually collecting vast amounts of data for analysis. François Arago was to write of these early practitioners, in volume 4 of his *Astronomie Populaire* (published in 1857), that they, "ended up putting a certain order in meteors whose inconstancy had been proverbial". Here was the great pay-off from the many mid-nineteenth century surveys and observations, and the hitherto piecemeal investigation of meteors, and meteor showers, was becoming a big-data-science, with the study of meteors becoming a recognized (and respectable) branch of astronomical research.

While counting meteors may seem like a straightforward exercise, it is an occupation that that requires much dedication, and in order to extract useful averages on daily sporadic meteor rates and meteor shower rates observations must be pooled over many years. For all this, by 1859 Coulvier-Gravier, in his monograph *Recherche's dur Les Meteore*s, was able to demonstrate that there was a distinct variation in the daily number of meteors recorded during the course of the year, with more meteors (on average) being seen per day in the later part of the year than seen in the earlier months. Meteor rates are at their highest around the time of the Autumnal Equinox (September 22nd), and at their lowest around the time of the Vernal equinox (March 20th)— see Fig. 5.9.

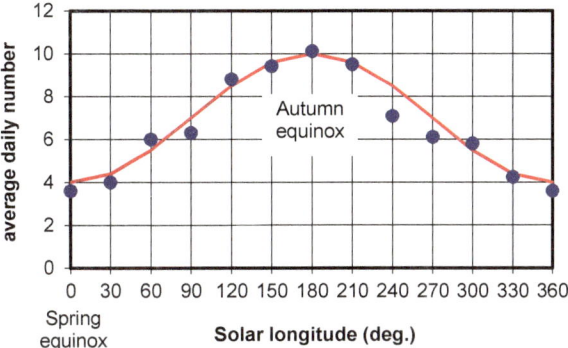

Fig. 5.9 The Annual variation in the daily-averaged number of visual sporadic meteors in the northern hemisphere

In addition to the average number of meteors seen per day showing an annual variation, there is also a sinusoidal, diurnal modulation (Fig. 5.10) in the number of meteors seen per hour over the course of a day. This latter variation relates to the direction of the apex of the Earth's way relative to an observer's location. At 6 p.m. local time an observer will be facing directly away from the direction in which the Earth is moving. In contrast, at 6 a.m. the exact opposite orientation holds true, with the observer facing directly into the direction in which the Earth is moving. In terms of present-day theory, this variation in meteoroid encounter speed means that in the morning (when head-on collisions occur), smaller mass meteoroids can produce slightly brighter, that is more luminous meteors, than they would otherwise produce during a catch-up (evening) encounter. Since, however, there are more smaller meteoroids in the sporadic background than larger ones, this velocity effect results in a greater visual meteor count in the morning relative to that which might be seen in the evening.

The sinusoidal variation in the number of meteors seen per hour during the course of a day was something that Schiaparelli was particularly interested in. Indeed, the idea that the variation in the hourly rate should be related to the position of the apex of the Earth's way was investigated by Heinrich Brandes in his book *Vorlesungen uber die Astronomie*, published in 1827, and by Herrick in 1838. Brandes had hoped that by understanding the variation in meteor counts that a direct measure of the Earth's orbital velocity might be obtained. While the program envisioned by Brandes bore no fruit, Schiaparelli, in 1866, was able to make more headway [9]. Specifically, Schiaparelli demonstrated that the frequency of meteors (that is, the number of meteors observed per unit time) F, should vary according to the Earth's orbital speed V, the speed of the meteoroids v, and φ the apparent height of the apex of

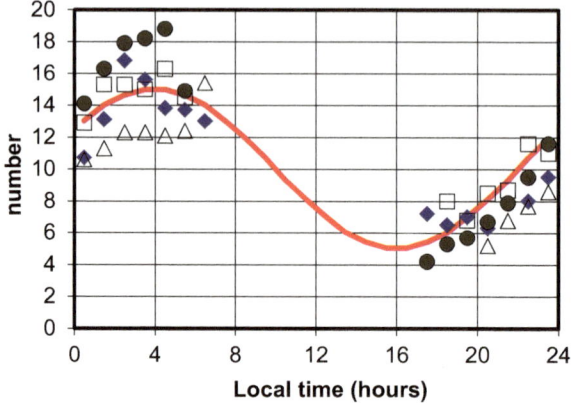

Fig. 5.10 The diurnal variation in the sporadic background of meteors. The meteor rates are at their highest at 6 a.m. local time, and at their lowest at 6 p.m. local time. The data points are from surveys conducted by: W.F. Denning (open squares); R. N. Coulvier-Gravier (blue diamonds); Cuno Hoffmeister (open triangles); and Johann Schmidt (black dots)

the Earth's way. In this manner he argued that:

$$F = K\left(1 + \frac{V}{v}\sin\varphi\right)$$

Using the data assembled by Coulvier-Gravier, Schiaparelli was able show that $K \approx 11$, and that $V/v \approx 0.7$. It is the latter ratio that turned out to be particularly important, since it indicated that the meteoroids producing meteors are characteristically moving faster than the Earth is in its orbit about the Sun. Indeed, the ratio deduced by Schiaparelli is close to what is known the parabolic limit with $V/v = 1/\sqrt{2} = 0.7071$. Accordingly, as Schiaparelli realized, his results implied that meteoroids must be moving about the Sun on highly elliptical, even parabolic, orbits—indeed, in a manner similar to those observed for comets. Simultaneously to, but independent of Schiaparelli, Hubert Newton at Yale came to the same conclusion, that meteor velocities were typically close to that of the parabolic limit in 1864 [10], and this implied that they must have comet-like orbits.[2] The point, at this stage, is that by the mid-1860's it was generally clear that the orbits of meteoroid streams were very similar to those deduced for comets, and that some form of commonality between the two phenomena was implied.

---

[2] It was also at this same time, 1864, that Newton introduced the word *meteoroid*, to describe the body responsible for producing a meteor prior to its encounter with Earth's atmosphere.

While figures such as 5.9 and 5.10 show the daily and hourly averaged variation in sporadic meteor counts, bringing order, as François Arago put it, to the proverbially inconsistent meteoric phenomenon, the actual sporadic plus shower meteor count presents us with a decidedly more complicated picture. For all this, such diagrams do bring out the times at which the more prominent annual meteor showers are active. For completeness in this section, Fig. 5.11 shows a relatively modern plot of daily meteor averages, as observed through radar techniques, over the time interval from 1958 to 1962. The annual variational trend is evident in Fig. 5.11, with a higher baseline count being evident between days 180 and 340 compared to days 10 to 160. Also evident in Fig. 5.11 is something that no visual observer can ever hope to see—that is, a daytime meteor shower (specifically the June Arietids = Ari), as discovered at the Jodrell Bank Radio Telescope Observatory, in 1947).

Over the time interval from 1830 to 1930 steady progress was made with respect to gathering data on the occurrence times of meteor showers, and in determining daily meteor rates. The data gathering process, of course, continues to this day, albeit now with instrumental techniques supplementing the visual observations. While observational data on meteor rates and showers was, and is, gathered by both amateur and professional astronomers, it was predominantly the professional astronomers who used the averaged meteor-count data to infer the properties of meteoroid orbits, and it was the professional astronomers who developed the theoretical models used to explain the observed variations in daily and annual meteor rates.

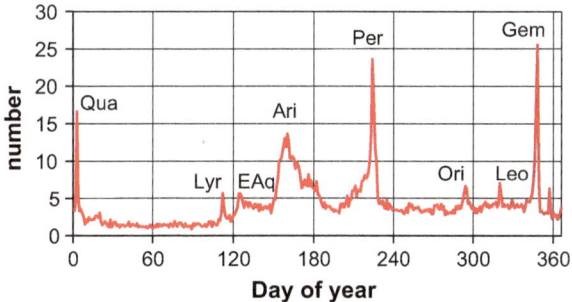

Fig. 5.11 Daily averages of meteors recorded during the course of the year. The data was gathered by a radar system (see Sect. 5.10.2), and only meteors brighter than zero visual magnitude are included. The key to the annual meteor showers is: Qua = the January Quadrantids; Lyr = the April Lyrids; EAq = the May Eta Aquarids; Ari = the daytime June Arietids; Per = August Persieds; Ori = October Orionids; Leo = November Leonids; and Gem = the December Geminids

5.5   Cosmic Streams

With the association between at least some comets and meteoroid streams being established by the close of the 1860s, the next important question to address was that concerning the origins of comets themselves. And, furthermore, it was questioned if every meteoroid stream must have a cometary companion. Indeed, might there be multiple sources of meteoroids? These very questions, along with their possible answers, were explored by Daniel Kirkwood in his 1867 treatise *Meteoric Astronomy*, and by Giovanni Schiaparelli in a series of letters to Angelo Secchi in 1866. Between them, these two researchers provided two distinctly different answers to the origins problem.

Schiaparelli adopted the view that comets and meteoroid streams were captured into the solar system from interstellar space. Specifically, he adopted the position that interstellar space contained distinct clouds or streams of meteoric material. If such a stream passed close to the solar system, he reasoned, and fell under the gravitational influence of the Sun, its component meteoroids would be drawn-out into a long parabolic arc, with the Sun situated at its focal point. If the Earth chanced to pass through such a parabolic stream, then a meteor shower would be observed, and perhaps meteorites might be seen to fall. According to the encounter conditions, Schiaparelli further argued that the material in a parabolic stream might be forced, by gravitational perturbations from the planets, into an elliptical orbit about the Sun. Within this framework, Schiaparelli saw comets growing by a process of accretion—that is, by being built-up from the interstellar meteoroids captured into an elliptical orbit. Joseph Norman Lockyer pushed the idea of colliding parabolic streams to its logical (but fantastical) limit. Indeed, Lockyer in his *The Meteoritic Hypothesis* (published in 1890) suggested that all self-luminous phenomena within the universe were the result of intersecting cosmic streams and the collisions between their meteoric components (recall Fig. 3.15).

In contrast to Schiaparelli, Kirkwood worked within the framework of the nebula hypothesis, positing that comets formed within the outer reaches of the solar system and were occasionally perturbed into elliptical orbits that brought them periodically close-in towards the Sun. In this manner, building upon the fact that comets were occasionally observed to split, and even disappear, Kirkwood saw comets as the primary bodies within a meteoroid stream, the meteoroids themselves being the small fragments produced during cometary decay. In this manner, over time, a meteoroid stream was built-up with each meteoroid having an orbit similar to that of its parent comet (Fig. 5.12).

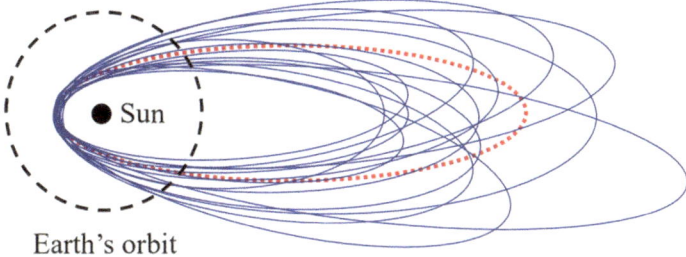

Fig. 5.12 A schematic diagram showing the orbits of meteoroids (in blue) and their associated parent comet (red dashed). A meteor shower will occur if the orientation of the stream is such that it cuts through the ecliptic at a heliocentric distance equal to the Earth's orbit

With the dawn of the twentieth century the idea of cosmic streams fell into disfavor. It became clear at this time, for example, that the Sun was not powered by the accretion of meteoritic material—the idea of nuclear fusion being articulated by Arthur Eddington in the mid-1920s. Likewise, it became increasingly clear that cometary nuclei were not formed through accretion from within meteoroid streams. Rather, meteoroid streams were the result of cometary decay. The contemporary picture of meteoroid stream development was first articulated by American astronomer Fred Whipple in the early 1950s. Indeed, Whipple introduced the dirty-snowball concept of cometary structure [11], with the nucleus being composed of water ice and embedded refractory grains. Accordingly, as a cometary nucleus nears the Sun, the surface ice begins to sublimate, forming distinctive jet-like features (Fig. 5.13). Any small grains that happened to be embedded within the sublimating ice would then be entrained by, and carried along, the jets, ultimately being ejected into space. These ejected grains (meteoroids) will have an initial orbit similar to that of the parent comet [12, 13].

It was the professional astronomers, ensconced in both national observatories and university halls, who developed the detailed theory of meteoroid stream structure—something that Charles Olivier, a product of the university system, took to heart [14]. Indeed, as seen earlier, the failure to predict the events surrounding the 1899 Leonids was seen by Olivier as a calamitous failure. The problem, however, was not one of fundamental ignorance, but rather one of being able to coax solutions from a complex set of differential equations. Time (and eventually electronic computers) would largely solve the dynamical and perturbative calculation issue. While mathematical difficulties were one problem, Olivier, along with many other practitioners, realized in the early years of the twentieth century that there were still observational practices that needed to be sorted out. In particular, the inordinately

Fig. 5.13 Comet 67P/Churyumov-Gerasimenko as imaged by the *Rosetta* spacecraft in 2015. The distinctive jet-like feature seen in the image is caused by the sub-surface sublimation of water ice heated by the Sun. The double-lobed nucleus is and about 4-km long. Image courtesy of ESA

large number of radiant points being report was becoming problematic. This problem, Olivier felt, was something that the observer's, rather than the theoreticians, were getting wrong—wrong in degree of course. What was needed was a re-think about how radiant points were identified (recall Chap. 2.1).

5.6   Radiants Galore

The essential geometry underlying the formation of a meteor shower radiant is illustrated in Fig. 5.14. To produce such a feature the meteoroids within a stream travel along parallel paths as they enter and then ablate in the Earth's upper atmosphere—a perspective effect will then result in the observed meteor trails appearing to radiate from a small, localized region of the sky (the radiant). If the geometry is such that a meteoroid chances to ablate in the line of sight of the observer, then it will appear as a temporary point of light at the location of the radiant itself. Meteors seen well away from the radiant will have much longer trail lengths than those seen to appear closer to it.

As discussed earlier, one of the important outcomes of the 1833 Leonid storm was the identification of its radiant point in the constellation of Leo. This was discernable then because of the great number of meteors appearing at any one instant. Furthermore, during the course of the storm, observers were able to see that the radiant point remained fixed with respect to the background stars. The observation of a radiant, and the fact that the radiant

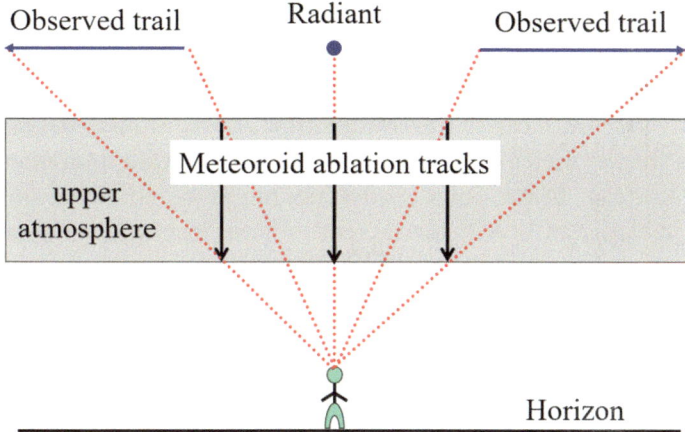

Fig. 5.14 A two-dimensional simplified diagram to illustrate the origin of a radiant point. Even though the meteoroids ablating in the Earth's upper atmosphere have similar path lengths and are moving along parallel paths, the observed meteor trails will have different angular path lengths on the sky

moved at a sideral pace, was strong evidence that the meteoroids responsible for producing the meteors must be entering the Earth's atmosphere from interplanetary space. In 1833, this result, in and of itself, was an important realization, since it was then still unclear, or perhaps we should say still open to debate, whether meteors were an atmospheric phenomenon (in some Aristotelian sense—recall Sect. 5.1, and see Sect. 5.10), or whether they were of an extraterrestrial origin. Chladni had argued in his 1799 thesis that meteors were derived from meteoroids having the same basic composition and structure as meteorites[3]; others, however, still questioned if some atmospheric, electro-chemical process might not provide a better answer to the question of meteor origins.

When seen by just one observer it is not possible to deduce a specific radiant point for a meteor—at least two observers are required (recall Fig. 5.2). For all this, however, if a single observer records the paths of multiple numbers of meteors, they can attempt to trace the paths backward on a star map to some intersection point. Accordingly, the question arises, must all meteors belong to some specific meteoroid stream, and therefore be traceable to some active radiant from which other meteors might radiate, or

---

[3] Only one meteorite has ever been observed to fall during an actual meteor shower. This was the Mazapil meteorite in Mexico, which fell during the Andromedid meteor storm of November 27, 1885. It is now clear that meteorites are collisional fragments derived from the main belt asteroid region, and accordingly there is no specific reason for them to fall at the times when (comet-derived) meteor showers are active.

are some (or all) meteors genuine loners, belonging to no specific stream, having, accordingly, no specific group radiant attachment. While Schiaparelli distinguished between sporadic and shower meteors in his letters to Secchi in 1866, it was not clear if this meant that sporadic meteors were simply unassigned shower meteors. This straightforward question took many decades to resolve. Indeed, before such a question might be answered one needs a detailed knowledge of how many meteor radiants might actually exist.

German mathematician Eduard Heis (Fig. 5.15), one of the great pioneers of meteor-mapping, began collecting meteor data in the mid-1830's [15]. Indeed, Heis along with his students and volunteers at the University of Münster gathered data on meteor trails, meteor brightness, and rates, for some 43 years, with a grand summary being published in 1877. This latter work, *Resultate der in den 43 Jahren 1833–1875 angestellten Sternschnuppen-Beobachtungen*, contained information on some 14,000 meteors.

Heis indicated that when conducting meteor observations at Münster, the sky was divided into 5 regions (north, south, east, west, and zenith), with two or more observers being tasked to each region. Any sightings were orally conveyed to a designated secretary, who recorded times, beginning and end

Fig. 5.15 Eduard Heis (1806–1877). One of the first practitioners to systematically study meteors and map-out meteor trails and radiant points

locations, and magnitudes. Details from an early catalog by Heis, covering the time interval from 1849 to 1860, were summarized by A. S. Herschel in the May 1864 issue of the *Monthly Notices*, with 69 radiants being reported. The *Sternschnuppen-Beobachtungen* listed 162 radiants based on those meteors observed between 1833 and 1875.

In 1860, Robert Philips Greg (Fig. 5.16) published his *Catalogue of Meteorites and Fireballs, from A.D. 2 to A.D. 1860*. This was an extension of an earlier 1854 catalog that brought together a vast quantity of data, and presented a detailed analysis of meteorite fall times and fall circumstances. Indeed, Greg constructed a series of tables in which he looked for specific meteorite fall epochs and/or favored fall dates—none were strongly indicated (just as present-day research informs us). While there are no favored fall dates, meteorites, as Greg found, do preferentially fall in the afternoon (as opposed to the morning or nighttime). Greg also concluded that fireballs associated with meteorites can be seen traveling in all directions—there being no predominance of one entry direction over another. Later, in 1868, Greg and A. S. Herschel published a catalog of meteor radiants in the *British Association Reports*. Based upon some 2000 meteor trails, garnered from the records of the Luminous Meteors Committee (LMC—see), Greg and Herschel deduced the presence of some 88 active radiants. Schiaparelli in his *Entwurf einer astronomischen Theorie der Sternschnuppen* (published in 1870) deduced the presence of 189 radiants from some 7000 meteor trails observed between 1867 and 1869. In the May 1872 issue of the *Monthly Notices*, Greg published, *A General Comparative Table of Radaint Positions and Duration of Meteor Showers*. This was both a compilation of previously cataloged radiants (by Schiaparelli, Hies, Herschel and Backhouse), and an analysis of materials collected by the LMC. In total 132 radiant points were described, and interestingly, Greg noted that, "meteoric showers having a duration of some six weeks appears to be still further confirmed by the comparative table or catalogue now presented". While Greg did not use the term stationary radiant, he essentially implied it, though importantly he also called for further observations to be made in order to confirm, or discount, radiant drift during the suspected times of activity. Denning would later draw attention to Greg's statement concerning long-lived activity. Indeed, writing in issue number 1 of *Popular Science Review* for 1880, Denning noted that not only Greg, but both Johann Schmidt (Director of the National Observatory of Athens) and Eduard Heis had described radiants that were active for thirty days and more.

In the February 1873 issue of the *Monthly Notices*, George L. Tupman published a catalog based upon 1960 meteors he had observed between 1869 and 1871, and deduced the location points of some 102 radiants,

Fig. 5.16 Robert Philips Greg (at age 79 years)

Henry Corder, writing in the December 1879 issue of the *Monthly Notices*, further indicated that he had found 230 active radiants from among the 5800 meteors he observed between 1870 and 1879. Also in 1879, Edwin Sawyer published, in the *American Journal of Science*, a list of 104 radiant locations based upon the observation of 1500 meteors made between 1877 and 1880. Indeed, in all of the catalogs published prior to the early years of the twentieth century, meteor observers were finding literally hundreds of radiant points, with a distribution, at least according to Schiaparelli, that was randomly distributed across the celestial vault. When Denning published his *General Catalog* in 1899, he, in effect, presented a vast compendium, a catalog of catalogs, containing his own observational work along with those of numerous others. In total 26 catalogs and surveys are combined in the *General Catalog* (hence its name) from observations collected over the 65-year interval between 1833 and 1898.

The transference of an observed meteor trail on the sky to a line drawn on a celestial chart, or chalked on to the surface of a globe, is an exercise fraught with possible inaccuracy. Accordingly, observers often placed great importance on recording, when chance made it possible, point meteors, or meteors with very short trails. Such meteors, as illustrated in Fig. 5.14, must be located exactly at, or at least very close to, their associated radiant points. Indeed, Denning published a catalog of 222 point (or stationary, as he called them) meteors in the April 1879 issue of the *Monthly Notices*. This particular work was a distillation of numerous observational catalogs (see Fig. 5.17). Indeed, this was an exercise in Big-Data-Set analysis, involving material from at least 12 catalogs of meteor observations. Included in his survey were the catalogs of: H. Heis, R. Greg, A.S. Herschel, J. Schmidt, E. Weiss, G. Zizioli, G. Schiaparelli, G. L. Tupman, M. Konkoly, J. E. Clark, H. Corder, E. F. Sawyer, T. W. Backhouse, and S. J. Johnson. Also included in the survey were his own observations, and data extracted from the reports published by the Luminous Meteor Committee of the British Association. Denning estimated that within the catalogs that he had surveyed there was information of some 59,000 meteors. To trawl through such vast quantities of data in the pre-electronic computer era shows Denning to be a remarkably focused (if not driven) researcher. Clearly, stationary meteors are only rarely seen, and from the raw numbers provided by Denning, they account for perhaps 0.5% of visually observed meteors. American astronomer Edwin Sawyer published another list of stationary meteors in the March 1881 issue of the *Monthly Notices*, with some 45 such meteors being identified from among the 1156 meteors recorded between 1879 and 1880)—suggesting that perhaps 1 percent of meteors appear as stationary.

Denning took these rare stationary meteors to be accurate markers of very low-density meteors showers—showers that produced perhaps only one or two meteors per night. For all this, he also believed that there was some structure within their observed distribution, noting concentrations in certain constellations. Denning allowed one point (or stationary) meteor to indicate the existence of an active radiant, on the supposition that all meteors must be linked to one active shower or another—an assumption, of course, that we now known to be untrue, but which, when Denning was working, was a genuine and accepted practice.

If every meteor is assumed to be traceable to some radiant, belonging to some active (perhaps at a low-level) shower, then the number of radiants is, almost by default, going to be enormous. Indeed, as seen in Chap. 2, Denning described some 4367 shower radiants in his *General Catalog* (published in 1899). Such a situation is fine enough, but requires that one must then also adopt the stance that not all meteors are derived from meteoroid streams directly associated with a parent comet. If, on the other hand, one believed that all meteoroid streams must have an associated parent comet, then the very large number of radiant points being identified was problematic. The problem being that the observed population of comets cannot readily account for the number of radiants being discovered (technical issues of meteoroid stream and cometary lifetimes aside).

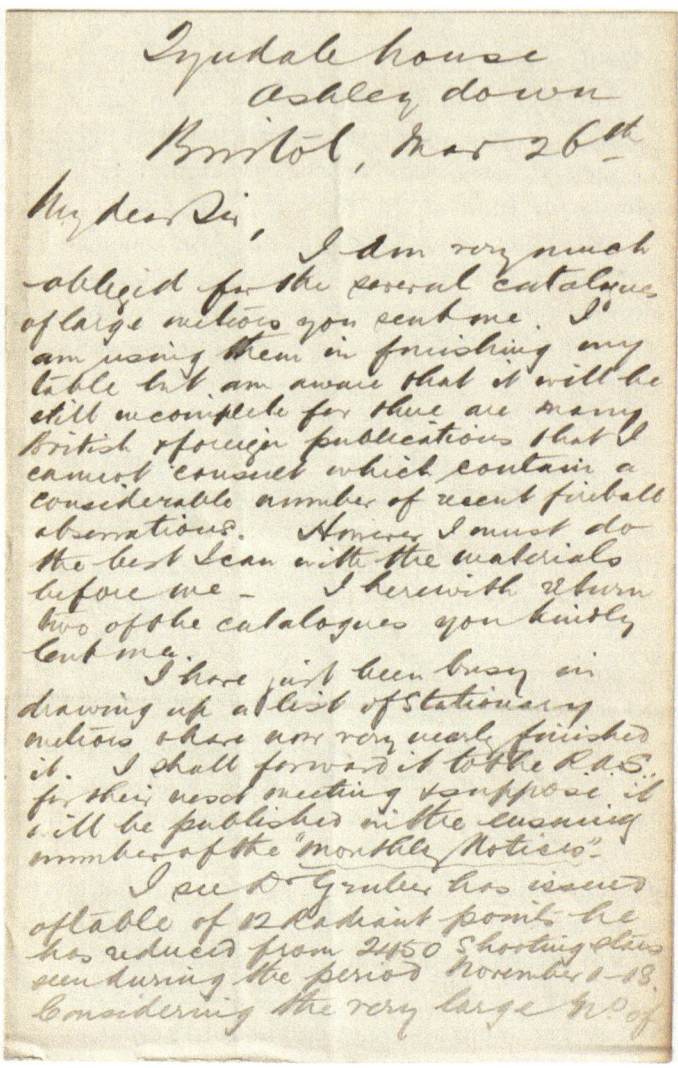

Fig. 5.17 First page of a letter from W. F. Denning to R. P. Greg for November 26 [1879]. In this letter Denning thanks Greg for catalog material, and indicates that he is very nearly finished compiling his "stationary meteor" catalog. Image from the authors collection

From circa 1910 onwards, Charles Olivier, as one of the key players among the new generation of meteor astronomers, was a firm believer that radiant reduction protocols (as used by virtually all previous observes) needed revising, and revising in such a manner that the majority of previously published radiant catalogs would be rendered all but useless [13]. The consequences of adopting Olivier's methods would be painful to all earlier practitioners—painful, that is, in the sense that vast quantities of data would need to be re-worked or completely abandoned. In 1948, J. P. M. Prentice bemoaned the great amount of time wasted on reducing radiant positions (and especially stationary radiant positions—recall Chap. 2.1) through the application of "unsound method" [16]. In this respect Prentice was entirely wrong. Observer's, such as Denning and, indeed, all his contemporaries, were applying accepted techniques. That those techniques eventually turned out to be inapplicable is not a sign of wasted effort or unthinking practices, it is a sign of scientists and the scientific method successfully at work. This is not to say, of course, that human nature is such that it will readily accept new methodologies and change.

For all the apparent consternation that surrounded Olivier's evangelism, he was, in fact, simply reverting to an extreme conservatism. It was not appropriate, Oliver argued, to co-add meteor trails recorded on different dates to deduce the existence of a group (that is active shower) radiant. Rather, radiants may only be deduced on the basis of meteors observed during a single observing session lasting no more than about ¼ of a day. This conservative stance will certainly allow for the capture all of the stronger annual meteor showers, but it will totally illuminate from the catalogs any stationary radiants. Olivier's radiant reduction protocol was not revolutionary or new, as such, but simply a means of riding radiant catalogs of the greater number of their false-positive entries. Not only this, Olivier implicitly adopted the stance that all meteoroid streams must have comet-like orbits. Indeed, as was seen in Chap. 2, Olivier had essentially ridiculed Denning's stance concerning the existence of stationary radiants, not on the principle of his inability to observer, but on the theoretical basis that such radiants could not be produced from a stream of meteoroids associated with a single comet.

Writing in 1935 Olivier reviewed the material that had been collected by the American Meteor Society (AMS) between 1919 and 1925. In total, society members had recorded data on some 8600 meteors, and from these Olivier deduced a total of 318 radiants. While this was a factor of 10 reduction in the number of radiants given by Denning in his *General Catalog*, many of the AMS radiants were still uncertain, and were presumed to be associated (given that they were actually real) with weak showers that were barely distinguishable under the condition of no co-adding over successive nights. This number of radiants was reduced by another factor 10 by Allan Cook (Smithsonian Astrophysical Observatory) in 1973 when he published a working list of radiants for 58 meteor showers (4 of these were daytime showers identified with the aid of radar observations) [17]. Furthermore, Cook noted that he would be more inclined to delete some of the radaints from his list, rather than consider adding new ones. The present working list of meteor showers, as maintained by the International Meteor Organization (IMO), includes 50 radiants, of which 11 relate to daytime showers [18].

5.7 The Luminous Meteors Committee

Founded in 1831, the British Association for the Advancement of Science (BAAS) embodied the organizational zeal of the British scientific community [19, 20]. Its aim was to transform science from a largely self-financed endeavor, conducted predominantly by the wealthy few, into a government-funded, professional activity for all able practitioners. To this end, the Association held grand annual meetings, at which there were not only public talks and scientific presentations, but committees were formed in order to study and make recommendations upon developing areas of study. During its founding years, the BAAS held gatherings at diverse locations across England, Scotland and Ireland, moving from city to city like some gigantic philosophical circus. It was an idealistic, even evangelical movement, looking to not only entertain and enthral the public, but to promote the development of science. For all of its grand, and well-intended bravado, however, the BAAS had its critics, and indeed, Charles Dickins in 1837–38 presented a wicked parodied of its aims and members in *The Mudfog Papers*, a series of stories published in *Bentley's Miscellany*. With pictures drawn by George Cruickshank, Dickins opened the first report on the inaugural meeting of the Mudford Society for the Advancement of Everything, by homing-in on the expected mayhem surrounding the accommodation of its members. "It is confidently rumoured", Dickens writes, "that Professors Snore, Doze, and Wheezy have engaged three beds and a sitting-room at the Pig and Tinder-box [a local Pub]". Later, we are informed that in Section C.—Statistics that, "Mr. Slug" presented some calculations concerning children's literature, noting that the popularity of Robinson Crusoe to Philip Quarll was, "four and a half to one". Later, we learn that in Section A.—Zoology and Botany, that, "Mr. Flummery exhibited a twig". The humor does not ascend to great heights in *The Mudfog Papers*, and Dickens, along with other haughty distractors were ultimately silenced by the continued growth in the Association's membership, its evident success in promoting science, and in its ability to enable the advancement of practical industry. Four BAAS meetings were held in Bristol during the period of study in this chapter—namely in 1836, 1875, 1898, 1930. Of these, Denning only attended the 1898 meeting (Fig. 5.18). There is an interesting snippet concerning the planning of the

1898 meeting in a letter from A.S. Herschel to Denning dated September 8th, 1898. Herschel writes, "wishing you a successful British Association week at Bristol in spite of the great calamity that befell the city and obstructed the great preparations for the meeting". It is not entirely clear what the great calamity referred to by Herschel was, but it may relate to a milk-borne cholera outbreak (in the Clifton area of the city) in 1897.

A review of the *Report* on the 1898 meeting indicates that Denning attended the meeting as the representative of the Toronto Astronomical and Physical Society. Indeed, the 1897 BAAS meeting had taken place in Toronto, and the Society was keen to continue the promotion of the idea for international time zones. The Society approached Denning, with a letter dated August 24, 1898, to ask if he would act as their representative (recall, Denning was elected a corresponding member of the society in 1891). *The Transactions of the Toronto Astronomical and Physical Society* for the year 1898 provides details of the correspondence:

Fig. 5.18 Advertisement for the 1898 British Association for the Advancement of Science meeting to held in Bristol. The corner images show statues that can be found in Bristol. Clockwise from the upper left are Queen Victoria (a statue unveiled in 1888); William of Orange seated on a horse (cast in bronze in 1733, and unveiled in 1736); Edward Colston (a Bristolian merchant and philanthropist. Unveiled in 1895 and toppled by demonstrators in 2022); Edward Burke (politician, economist and philosopher. Erected in 1894). At center are the Bristol City Coat of Arms, with its motto: *Virtute et Industria* (by virtue and industry—indeed, a motto worthy of being applied to and adopted by Denning, as well as the BAAS). Image courtesy of the BAAS Archive at the Bodleian Library, University of Oxford

Dear Mr. Denning,—After the meeting of the British Association in Toronto, the subject of becoming a Corresponding Society was mooted, and from time to time was under the consideration of this Society, its hope being that, at the meeting at Bristol, itself a seaport town, and for centuries a centre of ocean-going shipping, an opportunity would be afforded of presenting a paper or taking further action with respect to the Unification of Time at Sea, a matter which this Society, in connection with the Canadian Institute and the Royal Society of Canada, has been actively promoting for some years, and which it brought before the Association last year.

It transpired, however, that no positioning paper was produced, and the letter to Denning concludes:

Owing to the lack of time, the Society will not be able to furnish a paper to be read before the Association, or to suggest a line of action with respect to any matter which has occupied its attention. It may, therefore, prove that your duties as delegate, if you accept the position, will be perfunctory in character. Nevertheless, I can assure you that it will be no small matter of congratulation among the members of the Society if it should prove that you have taken charge of its interests before the Association.

By the time of the 1898 BAAS meeting, the Luminous Meteor Committee, which had been of fundamental importance in promoting meteor astronomy had ceased functioning. Denning did not present any material at the Meeting, and the only lecture relating to meteor astronomy was that by Irish mathematician/physicist George Johnstone-Stoney on the dynamics of meteoroid streams. Johnstone-Stoney did reference Denning's work on the Leonids, in his talk, but he focused mainly on the gravitational perturbations that a cloud of cosmic meteoroids might experience when encountering the planet Uranus. Indeed, Johnstone-Stoney suggested that the Leonid swarm was produced by an encounter between a compact [cosmic] cluster of meteorites and Uranus in March of A.D. 126. Furthermore, Johnstone-Stoney distinguished between two types of Leonid meteoroids—the "ortho and clino-meteors [sic]". The ortho-meteoroids, moved about the Sun in a coherent stream, with all the component meteoroids having essentially the same orbital parameters. The clino-meteoroids, in contrast, were distinguished by having slightly different orbits to the ortho-meteoroids as a result gravitational perturbations by the Earth. While Johnstone-Stoney's ideas concerning the origin of the Leonid stream have not stood the test of time, the paper itself was largely a theoretical disputation, and would have held only limited interest to Denning.

The first BAAS *Report* that includes a discussion of the status of meteor astronomy is that for the 13th meeting held in Oxford during the summer of 1847. At this meeting the Reverend Baden Powell, Savilian Professor of Geometry at Oxford, delivered a lecture at the Radcliffe Observatory on the topic of "shooting-stars". Powell began his lecture by agreeing with Chladni that there was good evidence to indicate a connection between the appearance of bright shooting stars [fireballs] and the fall of aerolites [meteorites]. With respect to meteorites, however, Powell argued, that they were likely formed when diffuse clouds of material encountered the Earth's atmosphere. These clouds were taken to be in orbit about the Sun, and were subject to orbital perturbations by the planets. When a cloud entered the Earth's atmosphere, the light phenomenon observed was caused by electrostatic discharge and combustion. The actual meteorite collected on the ground, Powell further argued, was the heated, reduced, fused, and consolidated material from the impacting dust-cloud. A meteorite therefore was a portion of newly-formed material, rather than the remnants of an old, solid meteoroid that had existed in space prior to encountering the Earth.

In addition to his lecture at the Radcliffe Observatory, the Reverend Powell also presented a summary paper on "Periodic Meteors". This latter paper was ostensibly a, "table of all the remarkable appearances of luminous meteors" that had been seen over the time interval from 1841 to 1846. The material was largely gathered from reports in various scientific journals, and is simply arrayed as a chronological account of sightings. Clearly evident in the material presented by Powell, however, are the July alpha-Capricornids, the August Perseids, the October Orionids, the November Leonids, and the January Quadrantids. Powell submitted annual reports to the BAAS from 1848 to 1859. With Powell's passing in 1860, a new committee, the Luminous Meteors Committee (LMC), was struck to continue the generation of annual reports on meteors and meteorites.

The first configuration of the LMC was composed of James Glaisher (recall Chap. 3), John Hall Gladstone, Robert Philips Greg, and Edward Joseph Lowe (see Table 5.1). These initial members had a diverse background of interests, ranging from meteorology, to chemistry, and botany, but Glaisher and Greg were the stalwarts, being involved with the committee's work for the full 21-years of its existence. The structure of the annual reports presented by the committee changed but little from those previously compiled by the Reverend Powell, being ostensibly chronological tables of meteor sightings, and reviews of relevant papers published within the scientific literature. In terms of costs, the LMC imposed very little burden on the BAAS. While, for example, grants amounting to over £1000 were made, between 1881 and

1910, to the committee in charge of standards for electrical resistance, and £500 was granted between 1874 and 1877 in order to compute tables of elliptical functions, with £899 being set aside for the purchase of rain gauges in 1836, the entire 21-years budget for the LMC amounted to £427 [19].

Robert Philips Greg was the son of a Robert Hyde Greg, a parliamentarian and well-to-do Mancunian merchant, and although groomed for a career in the family business, Greg had sufficient leisure time to indulge his interests in geology and meteorites. Indeed, Greg was not only a prominent member of the BAAS, he was a founding member of the Mineralogical Society, a Fellow of the Royal Astronomical Society, a member of the Society of Antiquaries, and a Fellow of the Geological Society. John Hall Gladstone was a chemist by training, and in addition to an active religious life, held various posts as a lecturer at St. Thomas's Hospital in London, and the Royal Institution. He was elected a Fellow of the Royal Society in 1853 (at the tender age of 26), and served on several Royal Commissions, including the Lighthouse Commission and the Gun Cotton Commission. In later life Gladstone was President of the Physical Society (1874–76), and the Chemical Society (1877–79). His research interests included optics and spectroscopy. Edward Joseph Lowe, was a polymath who conducted research in botany, conchology, meteorology, and astronomy. Independently wealthy, Lowe was

Table 5.1 Members of the BAAS Luminous meteors Committee

Name	Affiliation	Years on LMC	Background
J. Glaisher	FRS	1860–1881	Meteorologist
R. P. Greg	FRAS, FRGS	1860–1879	Geologist
J. H. Gladstone	FRS	1860–1861	Chemist (D)
E. J. Lowe	FRS, FRAS	1860–1861 1880–1881	Meteorologist/ Botanist
E. W. Brayley	FRS	1862–869	Geographer/ Meteorologist
A. S. Herschel	FRS, FRAS	1862–1881	Physicist (D)
C. Brooke	FRS, FRMS	1867–1872 1874–1879	Surgeon and inventor
G. Forbes	FRS, FRAS	1875–1881	Geologist
W. Flight	FRS	1875–1881	Chemist
R. S. Ball	FRS	1880–1881	Astronomer

Column 1 provides the names of the committee members, while column 2 indicates membership of national societies. Column 3 indicates the years over which the members served, and column 4 provides brief comments concerning specialities. A (D) in column 4 indicates that the person also served on the BAAS committee for the collection and identification of meteoric dust, which submitted reports from 1882 to 1889 active in many societies (including the RAS) and he assisted Baden Powell in his meteor-observing campaigns at Oxford.

A scan through the names included in Table 5.1 reveals that the members of LMC were all grand-gentlemen of science, and all fellows of the Royal Society. All were recognized experts in one scientific field or another; some were independently wealthy, free to work as they chose, others were members of formal centers of learning. Glaisher was a civil servant working at the Royal Greenwich Observatory, ostensibly in charge of meteorological and magnetic observations. Herschel was a professor of physics at Andersonian University (1866–71) in Glasgow (now Strathclyde University), and later (1871–86) Professor of physics at Durham University. Likewise, George Forbes was a professor of natural philosophy at Andersonian University. Edward William Brayley was librarian and professor of physical geography at the London Institution, while William Flight, a chemist by training, was an examiner at the Royal Military Academy in Woolwich. Robert Stawell Ball was Royal Astronomer of Ireland (1874–1892) at Dunsink Observatory, and later professor of astronomy and Director of the observatories at Cambridge University. Charles Brooke trained as a surgeon, and practiced at the Metropolitan Free Hospital, as well as the Westminster Hospital, in London. Brooke specialised in the invention and construction of self-recording instruments, and was an active member of the Royal Botanical Society, the Royal Microscopical Society, and the Royal Meteorological Society.

In the modern-era, apart from Glaisher, Greg and Herschel, most of the members of the LMC are better known for their work in areas other than meteor astronomy. Indeed, of all the members of the committee it is Alexander Herschel that was perhaps most influential in establishing, during the mid to late nineteenth century, meteor astronomy, at least in the United Kingdom, as a *bona fide* scientific enterprise. For all this, the LMC effused institutional weight, and their reports carried the seal of accepted wisdom. Furthermore, it can be argued that the LMC acted to distance the amateur observer from the practice of knowledge building, especially with respect to data analysis techniques and the development of physical theory. For all this, the amateur observer was an essential part of the data-gathering process, and it was in this somewhat subservient role that Denning began his observing career circa 1866.

As a final note in this section, there is no evidence that Denning 'officially' attended the Bristol meeting of the BAAS in 1875. This is perhaps surprising, since at that time Denning would, through his many contributions to the *Astronomical Register* (see Chap. 6), and his involvement with the Observing

Astronomical Society (see Chap. 7), have been a recognized persona amongst amateur astronomers. Other than personal, face-to-face networking, however, something that Denning never seems to have liked, there would probably have been little benefit in attending the technical sessions of the meeting. Furthermore, he may well have felt that the formal and regimented settings of a BAAS meeting were not the best places to promote the cause of amateur astronomy.

5.8   Meteor Spectroscopy

Unlike the majority of topics in the history of astronomy, the date upon which meteor spectroscopy began is known precisely. While making spectroscopic observations of the bright star Capella, on the night of 18 January, 1864 Alexander Stewart Herschel (recall Fig. 2.13) was pleasantly surprised to see the transient spectra of a bright meteor appear in his telescope's eyepiece [21, 22]. Inspired by this serendipitous detection, Herschel determined to establish a program to record more such events. His first step in this direction was to develop a direct vision meteor spectroscope, and this activity he described in a short publication to *The Intellectual Observer* for July, 1865. Inspired, we are told, by the way in which the two mirrors in a standard sextant work, Herschel looked for ways in which internal reflection might be used to limit the deviation of a light ray as it passed through a triangular prism. The design that he eventually isolated as being best was that provided by a right-angled prism. While there was still a small lateral displacement between the entrance and exit rays, Herschel went on to describe a binocular arrangement for a direct vision spectroscope (Fig. 5.19) in *The Intellectual Observer* for August, 1866. The binocular layout of Herschel's spectroscope is an unusual design, but it does provide for a wide field of view device that is easy to hold and point around the sky.

The Intellectual Observer article for August 1866 explains that the binary spectroscope, "which presents considerable difficulties of optical construction, has been very ably carried out by Mr. Browning; and the British Association Committee have ordered four of the instruments with a view to their employment in the coming November showers".[4] In a typical spectroscope it is necessary to place a thin slit in front of the entrance aperture in order that distinct absorption and/or emission lines can be observed, Herschel realized,

---

[4] At least one of these instruments has survived, indeed, it is labeled as belonging to A. S. Herschel (Observatory House, Slough), and forms part of the Royal Museums of Greenwich collection—ID: AST1030.74. The dimensions for the instrument are given as 3.9 × 13.7 × 8.8 cm.

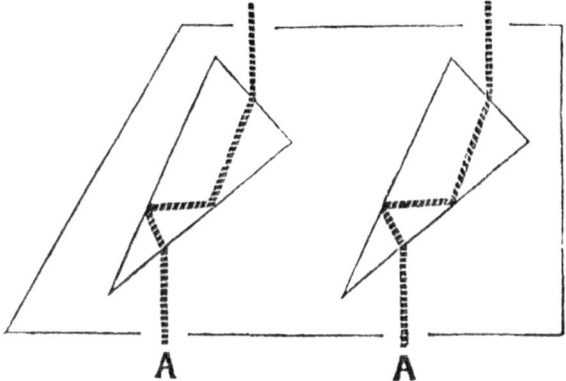

Fig. 5.19 Layout for the direct vision Herschel-Browning binocular spectroscope. Light from the meteor trail enters the device through the two slits labeled A. Each prism is angled so that there is only a small lateral deviation in the light ray path. No entrance slit is needed since the meteor trail is itself narrowly defined. Image from *The Intellectual Observer* for August, 1866

however, that in the case of meteor spectroscopy the meteor trail is already a long, and thin line source, and accordingly no slit is required. Herschel further noted that his spectroscope, "presents to the view a pretty considerable extent of the star-spangled surface of the sky". It was this combination of wide field of view, and low light-loss that made Herschel's spectroscope well suited towards meteor studies—particularly so since the device is hand-held, and can be directed, by pure reflex action, the instant a meteor is seen [23].

5.8.1   First Application

The first field-testing of Herschel's binocular spectroscope was conducted during the 1866 Perseid meteor shower on the nights of August 9, 10 and 11. Over these three nights Herschel managed to capture the spectra of 17 meteors—a remarkable achievement—with the results being presented in a detailed review published in the October issue of *The Intellectual Observer* (Fig. 5.20). Herschel provided brief notes on each spectrum that he observed, and with respect to spectra 1, observed at 8:40 pm on August 9th, he records that, "the meteor might be a solid body, heated to ignition". For spectra 2, he noted that, "its luminous substance was a gas". Meteor number 12, observed at 0:42 am August 11 was as bright as Sirius, and "the spectrum of the nucleus was red, green and blue…. the long endurance of the sodium line, after the rest had disappeared, was leisurely watched". Indeed, by the end of the observing session Herschel was prepared to conclude that, "in the majority of cases, a bright yellow line, having the unmistakable appearance of the sodium line, was clearly visible in the spectrum", to which he added, "the material of the August meteors is, therefore, probably a mineral substance, in which sodium is one of the chemical ingredients". With the spectroscopic results in place, Herschel further noted that, "the connection believed by the adherents of Chladni to exist between shooting-stars [meteors] and aërolites [meteorites] is now shown, at least in August, to extend itself in some measure to their chemical composition". This was an important result, and further strengthened the notion that the meteoroids responsible for producing meteor showers must be solid, and extraterrestrial in origin.

Two additional spectroscopes were deployed during the August 1866 field-testing campaign, one was used by James Glaisher at the Royal Greenwich Observatory, and the other was used by John Browning (Fig. 5.21), at Richmond-on-Thames. The British Association for the Advancement of Science report on meteor observations for 1866 sadly reveals, however, that there was a delay in the delivery of the spectroscope to Greenwich, and that on the night of August 10th the London observers were mostly clouded out anyway. Browning, who had constructed the spectroscopes, was also clouded-out during the display, and he made, on this occasion, no successful observations.

While Browning was an amateur astronomer, he was also a professional scientific instrument maker. Born into a family with a long-tradition of making navigational instruments, Browning originally hoped to train for a medical career, and although he graduated from the Royal College of Chemistry in the late 1840s, due to poor health turned to instrument making. Apprenticed under his father, and taking over the family business in 1856, Browning moved the product-base away from that of nautical instruments, to the construction of scientific instruments. To this end he began manufacturing telescopes, spectroscopes, microscopes, cameras and electrical equipment—all of these devices finding placement in both institutes of higher learning, and in the homes of well-to-do amateur scientists and enthusiasts. The quality of Browning's instruments was soon recognized, and he was an award recipient at the International Exhibition held in London in 1862. Browning was elected a Fellow of the RAS in 1865 and provided spectroscopic equipment to Joseph Norman Lockyer and William Huggins. Not just an instrument maker, however, Browning published a number of articles concerning the appearance of Jupiter's equatorial belt, and contributed numerous letters and articles to the *Astronomical Register* (see Chap. 6).

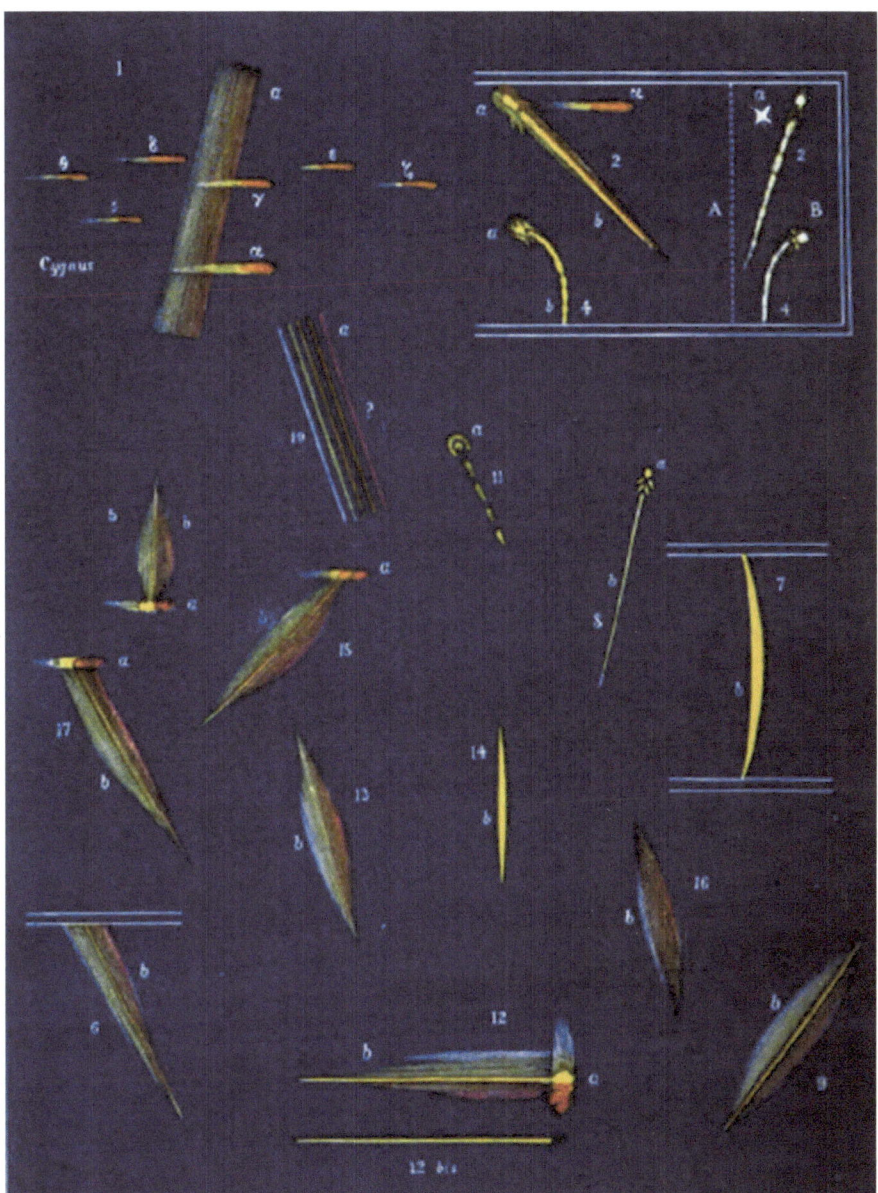

Fig. 5.20 Observations of Perseid meteor spectra as recorded by A. S. Herschel in August of 1866—see text for details. Image from the October 1866 issue of *The Intellectual Observer*

Fig. 5.21 Scientific instrument maker and amateur astronomer John Browning

With the success of the August meteor campaign behind him, Herschel began to turn his thoughts towards the November Leonids, promoting the event at meetings of the Royal Astronomical Society, in the *Times* newspaper, and within the pages of the *Astronomical Register*. Indeed, the 1866 Leonids delivered a spectacular storm, with many thousands of meteors being visible during the course of several nights. Strangely, however, it appears that only Browning specifically set out to make spectroscopic observations [24]. Observing from Upper Holloway in London, Browning watched the heavens for some 6.5 h and recorded a whole series (but unspecified number) of spectra. Browning noted, "from the rapid flight of the meteors rendering the spectra very difficult to catch, I cannot pretend to speak with confidence of the appearance of the spectra shown by the prism, but I saw a great difference between the spectra". Browning goes on, however, to identify four spectral types varying from continuous colour spectra (suggestive of a solid heated body), to monochromatic yellow and green streaks (suggestive of emission lines from sodium and magnesium).

Also observing the Leonids in London, from their Camden Square home, were Henry James Slack and Charlotte Walters. James Slack, at that time, was editor of *The Intellectual Observer*, Honoree-Secretary to the Royal Microscopic Society, and well known for his works promoting science and microscopy in particular. Writing in the December, 1866 issue of his journal, Slack graphically explains:

Catching meteors in a spectroscope is not an easy task. It is a sort of celestial snap-shooting in which there are more misses than hits. From the luminous glow which filled the whole atmosphere, and which was heightened by the London gas-light, it was hopeless to spectroscopize [sic] trains, except at their brightest. Out of a dozen or two I saw through the instrument, the best defined spectra were yellow and green from the trains, and with all the colours when the nucleus, or ball was in the field. My wife compared one she saw of a ball, to the spectrum of Sirius, but brighter.

Slack gives no specific details on the spectroscope that he and his wife were using, but it appears that they were highly successful in capturing a good number of spectra. The comparison of one meteor spectrum to that of the star Sirius (α Canis Majoris) is interesting since the color temperature of Sirius is equal to that of a blackbody radiator of 10,000 K, and this, from modern-day analysis, is the characteristic temperature expected of the plasma column immediately behind an ablating Leonid meteoroid.

Herschel was actively observing the Leonids on the nights of November 11–12th and 12–13th, traversing the length of England in the process. On the night of the 11th and morning of the 12th, Herschel was observing with his father, John Herschel, at Hawkhurst in Kent. On the night of the 12th and the morning of the 13th he was observing from the Observatory at the University of Glasgow. On the first night he counted meteor rates, while on the second he measured meteor trails (in order to determine the sky location of the shower radiant) and spent but a few hours using the meteor spectroscope, capturing just 4 spectra. Writing upon his observation in the January 1867 issue of *The Intellectual observer*, Herschel explained that the relatively small number of spectra observed, in spite of the otherwise impressive display, was due to his concentrating on the trail observations required to deduce the radiant point and the unexpected sudden drop in the meteor rates in the pre-dawn hours. Writing later in his report to the British Association for 1867, Hershel commented that spectroscopic observations of the November Leonids were conducted at Greenwich Observatory, but no spectra were recorded. Likewise, for the first time, we hear of observations being made by Robert Greg in Manchester. Greg was using a direct vision meteor spectroscope (possibly constructed by Browning, but not identified as such) and succeeded in observing three bright meteors, "their spectra all consisted of crimson, red, and blue".

With the 1866 Leonid shower behind him, Herschel apparently made no further attempts to systematically capture meteor spectra. This being said, he did continue a very active correspondence with other meteor observers, and he published numerous reports and reviews relating to meteors, comets and meteorites [25]. Indeed, in a letter to the journal *Nature* for 25 December, 1873 Herschel called attention to the peculiarities of the meteors emanating from the annual Geminid shower. "It may interest observers of shooting stars who attempt to obtain views of their spectra", he began, proceeding thereafter to set-down his naked eye-perceptions concerning the distinctive colors of the Geminid meteors, "a beautiful pale-green colour", contrasting them against the November Leonid meteors which typically produced "very ruddy streaks". Writing, again, to the journal *Nature* for 29 September, 1881 Herschel returned to the topic of meteor spectroscopy, but this time to mostly downplay his role in its initial development. Inspired to write this letter on account of the opening address by John Lubbock to the British Association, Herschel commented that his 1866, "easy recognition of the presence of sodium in meteor-streaks can only claim to be regarded as a slight and inconsiderable first-adventure in a province of spectrum analysis". Most of the major advancements in the field, he noted, had been due to others, and he particularly singled out the work by Nicolas de Konkoly working at the O'Gyalla Observatory in Hungary.

Konkoly began his observations in 1872, and in a paper communicated to the Royal Astronomical Society [26] in 1873 he noted that, "Last July I received a very fine meteor spectroscope from Mr. John Browning", and that meteor spectra had been recorded showing lines of sodium and magnesium. In a second paper [27] published in the *Monthly Notices of the Royal Astronomical Society* for December 1873, Konkoly described his observations of a long-lived meteor train—the spectrum showing sodium and magnesium lines in addition to several other unidentified emission lines. The train was observed for some 25 min (the first 11 min with his direct vision spectroscope, and the last 14 with a telescope and spectroscope attachment), and the event may well have related to a meteorite fall, although no actual find has ever been made. Konkoly also described the spectra of meteor trails observed in July and August of 1879, and in August of 1880.

It is probably reasonable to assume that many observers, over the later part of the nineteenth century, made, or attempted to make, visual observations of meteor spectra, but few details appear to have seen publication. One early success, however, was the serendipitous capture, on 14 November, 1868 of the spectra of a long duration Leonid meteor train by Italian astronomer and spectroscopy pioneer Angelo Secchi. Such trains are produced by extremely bright fireballs (see below), and in Secchi's case the train was visible for some ten minutes and revealed bright red, yellow and green coloured emission lines. We also learn in the very first volume of the journal *Nature* of attempts made by a "Mr. Meldrum", observing in Mauritius, to capture meteor spectra with a "small hand-spectroscope" in November 1869. In addition, a short note appeared in the *American Journal of Science* for January to June 1873, indicating that, "Professor Eastman [28] succeeded in catching the spectra of two small [meteors]"[5] during the Andromedid meteor storm of 27 November, 1872. No indications of what type and design of spectroscope Eastman was using are given, however.

On the 4th of November, 1898, Alexander Herschel wrote to Denning concerning his plans for that year's Leonid display [25]. "I would like to try and capture spectra of some of the bright trains", he wrote, continuing that, "in that attempt one must desist entirely from noting and recording path positions". No records of any spectra captured by Herschel during the 1898 Leonids survive, but the letter to Denning indicates that Herschel maintained his interests in meteor spectroscopy for at least 34 years. This letter to Denning probably goes a long way in explaining why so few visual meteor spectra were ever recorded. The work is exacting, requiring quick reflexes, and great patience of the observer, and to be successful it dictates that other work, such as recording the hourly meteor rate or trail plotting needed to be abandoned. For all this, Denning, who was certainly a capable and skilled observer, was never tempted to borrow a spectroscope (certainly, one feels, Herschel could have arranged this) in order to investigate meteor spectra.[6]

5.8.2 Further Refinements

A novel adaptation to the direct vision meteor spectroscope was described by Browning in 1868 with the purpose of reducing the perceived angular size of a meteor train. This compression, it was suggested, would enable the spectrum to be more readily captured and perceived [29]. The adaptation consisted of adding a doublet composed of two concave lenses to the spectroscope (Fig. 5.22). The doublet produced an extremely wide field of view, easing meteor train capture, but it also acted to compress the observed spectrum image. When testing the instrument Browning noted that he, "found it easy to obtain the spectra of balls shot from a Roman candle placed only a few yards off from the instrument". In revealing terms, no-doubt derived from personal experience, Browning concluded his *Monthly Notices* article with the words, "our opportunity of observing meteors being very few, I have

---

[5] John Robie Eastman (1836–1913) was working at the US Naval Observatory in Washington. He is perhaps best remembered today for his work on the Washington Double Star Catalog.

[6] It is also the case that Denning had no practical experience of laboratory spectroscopy, and/or detailed knowledge of the spectral lines that might be expected from specific elements.

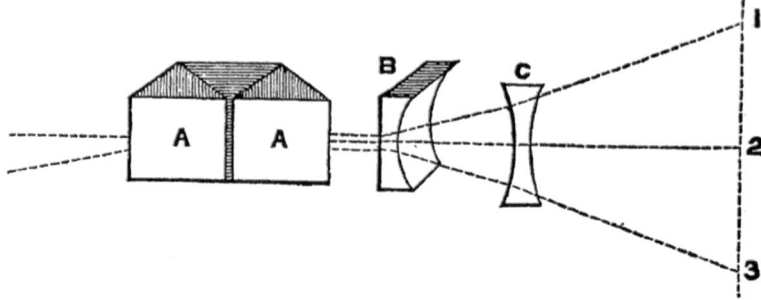

Fig. 5.22 The design for Browning's improved direct vision meteor spectroscope [26]. The meteor trail is shown to the far right and labeled 1, 2, 3. The doublet arrangement is composed of a plano-concave cylindrical lens B, and double-concave lens C. The arrangement of three prisms (labeled AA) constitutes the direct vision component of the spectroscope. Image from *Monthly Notices of the Royal Astronomical Society*, **28**, 50, 1868

made this adaptation, so that the same apparatus may be used for various purposes".

Additional adaptations to a direct vision spectroscope for meteor work were described by William Huggins [30] to the fellows of the Royal Society in December of 1867. Huggins explains that the instrument was designed during the summer of 1866 with the intent of observing meteors,[7] but that he was unable to actually use it during the November 1866 Leonid display. Undaunted, however, he recommended that the spectroscope might be used to observe the, "spectra of the lights which may be seen about the Sun during the total solar eclipse of [1868]". To this end, Huggins comments that, "four of these instruments, made by Browning, have been sent out by the Royal Society to India, to be placed in the hands of observers stationed at different places along the central line of the eclipse". After testing the spectroscope's design by observing "the spectra of fireworks seen from a distance of about three miles", Huggins comments that, "secret signals might be conveyed at night by means of the temporary introduction of certain suitable substances, as preparations of lithium, copper, strontium, &c., into the flame of a lamp giving a continuous spectrum; the presence of the bright lines due to these substances would not be perceived except by an observer provided with a

---

[7] Huggins's spectroscope has survived and resides in the scientific instruments collection at the Science Museum in London (Jon Darius, Curator of Astronomy – personal communication, 1987). The instrument was constructed by John Browning and the item label explains that, "it consists simply of a small telescope giving a magnification of about 2 ½ times, in front of which is a direct vision train of three prisms. As the instrument has no slit, it can be used only on small bright objects, or on objects at such a distance that they have only a small apparent size". The instrument (inventory number 1900–138) is on loan from the Royal Society.

spectrum telescope, to whom they might convey information in accordance with a previous arrangement". It is not known if Huggins ever used his device in an attempt to capture meteor spectra, or, indeed, send coded spectroscopic messages.

5.8.3 Later Developments

Konkoly's 1880s observations appear to have been the last serious attempt to systematically record visual meteor spectra—future advances would only be made through astrophotography. The great advantage of photography, of course, being that it produces a permanent record of the spectra, and that the wavelengths of the various emission lines can be measured with some accuracy. The photographic process still relies upon the serendipitous capture of meteor trails, but the fidelity of any data recorded is certainly superior to that acquired visually. The earliest record of a meteor spectrum being captured upon a photographic plate appears to date back to 27 November 1885, and was obtained during an outburst of the now dormant Andromedid meteor shower. The image was recorded by astronomer Jose Bonilla at the Zacatecas Observatory in Mexico with a 6-inch equatorial telescope fitted with a "Secchi spectroscope". Bonilla was a pioneer of astrophotography, primarily engaged in studying sunspot characteristics, and while no specific account of his observations seems to have been published, he did comment that the lines of sodium, carbon, iron, nickel and magnesium were present in the spectrum. The next meteor spectrum to be photographed, and in this case analyzed in detail, was that recorded on 18 June, 1897. This capture was made with the 8-inch Bache Telescope located at Arequipa in Peru—a telescope, in fact, constructed as part of the Henry Draper Memorial Survey.[8] Edward Pickering [31] explained that the spectrum consisted of six bright emission lines whose intensity varied as the meteoroid descended through the atmosphere. Pickering concluded his article by commenting that a special effort would be made to record Leonid meteor spectra in November, but no further reports were published for that year. Indeed, it was to be 1909 before the next meteor spectrum was to be captured by the Harvard survey.

The first dedicated program to capture meteor spectra photographically was established in 1904 by Sergey Blajko at the Moscow Observatory [32]. The first spectrum being recorded on 11 May, 1904 (apparently upon the

---

[8] The Henry Draper Memorial Survey began, under the supervision of Edward Pickering, at Harvard College Observatory in 1885. The aim of the survey was to obtain as many stellar spectra as possible and to thereby develop a systematic classification scheme. The HD catalog, containing information on 225,300 stars, was published between 1918 and 1924.

very first photographic plate to be exposed—but this could be an apocryphal story). A second meteor spectrum was also captured in 1904, during the Perseid shower on 12 August. The Moscow Observatory program ran, on and off, for about three years and recorded a total of 3 meteor spectra. From these, however, Blajko was able to clearly identify, for the first time, emission lines due to calcium and magnesium atoms.

Prior to the 1930s the detailed investigation of meteor spectra occurred on an ad hoc basis, and the topic was pursued by just a handful of individuals. This situation began to change, however, when Peter Millman, then a young Canadian graduate student working at Harvard University, set out to make meteor spectra the topic of his doctoral thesis.[9] Publishing his first research paper in the *Annals of the Harvard College Observatory* for 1932, Millman was able to bring together data on 9 spectra [34]. In an attempt to increase this number, the Harvard-Cornell Meteor Expedition was initiated in February of 1932. As part of this year-long campaign camera spectrographs were deployed at Flagstaff in Arizona, Fort Worth in Texas, and at various locations in Massachusetts. A total of 13 new spectra were obtained during this survey—remarkably, eight of these spectra were obtained during the November 1932 Leonid meteor display. Millman presented a detailed analysis of these spectra in 1935, introducing a new spectral classification scheme (which didn't, in fact, find much favor with other researchers, and is now obsolete), and he studied, for the first time, the detailed ionization state of the atoms responsible for generating the observed emission lines [35].

Writing 45 years after his 1935 review paper, Millman noted that, "meteor spectroscopy is a small field in experimental astrophysics but it has the unique feature of telling something about the chemistry of specific cometary fragments in the 0.1-to-0.001-g mass range" [36]. These statements are still true to this day, although thanks to the NASA *Stardust* and JAXA *Hyabusa* missions astrochemists now have direct access to meteoroids (dust grains) directly sampled from comet 81P/ Wild 2, as well as fragments from asteroid 25,143 Itokawa. Fragments from the surface of asteroid 101,995 Bennu have also now been retrieved. All of the above being said, it still remains the case that the inherent difficulty of meteor spectroscopy has not greatly changed since Herschel's very first attempts in 1864. It is interestingly the case that meteor spectroscopy is still an area of investigation largely conducted

---

[9] Peter Millman was a pivotal figure in the development of meteor astronomy in Canada. Indeed, during the 1930s to 1970s, Millman oversaw the establishment of a number of national programs concerning visual meteor observing, meteorite recovery, as well as radar and photographic surveys [33].

by amateur observers, although the techniques employed are now either photographic or video-capture in nature.

In parallel with the emergence of spectroscopy as a practical, laboratory-based science [37], the first direct-vision spectroscopes opened-up the field of spectroscopy to the amateur naturalist and the 'traveling' astronomer. Indeed, the direct vision spectroscope was an ideal instrument for all manner of meteorological work, both in the analysis of meteor spectra, and in the study of so-called rainbands.[10] Strangely, however, Browning makes no mention of meteor observations in his popular pamphlet [38] *How to work with the spectroscope*, published in 1878, although the spectroscopic study of the Sun, the planets, comets and nebulae are all described. This omission perhaps, again, says a great deal about the practical difficulties of studying meteor spectra, and in spite of Browning's otherwise encouraging words that, "these instruments [direct vision spectroscopes] are the easiest to use, and are therefore the best adapted for beginners", the topic is not even broached. That meteor spectroscopy is not mentioned by Browning is in stark contrast to the earlier enthusiastic discussion of the topic by Heinrich Schellen [39] in his *Spectrum Analysis* published in 1872. By the late 1870s, however, when Browning produced his pamphlet, the enthusiasm for and interest in rainband spectroscopy was in its early ascendancy, while interest in the visual study of meteor spectra was on the wane, with the field rapidly becoming, as Millman noted in 1980, a small and highly specialized branch of observational astronomy.

5.8.4 Meteor Trains

Systematic visual observations of long duration meteor trains (Fig. 5.23) were first conducted by Charles Trowbridge (Columbia University) during the early decades of the twentieth Century. Writing in 1907 Trowbridge noted [40, 41] that meteor trains typically have a yellow or greenish colouration, suggestive of the fact that the emission lines are due to sodium and magnesium respectively. Remarkably, and giving testament to the difficulty of obtaining data, fifty-two years later, in 1959, Robert Hughes [42] noted that up to that point in time, "we have not yet obtained a good spectrogram of a meteor train which is imperative for understanding the mechanism responsible for this luminescence". The first meteor train spectra to be recorded photographically appear to be those corresponding to several Leonid meteors,

---

[10] Charles Piazzi Smyth (Astronomer Royal for Scotland: 1846—1888) suggested in 1875 that the absorption lines due to aqueous vapor superimposed by the Earth's atmosphere on the solar spectrum might be used to study meteorological conditions.

all observed in November of 1965, by G. A. Nasyrov and L. A. Nasyrova at the Astrophysics Laboratory at the Turkmenian Academy [43]. These spectra displayed emission lines due to nitrogen, oxygen (these would be atmospheric lines) and magnesium as well as additional unidentified lines. Long-lived meteor trains, being both rarely formed, and intrinsically faint, are one of the least well understood of all meteoritic phenomena, and indeed, very little instrumental data exists on these objects even to this very day.

A study of American Meteor Society observations, by Charles Olivier in 1957, revealed that about 1 visual meteor in 5000 might produce a train that lasts longer than 5 min, with just 1 meteor in 125,000 yielding a train lasting longer than 10 min [45]. Train durations up to several hours have been reported, but these are exceptionally rare events (recall Fig. 3.19). Olivier deduced characteristic beginning and end heights of 100 and 75 km respectively for meteor trains, and from the time measurements of recorded distortions deduced characteristic upper atmosphere wind speeds of order 50 m/s. This latter measurement being particular interesting since it indicated

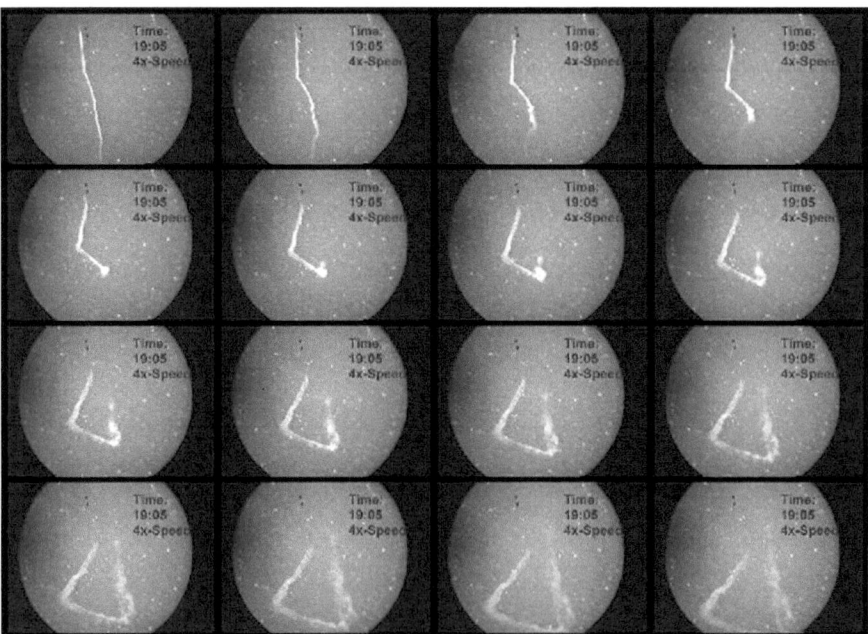

Fig. 5.23 This series of 16 images, taken from a video sequence, shows the time evolution of a Leonid meteor train. The images reveal the distortion of the initial train (upper left-hand corner) over a time span of about 30 s (lower right-hand corner). The evident distortions are due to variations in the wind speed at different atmospheric heights. The video sequence was recorded during an expedition to Mongolia, in 1998. Image courtesy of International Meteor Organization (IMO) [44]

an ability to study wind speeds at heights otherwise inaccessible to experimental study at that time. In this manner meteors became a 'natural' tool of aeronomy, enabling the means by which atmospheric density and dynamics could be studied.

The light generation mechanism for long lived trains has long puzzled astronomers, but it is now reasonably clear that it is driven by chemical reactions between atmospheric molecules, and metal atoms ejected during the meteoroid ablation process. One of the key processes responsible for generating light emission is the so-called Chapman mechanism,[11] which entails a two-step ozone dissociation reaction, using meteoroid-derived metal atoms as a catalyst. As the meteoroid ablates, so metal atoms, specifically sodium (Na), are injected into the atmosphere, and these can then interact with ozone, O_3, to produce sodium oxide NaO. According to the Chapman mechanism, the NaO can then interact with atomic oxygen, O, to produce an excited sodium atom Na^*, with the excited sodium atom rapidly de-exciting through the emission of a photon (of wavelength about 590 nm, corresponding to yellow light—this is the characteristic color of the sodium D line). Symbolically:

$$Na + O_3 \Rightarrow NaO + O_2$$
$$NaO + O \Rightarrow Na^* + O_2$$
$$Na^* \Rightarrow Na + h\upsilon$$

Similar reactions involving iron (Fe) and ozone can produce FeO in excited electronic states, leading to the formation of the so-called 'orange arc'. This latter process being verified for the first time with spectroscopic data obtained during the 1999 Leonid meteor storm [46].

5.9 The BAA Meteor Section

While the various members of the LMC can be viewed as the professional elite who took charge of directing meteor research in the United Kingdom, they were essentially compilers and reviewers, and as such, they controlled the dialog concerning accepted wisdom. Apart from A. S. Herschel, and R. P. Greg, most of the members of the LMC were not highly active meteor observers, and indeed, most of the data gathered in the LMC reports concerned serendipitously observed events. This began to change, however,

---

[11] This was first described by Sydney Chapman in 1939 as a mechanism to account for atmospheric airglow.

from circa 1870 onwards with the emergence of a core-group of dedicated amateur observers who made regular and coordinated observations.
Denning's first contributions, for example, to BAAS reports were received
in 1872.

While Denning did not try to organize a meteor observing sub-group in
the Observing Astronomical Society (OAS—see Chap. 7) in the early 1870s,
such a sub-group was immediately established upon the formation of the
British Astronomical Association (BAA) in 1890. Table 5.2 provides a list
of Directors to the Meteor Section of the BAA in the time interval 1890 to
1954.

The first director was David Booth. Born in 1860, the son of a wool-
spinner from Leeds, Booth made a living as a grocer and corn merchant.
He was a member of the Leeds Astronomical Society (see Chap. 7), and a
founding member of the BAA in 1890, although his membership lapsed in
1894. Booth was active as an observer in the late 1880s and is mentioned
in publications to the journal *Nature* by Denning in 1887 and 1888. His
first independent articles appeared in the *Observatory* between 1888 and
1891, and concerned the radiant of the August Perseid meteor shower. As
Director of the Meteor Section, Booth set straightforward observational goals.
Indeed, writing in the November 1890 issue of the *Journal of the BAA*,
Booth explained that, "our programme is necessarily a short one, as it is
chiefly involved in one sentence, namely, to ascertain the radiant points
of shooting stars". He expanded on these ideas in his Section Report for
1890–91, presenting suggestions for how observations should be recorded
and radiant points deduced. Some 8 observers submitted meteor observations
in the Section's inaugural year, including T. W. Backhouse (Sunderland), G.
T. Davies (Reading), W. F. Denning (Bristol), and W. H. S. Monck (Dublin).
Davies and Denning were, at that time, pursuing two station observations,

Table 5.2 Directors of the BAA Meteor Section: 1890 to 1954

	Birth—Death	Director	Work—Affiliation
David Booth	1860–1937	1890–1892	Grocer and corn merchant
Henry Corder	1855–1944	1892–1899	Market gardener
W. F. Denning	1848–1931	1899–1900	FRAS, FRMS
W. E. Besley	1877–1905	1900–1905	Civil servant—FRAS
C. O. Stevens	1865–1959	1905–1911	Independent researcher
Rev. M. Davidson	1880–1968	1911–1921	Clergy—FRAS
A. G. Cook	1877–1958	1917–1919	Acting Directors
F. Wilson	1864–1920	1917–1919	(Recall Chap. 1)
A. G. Cook	1877–1958	1921–1923	Independent researcher
J. P. M. Prentice	1903–1981	1923–1954	Lawyer—FRAS

with one real path being reported for a bright meteor seen on May 22, 1889. Appended to the first Meteor Section Report was a radiant catalog produced by W. H. Milligan (Dublin). In the time interval from early August to late November 1891 Milligan deduced a total of 16 radiants. Milligan, an amateur astronomer, was later to be one of the first ground investigators of the Crumlin meteorite fall on 13 September 1902.

Henry Corder took over the Directorship of the Meteor Section with Booth's retirement in 1892. Corder was, like Booth, a founding member of the BAA, but had been publishing reports on his meteor observations since 1875 (principally in the *Astronomical Register* and the *Observatory*). Born in Ipswich, Corder moved to Bridgwater, Somerset circa 1890, and was a market gardener by trade. Denning briefly took over the Directorship reigns in 1899, with the young, 23-year-old, Walter Ernest Besley becoming Director in 1900. Besley, a Civil Servant in Whitehall, London, certainly knew and communicated with Denning prior to taking over the Directorship of the Meteor Section, and began observing circa 1895. He was elected a Fellow of the RAS in January 1902, but resigned his Directorship, due to ill health, in April 1905, dying that same June at just 27 years of age.

When commenting on the first 50 years of the Meteor Section's work [16], then Director, J. P. M. Prentice, commented that its work, "collapsed with catastrophic suddenness in 1905", and that, "for six years very little meteor work was done in Britain". Such comments seem somewhat mysterious, and are, in fact, unnecessarily harsh, since the time interval indicated by Prentice covers the Section Directorship of Catherine Octavia Stevens (Fig. 5.24). Certainly, Stevens is deserving of more credit than Prentice provides, and she is certainly memorable as one of the pioneering grand amateur women of British astronomy [47]. Indeed, from the outset of its formation the BAA, unlike its RAS relative, not only allowed women to join its ranks, but also celebrated their work. Stevens was born into a relatively well-to-do middle-class family, her father being Dr. Thomas Stevens, Rector of Bradfield, in Berkshire, and from an early age was tutored in many areas of the arts and sciences. Having joined the BAA in 1891, her first article in the June 1896 issue of the *Journal of the BAA* concerned sunspot numbers. Other papers and notes concerned rainbows, mock suns, halos, aurora and the zodiacal light. Stevens travelled extensively during her lifetime, visiting New Zealand, and taking part in eclipse expeditions to Algiers (in 1900), Majorca (in 1905) and Canada (in 1932). Moving to Boars Hill, Oxford in 1910, Stevens had constructed a small observatory next to her house, and used a 3-inch aperture refracting telescope to make solar disk and sunspot observations. In the 1911 census, she described herself as an

'Astronomer' and her work as, 'British Astronomical Association'. In the 1939 census she described herself as a 'meteorological astronomer'. Stevens never married, and is an important and inspirational figure in the history of women in science, and especially so within the growing movement of amateur astronomy. Entering her house at Boars Hill was seemingly akin to visiting Aladdin's Cave, the rooms containing telescopic apparatus, a spectroscope, a barograph, and numerous neatly labelled specimens, "silver-bearing ore from the Rocky Mountains, coral from the Great Barrier Reef, monstrous insects from the Antipodes, …, and bundles of letters on scientific subjects from people all over the world" [47]. Stevens had a passion for meteorology, and developed a theory that attempted to link characteristics of auroral activity to local weather patterns and upper-level atmospheric winds. Three years into her Directorship, Stevens published an "interim report' on the Meteor Section in 1908, indicating with some reticence that the role of Directory was "onerous", and that, "the duty of the Director is not merely to sweep together the records" of section members. In this respect Stevens did not feel that every single meteor seen in the sky should be reported, but rather that section members should organize two-station observation campaigns. In this regard, Stevens gave praise to the Société Belge d'Astronomie for their highly successful two-station campaign during the Perseid meteor shower in August 1907. Contrasting the Belgian results against the relatively lack-luster observations of BAA section members, Stevens warned that, "our work must fall short of usefulness so long as we continue to work by *luck* rather than by *cunning*". Indeed, Stevens was right, and while her comments may well have alienated and even upset some section members of long-standing (Denning, specifically), a need for change was warranted. "It is the ambition of the present Director that the method of cunning should be forthwith and permanently adopted as the only method to be credibly pursued by Members of this Association", wrote Stevens in her 1908 interim report, but it would appear that the section members were unable to adapt. Stevens produced no further reports, although she remained Director for a further 3 years. Stevens should be remembered as an important instigator of change within the BAA Meteor Section. She realized that past methods and procedures needed to be revised, but was not able to bring the required changes about herself.

The Reverend Martin Davidson took over as Meteor Section Director in 1911, and he appears to have had much greater success in bringing about the changes that Stevens had called for. Born at Armagh in Norther Island, Davidson studied at Queen's University in Belfast, and was ordained in 1904, thereafter working in the East End of London for many years. He joined the BAA in 1911 (serving as Treasurer from 1933–36, and President from

Fig. 5.24 Catherine Octavia Stevens. Director of the BAA Meteor Section from 1905 to 1911. Image courtesy of the British Astronomical Association

1936 to 1938), and became a Fellow of the RAS in 1914 (serving as Vice-President from 1946 to 1948). Indeed, Davidson was a keen observer, and a good theoretician, writing many papers for the BAA journal and publishing numerous books on topics relating to astronomy and physics. Davidson was appointed Chaplin to the Forces with the outbreak of WWI, and his friends, and stalwart observers, Alice Grace Cook and Fiammetta Wilson took-on the role of running the Meteor Section until his return in 1919. In a sad report, returned from the Western Front, in 1917, Davidson noted that he had not been able to pursue any meteor related work on the front since, "there was no adequate means of distinguishing between a fireball and a star-shell" [16]. Unlike Stevens, Davidson assembled and published many reports on the activity of Meteor Section members, and was able to initiate a some-what loosely organized program of two-station observations. While Davidson stepped down as the Meteor Section Director in 1921, he remained a strong presence within the BAA. In addition to taking-on many administrative roles, he acted as Director of the BAA Comet Section from 1939 to 1945.

During Davidson's term as Meteor Section Director, a long running, but Herculean, project by T. W. Backhouse came to fruition in 1911. An experienced meteor observer, Backhouse announced his plans to produce a series of star maps, suitable for meteor work, in the *Astronomical Register* for February, 1886 (with the same article being reproduced in the *Observatory*, *Astronomische Nachrichten*, and the *Sidereal Messenger*). The maps were to be produced using a gnomic projection (recall Fig. 2.4), in which great circles are projected as straight lines, and would only show those stars that are visible to the human-eye (that is to magnitude +6). A total of 14 maps were to be produced and an accompanying star catalog would provide details concerning star positions and magnitudes. The star catalog would be compiled from pre-existing star catalogs with the star magnitudes being taken from the Harvard Photometry Catalog (published in 1908). The aim of the star maps was to make plotting meteor paths easier. By cutting out unnecessary detail (i.e., stars not visible to the human eye) the idea was to simplify the process of transfer. Four years after announcing the project, Backhouse explained in the February (1890) issue of the *Observatory*, that the compilation was, "taking a great deal more time than I expected", but that the star catalog was nearing completion. A further eleven years were to pass before the *Catalogue of 9842 Naked Eye Stars* (Hills & Co., Sunderland, 1911) was to appear. And, it was to be a further 2 years before the first star charts were engraved and printed by George Pearson in London. It was Fiammetta Wilson who introduced, and explained the purpose of the maps, to the assembled audience, at the May 1913 meeting of the BAA. Concerning the maps and catalog, Wilson commented, "it is with feelings of, may I say, reverence as well as gratitude that one examines this work". In spite of being well received, it would appear that few maps were ever produced and/or used—and the author is not aware of any marked maps surviving to the modern era, although the star catalog is still in print. Indeed, several members at the May 1913 meeting commented that the maps were too well produced (and to expensive) to consider marking them up with meteor tracks. When first proposed, in 1866, the maps would probably have found greater usage (indeed, the Luminous Meteors Committee of the BAAS had produced a limited selection of similar such maps), but by 1913, interest was beginning to shift away from the process of plotting meteor tracks.

Alice Grace Cook assumed full Directorship of the Meteor Section, upon request of the BAA Council, in 1921. In her first report, she interestingly reflects upon advancements made over the past ten years, but cautions that much more work needed to be done—especially with respect to establishing

the actuality of stationary radiants. Cook also outlined her desire to organize dedicated observing programs, with section members coordinating their
efforts to detect meteors from a specific shower radiant—the aim being to
refine radiant positions and structures, and to look for night-by-night radiant
motion. Towards this goal, Cook added the plea, familiar to all Section Directors, that any future successes would be dependent upon the recruitment
of new members. Furthermore, Cook also noted the importance of gathering data on telescopic meteors, and she encouraged section members to
experiment with photographic techniques. While Cook was encouraging of
change with respect to observing strategy, she continued to practice the same
methods of radiant reduction as performed by earlier Directors.

John Philip Manning Prentice (Fig. 5.25) was appointed Director of
the BAA Meteor Section in 1923, and, living in Stowmarket, was a close
neighbor of Alice Grace Cook. A lawyer by training, Prentice joined the
BAA in 1919, and was, in many ways, the enthusiastic new-blood that
Stevens, Davidson, and Cook had all been looking for. Indeed, under Prentice the Meteor Section was thoroughly restructured. Prentice was keen to
modernise practices, and is particularly remembered for abandoning the old
radiant reduction protocols, adopting the more conservative approach to
radiant identification advocated by Olivier and others. He was also keen
to improve methods of record keeping and observing, specifically asking
observers to change the way in which beginning and end points of meteors
were documented. Furthermore, with the help of John Guy Porter (Director
of the Computing Section, 1938–1959; BAA President 1948–1950), Prentice
introduced improved analysis techniques for reducing two-station observations, moving away from the graphical approach that had been previously
employed, and including a proper accounting for observational errors. A
skilled observer himself, Prentice is credited with the discovery of nova DQ
Hercules on 12 December 1934, and earlier, in 1926, he was one of first
observers to deduce a good radiant location for the October Draconid meteor
shower (Denning, however, was the first to record the shower in 1915).
Indeed, Davidson had predicted that such a shower might exist on the basis of
the orbital characteristics derived for the then newly discovered comet, comet
21P/Giacobini-Zinner. This outburst shower, typically delivering just a smat-
tering of meteors each year, produced dramatic storms in October 1933, and
October 1946. The 1946 display was particularly important, since it established a clear relationship between the appearance of visual meteors and the
occurrence of radar transient signals (see below). Indeed, in 1946 Prentice had
traveled to the Jodrell Bank Radio Telescope Observatory, to assist A. C. B.

Lovell, in anticipation of a good Draconid display. With the continued developments of both photographic and radar techniques applicable to meteor studies, it would seem that Prentice lost faith in the value of visual observations, and he began to pursue other projects and activities. When Prentice stepped down in 1954, Harold Ridley, the new Director, inherited a very different Meteor Section to that which existed at the beginning of Prentice's tenure.

By 1950, the Meteor Section of the BAA had undergone a thorough transformation. The old-school observers had now mostly retired (or died—as in the case of Denning), and new observational protocols, as well as reduction techniques, had been introduced. This is exactly as one would expect from a vibrant and developing science, and from a vibrant and active society. Furthermore, while in 1890, at the time of the Meteor Section's formation, all observations were made visually, by 1950 instrumental techniques had far outstripped the human observer in terms of accuracy and ability. This is not to say that the visual, and specifically the amateur, observer was obsolete. Far from it. Rather it forced the roles of the amateur and the professional observers to change. Instead of using visual observations to deduce radiant locations and meteor trail characteristics, the emphasis now shifted towards recording shower activity. That is, visual observers directed their attention towards counting the number of meteors visible per hour, over the duration of a meteor shower (the so-called activity profile -recall Fig. 2.6), and they additionally gathered counts upon the magnitude distribution of

Fig. 5.25 John Philip Manning Prentice. This image, from 1934, shows Prentice using an aircraft navigation machine for converting declination and right ascension into altitude and azimuth coordinates. Image courtesy of the British Astronomical Association

meteors, from a specific shower, as a function of time. In contrast, the professional astronomers turned their attention towards improving instrumental techniques, the determination of meteor velocities, and the study of ablation characteristics. Not only this, the professional astronomers used their successes in the understanding of the interaction of meteoroids with the atmosphere to ask new questions about its structure. Indeed, Geoffrey Nigel Gilbert (University of Surrey, UK) argued in 1977, that in terms of professional research activity, the direct study of meteors had ceased to exist by 1960 [48, 49]. This is not to say that meteors were no longer being observed, but rather that meteor studies had been subsumed into the broader arena of atmospheric research. Gilbert is partly correct in his claim—but he somewhat misguides his readers. There was no loss of interest in meteor physics in the 1950s, nor, indeed, has there been a loss of interest right up in the present day. Rather in order to sustain their research interests the professional astronomers, in the 1950s (and after), had to broaden their outlook, and chase money wherever it was available, and specifically where it was being directed to by funding agencies. In the 1950s (specifically) it was the opening-up of the space age (along with ballistic missile defence programs) that resulted in funding agencies favoring research in atmospheric physics. In order to capture research funding, therefore, researchers had to couch their projects in terms of what meteors might reveal about the structure of the upper atmosphere, rather than what their observations might reveal about meteors and meteoroid ablation.

The Meteor Section (MS) of the BAA, founded in 1890, was the first instance of an organized amateur group being established with the specific aim of studying meteor activity. This was followed by the formation of the American Meteor Society (AMS) in 1911 [14]. While ostensibly an amateur observing society, the AMS, unlike the MS, was founded and run, for its first 60 years, by a professional astronomer—that is, by Charles P. Olivier. It was Olivier who directed the AMS agenda, and who organized its specific observing activities. This was in definite contrast to the MS, where observers were largely left to their own devices, pursuing their own specific goals, with only limited organized (or Director directed) activity being evident. Next to organize were the professional astronomers, with the establishment of the International Astronomical Union (IAU) in 1919. From the outset the IAU established Commission 22, *Etoiles Filantes*, and Denning (recall Chap. 1) was appointed its first President. The first General Assembly of the IAU was held in Rome, in 1922. Denning did not attend, but his report was read and adopted. Only a very brief summary of Denning's report has survived, with the *Observatory* magazine for June 1922 providing the following:

The report of the Committee on Shooting Stars, drawn up by Mr. Denning (Gt. Britain), was adopted. It dealt mainly with recent observations, with the question of stationary radiants, and the addition of a new periodic shower associated with Pons-Winnecke's Comet. The publication of a new catalogue of radiant-points incorporating the results of the past 20 years was under consideration

Denning resigned from Commission 22 in 1925, and Charles Olivier was approached to take over as President. One cannot but feel that Denning was probably unhappy about the situation. Indeed, Denning had long passed into to a dogmatic stance with respect to stationary radiants, and Olivier was his most vocal critic. It was certainly a victory for Olivier, and it established his position as the new champion of meteor astronomy. Indeed, Olivier on assuming the Presidency of Commission 22 from Denning symbolically marks the end of the formative era of meteor astronomy. While Olivier publicly praised Denning in the October 1931 issue of the *Observatory*, writing upon Denning's death that, even though he did not know him personally, "his name deserves to be remembered for all time as one of the pioneers, and his example of enthusiastic work up to the very end of a long life … should be an inspiration", this was a far cry from his writings, a few years earlier, when he accepted the Commission 22 Presidency. Indeed, Olivier set very specific conditions on his acceptancy, and wrote to each Commission 22 member requiring them to answer a series of pointed questions. "What we want is the truth; and if some of us have to acknowledge that a pet theory is wrong or needs modification, the quicker we become convinced and admit our error the better", wrote Olivier on 19 November 1925 [14]. Effectively, Olivier wanted obedience to the new methodologies, and he specifically wanted to quash the older ideas championed by Denning (especially that concerning stationary radiants). By the 1928 General Assembly Olivier had developed a set of recommendations concerning future actions; these included the possibility of compiling a new radiant catalog based entirely upon the new methods of reduction (this effectively killed-off the stationary radiant debate). The Commission (that is, Olivier) also called for the thorough investigation of the accuracy of visual observations (a topic enthusiastically taken up by Prentice, and the Meteor Section of the BAA), and the continued development of meteor photography. Circa 1930, the old guard had effectively been replaced—indeed, it had been a rout, with a new regime of enthusiastic, young observers and theoreticians, with a new outlook, coming to the fore.

5.10   The Development of Instrumental Techniques

One of the main reasons why it had been difficult to rule-out a cosmic origin for at least some meteoroid streams was the observational problem of determining meteor speeds. Almost invariably, the speeds deduced by visual observers yielded values well in excess of the parabolic limit—indeed, meteoroids appeared to be moving so fast that they had hyperbolic velocities. This was problematic since, a hyperbolic velocity (that is a velocity greater than the Sun's escape velocity of 42-km/s at the Earth's orbit) indicates that the meteoroid could not be moving along a bound (that is elliptical orbit) about the Sun. While dynamically speaking a meteoroid having a hyperbolic velocity is not problematic, what is problematic is that such a meteoroid could not be associated with, or derived from, a comet moving along an elliptical orbit. Accordingly, the critical question pivoted on the accuracy with which meteor velocities could be determined.

An estimate of a meteor's speed is in principle a straightforward calculation. Using two-station observations to determines a meteor's trail length L, the characteristic velocity is essentially $V = L/D$, where D is the duration of the meteor.[12] Visual observations, however, provide inherently inaccurate values for both the meteor's trail length, and its duration, with the deduced velocities being 'doubly' uncertain. No amount of human experience, and or skill, will yield accurate meteor velocities, and accordingly the only solution is to turn to instrumental techniques. In this latter respect, however, the astronomer is entirely dependent upon the actual technology available for use in their specific era.

5.10.1   First Instruments

One of the great attractions of meteor observing to the amateur observer is that, at least in principle, no great investment in instrumentation is needed— one just needs a good pair of eyes, and a note book. Given, however, unavoidable limitations with respect to human acuity, timing ability, and reflex responses, high fidelity data, that is data of greater scientific value, can only be obtained through the development and deployment of special and appropriately designed instruments [50]. While photographic and later radar observations were to revolutionize meteor astronomy, the first meteor

---

[12] This is really an average velocity, since a meteoroid undergoes dramatic and rapid deceleration once rapid ablation begins.

observing instruments were largely developed along the lines of visual aids. Such simple devices helped to focus the observer's memory and/or offered a simplification of otherwise tedious data reduction. The first instrument to see wide usage was that of the meteor-wand, which seems to have seen usage since at least the mid-nineteenth century. Denning is seen with one such wand in Fig. 1.1—likewise, Alice Cook is seen with an observing wand in Fig. 1.9. This device is very simple to use. Upon seeing a meteor, the observer aligns the wand along its trail, and holds the position while estimates are made of the beginning and end points on the sky. Accordingly, it is a device to help focus the observer's mind on sky locations and the direction of a meteor's flight. A variant on the meteor-wand theme was described by J. P. M. Prentice in volume 36 of the *Memoirs of the British Astronomical Association* for 1948, where it is suggested that a flexible string or thin wire might be employed. In this case the string is held in both hands stretched out at arms length, with the right and left hands indicating the beginning and end points, with the string running along the meteor trail. Such a calibrated string could be well-adapted to determine the degree arc length of a meteor trail. A further variant on the meteor-wand was introduced by Ernst Opik in the early 1930's, where a fixed, thin-wire grid (or reticle) was placed in front of the observer. Indeed, during the Arizona Meteor Expedition conducted in the early 1930s (see next section), the observers were fully isolated from their surroundings, with the only view of the sky being through a square wire grid (Fig. 5.26). In this manner, the path of any observed meteor was described according to the XY-grid locations of its beginning and end points. While such aids and isolation did help observers focus their attention on the sky, there is no clear or specific indication that they significantly improved the accuracy with which meteor paths were recorded.

Estimating the duration of a meteor was recognized as being especially problematic from the very earliest times. Indeed, human beings are just not good at accurately estimating time intervals. As seen in Chap. 2, Denning and Herschel advocated a simple counting method to determine meteor durations. In this manner they repeated a string of numbers or letters and used this as a guide to how long a train was visible. Other observers attempted to use a stop-watch, but again this hardly improved duration estimates since the reflex delay in starting and then stopping the watch made for uncertain errors. Writing in the June-July 1916 issue of *Popular Astronomy*, 19-year-old Lincoln LaPaz[13] attempted to avoid the reflex delay by adapting an alarm clock to provide a series of audible clicks (Fig. 5.27). The adaptation involved

---

[13] Lincoln LaPaz (1897–1985) was to become a prominent researcher in the field of meteoritics, founding the Institute of Meteoritics at the University of New Mexico in 1944, being its first director

Fig. 5.26 One of the meteor 'houses' deployed during the Arizona Meteor Expedition. The wire reticle for estimating the sky coordinates of a meteor's path is visible in the observer's window on the sloping roof. Image from *PNAS*, **18**, 1933

soldering a thin piece of wire to the clock's escapement, with the clock tilted so that the wire would periodically dip into a small mercury bath. With the wire immersed in the mercury, an electrical circuit was completed, and a click was heard in a set of headphones. LaPaz wanted to use headphones specifically so as to free-up the observer's hands for recording the information about the visual appearance of the meteor. The 'meteor durimeter', so called, produced 4 audible clicks per second. LaPaz further indicated that when in use he found it useful to beat time with his fingers "in the unoccupied moments between meteors" so that the rhythm of the clicks became familiar. Charles Olivier provided a testimonial note to LaPaz's article, suggesting that all members of the American Meteor Society should make their own versions of the durimeter, although there is no evidence to indicate that this was strongly heeded.

Some attempt to investigate the problem of determining meteor durations were made in a series of studies by Bemrose Boyd (University of Iowa) in 1935–6. Writing in the November 1935 issue of *Popular Astronomy*, Boyd described a series of experiments concerning the ability of observers to estimate how long a lamp (situated on a distant radio mast) was illuminated for. The lamp was operated at random, without warning, for various duration times. The observers were divided into several groups. One set was given no coaching, while other groups were coached to employ a steady counting

through to 1966. He is also known for his investigations (on behalf of the American military) into the 1947 Roswell UFO incident.

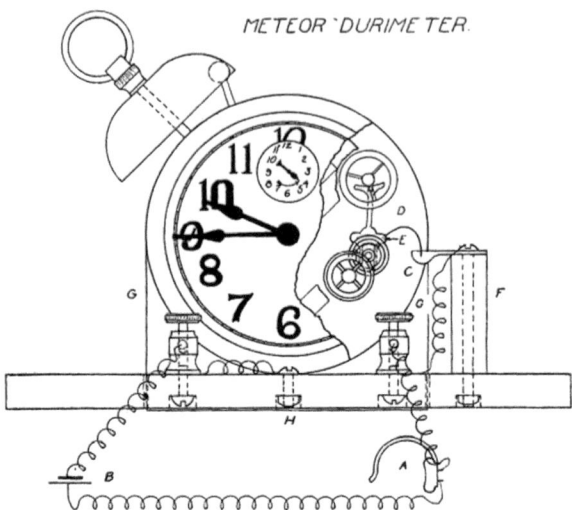

Fig. 5.27 The 'Meteor Durimeter' constructed by Lincoln LaPaz. The clock's balance wheel is labeled D, and the small contact wire is labeled E, with C being the small mercury reservoir. When the contact wire is emersed in the mercury bath the circuit is closed and battery (B) will produce a click in the headphones (labeled A). Image drawn by Merle Myers

method while the light was on. The results were quite clear. For the 26 observers in the un-coached group, the average error in estimating the lamp duration was 147 percent. For the 13 observers coached to employ a steady counting method while the lamp was lit, the average timing error was 14 percent. Boyd concluded that it made sense to, "advise amateur astronomers and other observers to count during the appearance of a bright meteor". Boyd took his experiments even further by developing an artificial meteor—these results being described in the January 1936 issue of *Popular Astronomy*. In this second experiment he attached a small electrical lamp on a moving belt—the belt being able to move at various speeds, with the light bulb being illuminated, over a 6-foot track, for intervals in the range from 0.4 to 1.8 s. To the observer's eye, the artificial meteor would traverse an angle of about 30-degrees. Eleven observers (college students) were used in this investigation. Of these, Boyd found that 5 always underestimated durations, 1 observer always overestimated durations, and 5 made as many overestimates as underestimates in their determinations. Furthermore, for all observers, Boyd found that the timing errors increased in step with the actual duration of the meteor—the longer the duration, the greater the error in the time estimate. In concluding

his article Boyd essential argued that progress in determining meteor durations will only be made through the development and deployment of purely
instrumental methods.

An early attempt to improve the determination of the beginning and end
points of a meteor's path on the sky was introduced in 1866 by James Challis,
Directory of the Observatory at Cambridge University. He had actually
invented the device, which he called the meteoroscope, nearly 20 years earlier,
in 1848, for a study concerning the heights of auroral displays. The meteoroscope, however, was hardly original, being essentially a pointer attached to a
tripod, with a base that enable the observer to read-off altitude and azimuth
positions (essentially it was a form of telescope-less theodolite). By recording
the time of the observations, the measured altitude-azimuth positions could
be converted to right ascension and declination coordinates, and these could
then be compared with the results from a second observer, located at some
distance away (recall Fig. 5.2), to deduce a radiant location and potentially
real path characteristics.

An improved meteoroscope (Fig. 5.28) was developed by the reverend
Martin Davidson, while Director of the BAA Meteor Section, in the early
1920s. Again, this device was hardly new, being a variant of the armillary
sphere instrument (as used by Tycho Brahe, for example, in the sixteenth
century). With reference to Fig. 5.28, Davidson's meteoroscope consisted of
a 32-inch diameter meridian hoop A mounted on a stand. The device was
accordingly orientated so that the plane of the circular hoop coincided with
the great circle running from the north celestial pole P, through the observer's
zenith Z, and down to their southern horizon. The circle is also graduated
running from +90° at pole P, through to 0-degrees (corresponding to the
celestial equator, and then on down to −90° at the southern celestial pole.
Attached at right angles to the meridian hoop A is a semi-circular hoop B—
this hope is also graduated from +90 at the north celestial pole P down to
−90° at the south celestial pole. At the center of the meridian hoop is a
swivel bracket to which is attached a pointer R. This pointer can be moved up
and down in elevation, and rotated in azimuth by twisting about spindle H
(which in turn is attached to a horizontal table T). Attached to the end of the
pointer R is a semi-circular arc K. Arc K in turn rotates about the end of the
pointer. Each of the moveable components are friction tight or have clamps
to secure them in position. In operation, the observer places their head at the
center of the meridian hoop A and sights along rod R. On seeing a meteor,
the observer adjusts the elevation and azimuth of rod R and rotation angle of
moveable arc K so that the latter runs along the meteor's trail on the sky. The
observer also records the time of the event. Having aligned arc K along the

meteor's path on the sky, the observer can read the declinations (correspond to the arc of the meteor trail) from hoops A and B. Since semi-circle hoop B is attached at 90° to circle A, its right ascension is automatically 6 h greater than the right ascension at A. The right ascension at A is determined form the meteor's logged time (by converting from local, to sideral time, and then calculating the right ascension of the meridian). The great circle arc of the meteor's path is now defined by the four coordinates: (RA and dec.) at A and (RA and dec.) at B, and this is the key information required to determine the radiant point of the meteor once combined with a similar data set obtained by a distant observer (recall Fig. 5.2).

Davidson's meteoroscope did not see wide usage—indeed, one suspects that, even if well constructed, it was a decidedly unwieldy device to use *en plein air* in the dark. Furthermore, one suspects that it was probably no more accurate at determining the path of a meteor than a straight-wand held at arms length. No doubt, however, it would aid in the relatively rapid evaluation of the right ascension and declination of two points along the arc of the meteor trail. The meteoroscope was not the first of Davidson's mechanical inventions, and, he additional constructed a meteorometer (Fig. 5.29). This

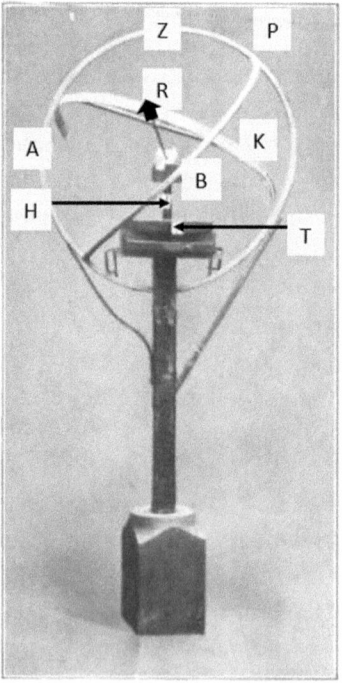

Fig. 5.28 Davidson's *Meteoroscope*. Image adapted from the April 1920 issue of the *Journal of the British Astronomical Association*. See text for details

was a pure analog computing device (not an outdoor instrument) consisting of two, one-foot-long, pointers mounted on moveable plates positioned upon a scale map. The idea here was to compute beginning and end heights of a meteor from the observed altitude and azimuth determinations provided two observers a known distance apart (recall Fig. 5.2). The advantage of this analog device was that the beginning and end heights could be determined even if one observer only sees a portion of the meteor's trail. There is no indication that this device ever saw much usage.

Showing great ingenuity, Davidson further developed, in 1930, an analog computer for deriving the parabolic orbit of a comet based upon three sets of observations. The device was constructed according to a scale of 1 m corresponding to 1 astronomical unit, and as seen in Fig. 5.30 filled a good fraction of a room. While Davidson acknowledged that the arrangement was far from ideal, it had utility in being able to deduce an approximate orbit, and it illustrated the geometry underlying the mathematical equations—again, this was a device developed entirely in order to save time with respect to what would otherwise be tedious hand calculations. Once again, there is no indication that this device ever saw much general (if any) usage.

Davidson's 'inventions' are very much of their era. Being analog devices, they were constructed according to the dictates of well-known mathematical theory, but saved the practitioner from having to make a long series of calculations. Recall this is the era before the mass production and general

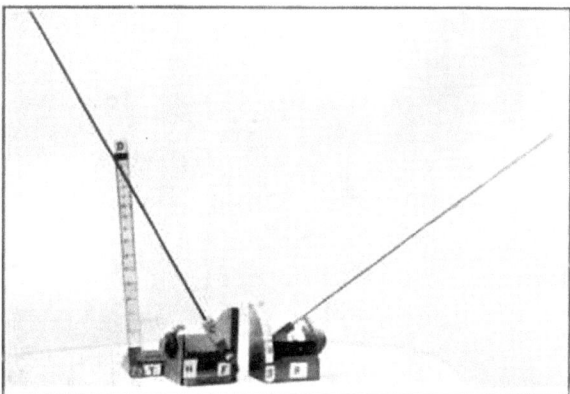

Fig. 5.29 Davidson's meteorometer – an analog computer for determining the real path of a meteor. Image from the December 1919 issue of the *Journal of the British Astronomical Association*

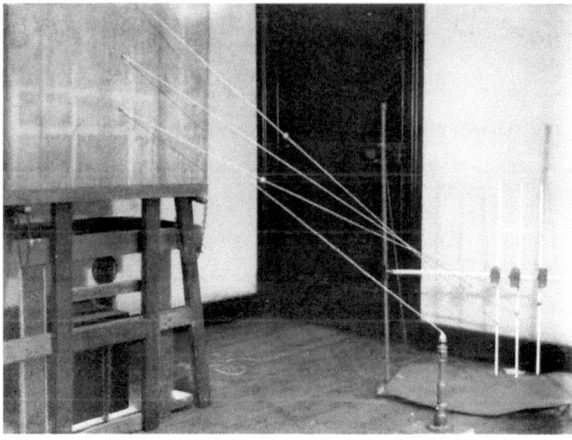

Fig. 5.30 Davidson's analog computer for determining the parabolic orbit of a comet. The Sun is located atop of the wooden pillar (to the lower righthand corner). The Earth's orbit has a radius of 1-m, and the three vertical bars (at the far-right center) indicate the position of the Earth when the comet was observed. Image from the March 1930 issue of the *Monthly Notices of the Royal Astronomical Society*

availability of mechanical calculators and even slide rules.[14] Furthermore, the analog devices tap into the zeitgeist of the era, and fully endorses the famous comments made by William Thomson (Lord Kelvin) that he could only understand a physical problem if he could make a model of it.

5.10.2 Photography

The word 'photograph' was coined by German astronomer Johann Heinrich von Madler in 1839. Predicated upon the pioneering works of Louis Daguerre in France, and William Fox Talbot in England, it was soon realized, during the early to mid-1830's, that photography offered all branches of science, not just astronomy, a new and valuable observational tool for recording precise images. In principle it removed the human, along with their inherent biases and limitations, from the data recording process. The photograph captured an impartial view of events, and it held those events within a permanent image that could be measured in the laboratory with precision instruments.

---

[14] Although, recall Fig. 5.25, where Prentice is pictured using a specialised analog computer for the computation of altitude and azimuth angles from celestial right ascension and declination measurements. Such instruments, however, were not mass produced and/or generally available to the public.

The first image of the Moon was obtained by Louis Daguerre on 2 January 1839, although the oldest known surviving image of the Moon is that taken a year later, from a rooftop in New York, by medical doctor and chemist John William Draper. These first images, both Daguerreotypes, required long exposures to be made, even for something as bright as the Moon, and it was to be nearly 20 years before the technology had progressed to a level that a comet could be photographed. William Usherwood, in England, and George P. Bond at Harvard in the U.S., obtained the first such images of a comet, comet Donati, in September of 1858. It was to be a further 27 years, however, before the first meteor trail was to be captured—this being achieved by Ladislaus Weinek, Director of the Klementium Observatory in Prague, on 27 November 1885. The meteor imaged was from the Andromedid meteor shower, which in 1885 produced a strong and dramatic display. Unfortunately, while Weinek had set up two cameras, at separate locations, so as to enable height determinations, the Andromedid trail was only recorded at one station.

The first dedicated meteor photography programs were established at Yale and Harvard University observatories in the 1890s [51]. From 1893 to 1909, William Lewis Elkin, whilst Director of the Yale Observatory, established a two-station photographic patrol system in the hope of capturing meteor trails. The first photographs of Leonid meteor trails were obtained in 1899. One of the important innovations introduced by Elkin was a rotating occultation disk [52]. This disk would periodically block, several times per second, any light from reaching the camera lens, and accordingly the trail of any moving object (i.e., that of a meteor) would be broken into a series of segments (Fig. 5.31). The reason for this shuttering is that each segment recorded would correspond to a precisely known time interval, and accordingly, not only could the velocity of a meteor be determined, so to could its deceleration. Elkin's original occulting disk was constructed out of an old bicycle wheel (which gives new meaning to the term heavenly cycles), but the main mechanical problem centered on the issue of maintaining a constant rotation speed. Any drift in the rotation rate of the wheel would cause uncertainty in the time interval corresponding to each meteor trail segment, and accordingly introduce errors into any velocity estimate. For all this, the timing uncertainties were significantly better than those associated with human, visual observations.

While photography was most definitely a new and popular pastime from the late nineteenth century onwards, its application to astronomy, by amateur astronomers, was slow to develop (no pun intended). This was not due to any lack of skill and/or interest, but the practical limitations of plate (that is, film/

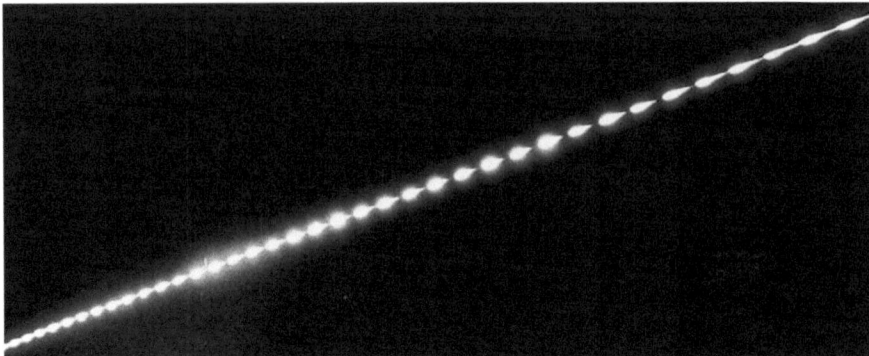

Fig. 5.31 A bright meteor trail 'chopped' into a series ¼ of a second long segments by the use of a rotating shutter. This particular trail presaged the fall of the Innisfree meteorite in Alberta on 5 February, 1977. Image courtesy of the Geological Survey of Canada

emulsion) sensitivity. The Photographic Section of the British Astronomical Association was established under the Directorship of William Schooling, with the intention of forming a collection of lantern slides for the use of Association members. It was not until well into the twentieth century, however, before BAA members began to produce respectable astrophotography results. Experiments attempting to capture meteors were made early on, but none were particularly successful until the early 1920s. This success was partly driven by fast, wide-field, war-surplus lenses becoming available for general use, and in the improvement of film sensitivity. James Hargreaves (Fig. 5.32) described his meteor photography plans in the January 1921 issue of the *Monthly Notices*, highlighting the importance of such studies in terms of what photographic results might resolve. "The long-continued controversies on certain questions in the study of meteors make it evident that more accurate methods of observing are required", writes Hargreaves, and by this he most probably means the issue of stationary radiants. In the same article Hargreaves indicates that he proposes to construct a camera with several lenses of, "clear aperture 5½ to 6 inches", with which he estimates that meteors brighter than magnitude +2 should be recorded. Hargreaves also notes that he did not intend to use a rotating shutter since this would hinder the capture of faint meteors. In this respect Hargreaves was concerned with the determination of sky location information rather than meteor speeds.

In the March 1929 issue of the *Journal of the BAA*, Edward H. Collinson described the design and construction of an automatic meteor camera [53]. The mechanical innovation introduced by Collison was a rotating mechanism that enabled the exposure of 6 individual plates (Fig. 5.33). Driven by

Fig. 5.32 Meteor photography (circa 1920). This un-accredited image was used by Denning in his article contributed to *Splendors of the Heavens* (edited by T. E. R. Phillips and W. H. Stevenson, 1923). It is suggested here that the seated observer (stopwatch and notebook in hand) is James Hargreaves (see ref. 47, Chap. 1)

Meccano[15] gears and solenoid actuators, the system advanced the plates at a rate of one per hour. The article by Collison indicates that a second system was being constructed for J. P. M. Prentice, with the aim of performing two-station observations. Collison indicates that, "the camera measures 10 in. by 10 in. by 12 in. and is fitted with an Aldis F/3 anastigmat lens of 6 inches focal length 6".

Rather than use a rotating occultation disk to determine meteor speeds, Ernst Opik developed a rotating mirror system for visual observers in 1935 [54]. In this case the mirror was made to vibrate ten times per second, and this turned the otherwise straight meteor trail into an epicycloid. The number of epicyclic loops associated with a given trail could then be used determine the time of flight as well as the angular velocity of the meteor. This system was later developed, in the 1950s, for photographic use. Where the rocking mirror system for visual observations came into its own, however, was in the Harvard run, Arizona Expedition for the Study of Meteors. Operational between October 1931 and July 1933, Opik organized a whole cadre

---

[15] Meccano is a construction system that was invented in 1898 by Frank Hornby. Still sold to this day, the system consists of reusable girders, plates, brackets, wheels, pulleys, and gears that can be bolted and meshed together in order to create articulated mechanical models.

Fig. 5.33 E. H. Collinson's automatic plate changer for a meteor camera. The clock-work Meccano motor (M) is seen to the right in the image, while the arrows labeled A, E and F indicate the rotation advancement and timing mechanism. The wooden base of the changer measures 9 by 7-inches. Image from [53]

of observers, who eventually recorded details on some 22,000 meteors, of which 3540 had their beginning and end heights determined. Following the close of the Arizona expedition, in 1936, Fred Whipple and Fletcher Watson initiated a two-station camera patrol system in Massachusetts. One camera was located a Harvard University, and the other some 38-km away in Oak Ridge. Many meteor trails were successfully captured with this system, and technical details concerning the stabilization of occultation disk rotation rates were investigated and largely resolved. The patrol system ran until 1951, after which time the Harvard Super-Schmidt Photographic Program was initiated in New Mexico. This latter program used specially designed widefield cameras designed by James Baker, and was funded by the U. S. Naval Bureau of Ordinance [55]. It was through the use of these cameras that the first highly accurate data set of meteor heights, velocities, and ablation characteristics was assembled.

5.10.3   Radar

The study of meteors via radar techniques was first begun in the post WWII era [56]. Indeed, the first meteor trails were detected and identified in step with the wartime development of radar warning systems. While the development of this technology falls well outside the time interval set for this chapter, a few very brief comments will be useful. Much of the early work on radar meteor studies was conducted, with army surplus equipment (Fig. 5.34), outside of Manchester, at Jodrell Bank, under the Directorship of A. C. B. Lovell. The initial projects were concerned with the detection and study of cosmic ray[16] interactions with the atmosphere, along with the identification of ionization trails associated with meteoroid ablation. Indeed, the very first time that the radar equipment at Jodrell Bank was turned on was on the peak night (December 14th, 1945) of the Geminid meteor shower.

The radar detection of meteors is predicated on the development of a typically short-lived, highly ionized trail being left in the region behind an ablating meteoroid. It is the interaction and reflection of the radar signal with this ionization trail that enables a detection to be made [57]. The connection between meteors and the appearance of transient radar echoes was clearly established by J. S. Hey and G. S. Stewart, who, writing in the 5 October

Fig. 5.34 Army surplus GL II (gun laying, 4.2-m wavelength) radar installed at Jodrell Bank in 1945. Image courtesy of the University of Manchester

[16] These are atomic nuclei that travel through space with relativistic speeds. They originate from the Sun and from outside of the solar system. Cosmic rays were first detected by Victor Hess in 1912 with balloon borne instruments.

1946 issue of *Nature* reported a distinct increase in radar echoes during the January 1946 Quadrantid meteor shower. Further proof of the connection between transient radar echoes and meteors was made during a spectacular outburst of the Draconid meteor shower on 9 October, 1946.

5.10.4   Taming the Hyperbolic Hare

As new photographic and radar technologies became available, so the question concerning characteristic meteor speeds was gradually addressed. Recall, the problem of interstellar meteoroids pivoted upon the determination of meteor velocities. If the velocity was higher than 42 km/s, so called hyperbolic velocities, then the parent meteoroid was not gravitationally bound to the Sun, and must have entered the solar system from interstellar space. The issue of hyperbolic velocities came to prominence with the 1925 release of a new catalog of fireball orbits as determined by Austrian astronomer Gustav Niessl von Mayendorf and German astronomer Cuno Hoffmesiter. Of the 611 orbits studied 79% were deemed to have hyperbolic orbits. Such a high percentage of interstellar meteoroids was remarkable, and in short order, in *Harvard College Observatory Circular* 331 (1928), Willard James Fisher (described by Fred Whipple in 1972 as a "professional amateur") reviewed the catalogs findings and implications. Fisher's review of the *Katalog der Bestimmungsgroßen fur 611 Bahnen großer Meteore* was critical of the analysis and selection methodologies employed by its authors, and his final sentence is telling, "there is probably a systematic error in the observations, leading to an overestimate of geocentric velocities". In short, the data and reduction methods were not to be trusted. The issue of hyperbolic meteors, however, refused to go away, and the Arizona Expedition for the Study of Meteors, conducted by Opik from 1931 to 1933, was implemented (in part) with the aim of resolving this very problem. As discussed earlier, Opik employed visual observers (who used special grids to determine meteor positions—recall Fig. 5.26) and a rocking mirror device to determine angular speeds. Opik's conclusions, however, indicated that some 57% of meteors observed (mostly sporadic meteors) had hyperbolic speeds. This was a smaller percentage than that deduced by Niessl and Hoffmeister, but still a remarkably high number. Through the 1940s, both Fred Whipple in America, and J. G. Porter in England [58], continued to review the problem, and both became convinced that systematic observing errors were the reason for the large proportion of hyperbolic meteors being reported. Indeed, they argued that no more than perhaps 10% of meteors had hyperbolic speeds. Furthermore, in 1940 Whipple, using observations gathered with a two-station camera system run by Harvard College Observatory, concluded that the October–November Taurids meteor shower, long touted as being of interstellar origin, was, in fact, a 'normal' meteor shower, with meteoroids moving along closed elliptical orbits—and, indeed, moving along orbits similar to that of comet 2P/Encke. For Whipple this latter discovery was a complete reversal of earlier expectations. Indeed, as a young post doctoral student at Harvard, he had been so impressed with Opik's ideas concerning meteoroid origins, that he had set out to determine the theoretical radiants that meteoroids associated with the star Sirius might make at Earth's orbit during the course of a year [59]. This "Bold Hypothesis" as Whipple later called it, was predicated on Opik's idea (developed in the early 1930s) that the outer regions of our solar system, and indeed the outer regions surrounding all stars, contained massive swarms of meteoroids and comets, all moving in long and looping elliptical orbits. In Opik's view not only were there free-moving interstellar meteoroids, but there were potentially interstellar meteoroid streams that could intersect the Earth's orbit, but which were gravitationally bound to larger (that is more massive) near-by stars. While Opik's ideas have not survived the test of time, he did effectively pre-empt Jan Oort, by some 20 years, in positing a vast swam of comets delineating the outer boundary of the solar system.

Moving into the post WWII era, the Harvard Super-Schmidt photographic program (Fig. 5.35), with its refined instrumental precision and control, revealed, under the analysis directed by J. G. Jacchia, that no more than 1% of visible (that is bright enough to be photographed) meteors had hyperbolic speeds [60]. Likewise, as the study of meteors via radar techniques improved, so in the June 1950 issue of the *Observatory*, Mary Almond, John G. Davies and A. C. B. Lovell, at the Jodrell Bank Observatory, were able to report that less than 1% of the meteors they detected had hyperbolic velocities. Peter Millman and Donald McKinley, working in Canada, came to the same conclusion. Opik, at this stage, however, was still not convinced, and continued to hold the opinion that most meteors had hyperbolic speeds. For all this, Opik's comments at this time were more bluster than reasoned argument. Indeed, at the 9 December 1955 meeting of the Royal Astronomical Society, a talk was presented by John G. Davies and J. C. Gill, both of the Jodrell Bank Observatory, on the radio-echo method of meteor orbit determination. During the talk, Davies made the comment that no hyperbolic orbits had been recorded. When questioned by R. O. Redman (then Director of the University of Cambridge Observatories), in the after-discussion, on this particular point, A. C. B. Lovell commented, "I am rather concerned in case Professor Redman's remarks should start a hyperbolic hare running.

I would, therefore, like to emphasize that these new measurements of individual meteor [sic] orbits made by Dr. Davies does not show a single example of a hyperbolic meteor" [61]. Opik picked-up upon the comment by Lovell, and using his privilege as Editor of the *Irish Astronomical Journal*, noted in its June 1956 issue that, with respect to Lovell's hyperbolic hare, "the danger is obvious: the hare may turn and chase the greyhounds! Besides, the hyperbolic hare has never been really on the run" [62, 63]. Thirteen years after this editorial jibe, however, Opik, then the sole hardliner, finally conceded that his analysis of meteor velocities had unaccounted for errors. Writing in the December 1969 issue of the *Irish Astronomical Journal*, and using the self-deprecating heading of "The Failures", Opik reminded his readers that in pursuing science, "failures are inevitable, but usually not in vain". Indeed, Opik insisted that the methods employed, and the observations gathered during the 1931–33 Arizona Expedition were not the problem, but rather, the problem lay in the assumptions made before the analysis began—these assumptions being that shower and sporadic meteors were physically similar, and that the systematic and random errors were the same for each observer on every night of the campaign. It was these, Opik concluded that, "led to the unfortunate misinterpretation[s]" concerning hyperbolic meteors.

There is a philosophical parallel that can be drawn between the apparently high percentage of hyperbolic meteors determined by Opik, and the apparent existence of stationary radiants as described by Denning. Both Opik and Denning were correct to insist that their observations were not at fault—they were both experienced observers and experts in the field. Indeed, we may safely take it as read that their observations were as accurate as could be achieved by any human observer. Accordingly, they stuck to their controversial deductions partly because they were rightly confident in their observational abilities, and partly (no doubt) out of shear belligerence towards their critics. In both cases of hyperbolic meteors and stationery radiants, it was not the practitioners that were at fault. Rather it was the faulty assumptions that went into the practitioner's data analysis. These mistaken assumptions, in effect, being put in place before even a single meteor was observed. Indeed, no scientific observation or experiment is ever conducted in a fully bias-free manner—there are always underlying theoretical assumptions, and it is these that can be (and indeed, often are) wrong. Denning applied the reduction methods that had been put in place by the pioneering observers—observers that he respected and learned his trade from. Opik, in contrast, applied massive statistical corrections to his meteor observations, but was led astray by assumptions relating to mathematical rigor and observer inconsistency.

Fig. 5.35 Fred Whipple, circa 1952, standing in front of a Baker Super-Schmidt meteor camera. Designed by James Baker and manufactured by Perkin-Elmer Corporation, the camera had an aperture some 12 ¼—inches in diameter, a 50-degree field of view, and could capture images of meteor trails to a faintness limit of about magnitude 3.5. Image courtesy of the AIP Emilio Segrè Visual Archives

While Opik acknowledge the errors inherent to his analysis of the meteor velocities just as the swinging 60s were about to end, it transpired that some 23 years later, starting in the early 1990's that the hyperbolic hare did, in fact, turn upon the hounds. Dust detectors carried aboard the *Ulysses*, *Galileo* and *Helios* spacecraft, for example, unambiguously detected very small dust grains (with masses in the range 10^{-18} to 10^{-13} kg) moving at speeds well in excess of the local solar system escape velocity, and in a direction parallel to the local flow of interstellar gas [64]. These are truly interstellar meteoroids, but they are far too small in mass to produce visible meteors in Earth's atmosphere. Indeed, to find a hyperbolic hare one has to physically go into outer space.

5.11 Atmospheric Interactions

While Aristotle had argued that luminous meteors were a fiery phenomenon of the upper atmosphere, it was Edmund Halley, in 1714, who first used the then known properties of the atmosphere to distinguish between ideas suggested for the physical makeup of meteors. Halley's arguments were constructed upon his earlier investigations into the Earth's atmosphere, and specifically the observations he presented to the fellows of the Royal Society in 1686. Specifically, Halley inferred that the upper limit to the height of Earth's atmosphere was about 50 miles altitude [1, 2]. With this result in place, Halley correctly reasoned that any terrestrial meteors must accordingly occur at heights lower than this upper limit. Furthermore, he questioned what sort of mechanism must be in play in order that terrestrial vapors might rise to the top of the atmosphere, and then apparently assemble in straight lines. Accordingly, in 1714 Halley suggested that bright fireballs "must be some collection of matter form'd in the aether [i.e., interplanetary planetary space], as it were by some fortuitous concourse of atoms, and that the Earth met with as it past along its orb" [65]. This was a bold hypothesis, not least for its suggestion of an extraterrestrial origin for meteors, but largely because Isaac Newton, then President and dominant force in the Royal Society, had earlier argued that interplanetary space was devoid of all matter. In later publications, Halley reverted to the idea that meteors had a terrestrial origin, and it was left to Ernst Chladni to reinvent the extraterrestrial hypotheses at the end of the nineteenth century.

While observations of meteor heights, via two-station observations, would ultimately reveal that meteoroids must originate from beyond the Earth's atmosphere, it was still far from clear, throughout most of the nineteenth century, exactly what was physically happening to produce the luminous (that is light) part of the meteor phenomenon. Certainly, it was clear, that it must be an interaction between a small solid body and the Earth's atmosphere, but exactly how was the light of the observed meteor produced? British physicist James Joules opened-up the debate on the physical interaction in the mid-nineteenth century. By applying the then new ideas of the conservation of energy and the equivalence of heat to mechanical energy [66], Joules reasoned that a meteoroid was ignited by violent collisions with atmospheric molecules. It was clear by the mid-nineteenth century that the speed (while poorly constrained) with which a meteoroid encountered the atmosphere was well in excess of the sound speed, and that any collisions between the meteoroid and an atmospheric molecule would be highly energetic—the energy available in the overall process being specifically the kinetic energy of the meteoroid $E = \frac{1}{2}\, m\, v^2$, where m is the meteoroid mass and v is its initial velocity.[17]

As to the typical mass of a meteoroid, one of the first attempts to constrain this property was given by A. S. Herschel in 1863 [67]. Indeed, Herschel's calculation was a heroic application of standardized experimentation and unit conversion. He began by stating that the light of the full Moon was equivalent to that produced by a "flame consuming 4 ½ cubic feet per hour of ordinary coal gas at a distance of fifteen yards". He then calculated the mechanical equivalent energy of burning a single cubic foot of coal gas, concluding that such a volume of gas would suffice to arrest the motion of 9.378 grains weight of matter moving at 30 miles per second. With these equivalents, it was possible to determine a relationship between meteor magnitude and meteoroid mass. A meteor as bright as Jupiter (magnitude -2), for example, would be produced by a meteoroid of mass 4×10^{-2} g; a first magnitude meteor would be produced by a meteoroid of mass 4×10^{-3} g, and a fifth magnitude meteor would be produced by a meteoroid of mass 2.6×10^{-4} g. These numbers are in the right sort of ball park, but it is now known that the maximum brightness of a meteor is a complicated function of the initial meteoroid mass, zenith angle, and entry velocity. The chief uncertainty in establishing a calibration equation being that of the efficiency with which the kinetic energy of the incoming meteoroid is converted into light. A formula presented by Whipple, based upon Super-Schmidt Camera observations, for example, gives:

$$\mathrm{Log}\, m = 27.2978 - 4.25\, \mathrm{Log}\, V - 0.3448\, M_V$$

where m is the meteoroid mass in grams, V the initial velocity in cm/s, and M_V is the maximum visual magnitude. This formula indicates, for example, that at $V = 30$ miles per second (48 km/s), a 4×10^{-2}-g meteoroid will produce a first magnitude ($M_V = 1$) meteor; a 1.4×10^{-3}-g meteoroid producing a 5th magnitude meteor[18]—indicating that Herschel's mass estimates are a factor 5 to 10 too small.

The first attempts at formulating a full theory for the physical interaction of a meteoroid with the Earth's upper atmosphere were presented in the second decade of the twentieth century. At this time, in 1922, Ernst Opik, then working largely in isolation at the Tartu Observatory in Estonia,

---

[17] In terms of a 1-g Leonid meteoroid, traveling at some 70 km/s at the top of the Earth's atmosphere, the energy budget amounts to some 2.45 million Joules – this is equivalent to about 0.6-kg of TNT explosive.

[18] For our 1-g Leonid meteoroid of footnote 17, the maximum brightness according to Whipple's formula will be magnitude -5. This is about the same brightness as planet Venus.

applied a conservation of energy argument, to determine the efficiency with which the kinetic energy of a meteoroid was converted into light. At the same time Frederick Linderman and Gordon Dobson (both meteorologists at Oxford University), expanded the theoretical discussion to include not only the heating effect of direct impacts upon the meteoroid's surface, but also the heating effect due to a hot cap of compressed air in front of it [68]. Importantly, Lindemann and Dobson (in line with the spectroscopic observations) identified atomic line emission as the process by which the light from a meteor is generated (Fig. 5.36). Specifically, impacts with atmospheric molecules will result in the ejection of atoms from the surface of a meteoroid, and these will further interact with surrounding atmospheric molecules, leading to their ionization. The excited atoms will then lose their excitation energy through the emission of a photon. The light phenomenon (the meteor) associated with the passage of a meteoroid through the atmosphere, therefore, is not produced via a chemical reaction, nor a burning process, but is rather the result of atomic, line-transition processes governed by the rules of quantum mechanics. Given this complex chain of interactions, it is very difficult to determine the efficiency with which the kinetic energy E of a meteoroid is converted into visible light, but most (present-day) estimates place it at 1% or less.

Having developed a physical description of the meteoroid ablation process, Lindemann and Dobson then combined their equations with a standard atmosphere model, and compared the results against the observed beginning and end heights of meteors (the data they used being supplied by Denning). This model—observation comparison indicated that the atmospheric density, at heights of order 100-km, must be significantly higher than was then thought. Writing in the March 1926 issue of the *Astrophysical Journal*, C. M. Sparrow (University of Virginia) criticized, and refined several of the physical assumptions introduced by Lindemann and Dobson, and specifically showed that it was direct impacts upon the surface of a meteoroid that drove the ablation process (this reduced the apparent need to enhance the density of the upper atmosphere as found in the 1922 analysis by Lindemann and Dobson). Further refinements of the theory of meteoroid ablation were developed, by many researchers, from the late 1940s onwards, including significantly F. Whipple, L. Jacchia, A. F. Cook, and R. N. Thomas in the US, and B. J. Levin in the USSR. The first monographs describing the detailed physics and theory of meteoroid orbits, observations and ablation were published by A. C. B. Lovell (Manchester University) in 1954 [69], and E. Opik (Armagh Observatory) in 1958 [70].

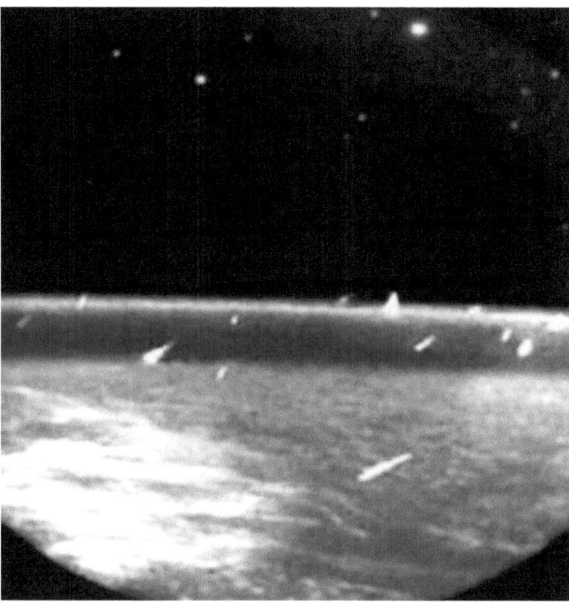

Fig. 5.36 Leonid meteoroids (bright streaks) ablating in Earth's upper atmosphere. The image was obtained by the *Midcourse Space Experiment* Satellite at the time of the shower's peak activity in November 1997. A patch of reflected Moonlight is visible to the lower left of the image, while stars are visible in the upper righthand corner. Image courtesy of NASA

5.12 Chapter Summary

In the time interval from 1830 to 1930 the study of meteors evolved from being a minor branch of astronomy, to a major component of atmospheric research initiatives. It was effectively turned into a field of practical observation and measure in the first half of the nineteenth century, and then, in the first half of the twentieth century, it saw significant advancement as a theoretical and instrument-based science. Both professional and amateur astronomers were involved in this transformation, and both professional and amateur astronomers had to correct and re-asses cherished ideas and understandings. The first significant steps were made through two-station observations of meteor heights, revealing that meteoroids must be extraterrestrial in origin. Such observational results established the ideas that had been introduced by Chladni in the late eighteenth century. The fortuitous observation of several spectacular meteor storms (predominantly from the Leonid shower), focussed attention on stream properties, and this ultimately resulted in a physical connection being made between meteoroid streams and comets. As with all new sciences, there were instances where pioneering ideas had to be abandoned. Three key observational problems can be identified: (1) the apparent existence of many hundreds, even thousands, of radiants, (2) the apparent existence of long-lived, stationary radiants, and (3) the apparent predominance of hyperbolic meteors. Denning, an amateur observer, was directly involved in the first two of these issues, although he is identified here as the last (vocal) hold-out, rather than the sole instigator of the problems. New, stringent rules for the identification of shower radaints were recommended by Tupman in the late 1870s and championed 50 years later by Olivier and Prentice (among others) during the first several decades of the twentieth century [14]. Opik, a professional astronomer, was the major protagonist in the hyperbolic meteor debate, but advancements in instrumental techniques ultimately showed that his dogged promotion of the concept was misplaced. Indeed, meteor astronomy was thoroughly revolutionised by the development of photographic and radar techniques from the mid-twentieth century onwards. The former of these techniques only became practically available in the first several decades of the twentieth century, while the latter came to prominence in the post WWII era. Likewise, a detailed physical theory of meteoroid ablation only appeared in the post WWII era, and this played into the development of a detailed understanding of the Earth's upper atmosphere. Meteor spectroscopy began early, in the mid- to late-nineteenth century, but was purely a visual exercise in nature, with only the most prominent emission lines being identified. The first photographic meteor spectra were captured serendipitously, as a result of other astronomy work, and these were first studied in detail by Peter Millman in the early 1930s [34, 35].

Bibliography

1 Beech, M. (2021). *A Cabinet of curiosities—The myth, magic and measure of meteorites*. World Scientific.
2 Beech, M. (1990). Halley's meteoric hypothesis. *The Astronomy Quarterly, 7*, 3–18.
3 Burke, J. G. (1986). *Cosmic debris—Meteorites in history*. University of California Press.
4 Hughes, D. W. (1982). The history of meteors and meteor showers. *Vistas in Astronomy, 26*, 325–345.
5 Beech, M. (1995). The makings of meteor astronomy: Part X. WGN. *Journal of the International Meteor Organization, 23*, 135–140.
6 Littmann, M. (1998). *The heavens on fire—The great Leonid meteor storms*. Cambridge University Press.

7 Olmsted, D. (1834). Observations of the meteors of November 13th, 1833. *American Journal of Science and Arts, 26*, 132–174. See also

8 Twining, A. C. (1834). Investigations respecting the meteors of Nov. 13th, 1833. *American Journal of Science and Arts, 26*, 320–352.

9 Buffoni, L., Manara, A., & Tucci, P. (1990). G. V. Schiaparelli and A. Secchi on shooting stars. *Memoire della Societa Atronomia Italiana, 61*, 935–959.

10 Newton, H. A. (1864). Abstract of a memoir on shooting stars. *American Journal of Science and Arts, 39*, 193–207.

11 Whipple, F. (1950). A comet model I: The acceleration of comet Encke. *The Astrophysical Journal, 111*, 375–394.

12 Williams, I. (2011). The origin and evolution of meteor showers and meteoroid streams. *Astronomy and Geophysics, 52*, 2.20–2.26.

13 Jenniskens, P. (2017). Meteor showers in review. *Planetary and Space Science, 143*, 116–124.

14 Taibi, R. (2017). *Charles olivier and the rise of meteor science*. Springer.

15 Sperberg, U. (2021). Eduard Heis, an early pioneer in meteor research. *History of Geo- and Space Sciences, 12*, 163–170.

16 The British Astronomical Association—the first fifty years. *Memoirs of the British Astronomical Association, 42*, 1989.

17 Cook, A. (1971). Evolutionary and physical properties of meteoroids. In *Proceedings of IAU Colloq. 13, held in Albany, 14–17 June 1971*. National Aeronautics and Space Administration SP 319, 1973.

18 See: http://www.imo.net/members/imo_showers/working_shower_list

19 Howarth, O. J. R. (1931). *The British association—A retrospective*. British Association.

20 Morrell, J., & Thackray, A. (1981). *Gentlemen of science—Early years of the British association for the advancement of science*. Clarendon Press.

21 Herschel, A. S. (1865). *Les Mondes, 7*, 139–141.

22 Millman, P. (1980). The Herschel dynasty III. *Journal of the Royal Astronomical Society of Canada, 74*, 279–290.

23 Beech, M. (1990). A simple meteor spectroscope. *Sky and Telescope* magazine, November, 554–556.

24 Browning, J. (1867). On the spectra of meteors of Nov. 13–14. 1866. *Monthly Notices of the Royal Astronomical Society, 27*, 77–79.

25 Beech, M. (1991). The Herschel-Denning correspondence: 1871–1900. *Vistas in Astronomy, 34*, 425–447.

26 de Konkoly, N. (1879). Spectroscopic observations of meteors. *The Observatory, 3*, 157–158.

27 de Konkoly, N. (1880). The August meteors. *The Observatory, 3*, 577.

28 Tucker, R. H. (1914). *Publications of the Astronomical Society of the Pacific, 26*, 41.

29 Browning, J. (1868). On a contrivance for reducing the angular velocity of meteors, so as to facilitate the observation of their Spectra. *Monthly Notices of the Royal Astronomical Society, 28*, 50–51.

30 Huggins, W. (1967). Description of a hand spectrum-telescope. *Proceedings of the Royal Society of London, 16*, 241–242.

31 Pickering, E. (1897). Spectrum of a meteor. *Astronomische Nachrichten, 145*, 77–80.

32 Blajko, S. (1907). On the spectra of two meteors. *The Astrophysical Journal, 26*, 341–348.

33 Tors, S., & Orchiston, W. (2009). Peter Millman and the study of meteor spectra at Harvard University. *Journal of Astronomical History and Heritage, 12*, 211–223.

34 Millman, P. (1932). An analysis of meteor Spectra. *Annals of the Harvard College Observatory, 82*, part 6.

35 Millman, P. (1935). An analysis of meteor spectra: Second paper. *Annals of the Harvard College Observatory, 82*, part 7.

36 Millman, P. (1998). One hundred and fifteen years of meteor spectroscopy. In I. Halliday and B. A. McIntosh (Eds.), *Solid particles in the solar system*, D. Reidel Publishing, Dordrecht (1998)

37 Hearnshaw, J. B. (1987). The Analysis of Starlight: Some comments on the development of stellar spectroscopy, *1815–1965. Vistas in Astronomy, 30*, 319–375.

38 Browning, J. (1878). *How to work with the spectroscope*. W. J. Johnson, Printer.

39 Schellen, H. (1872). *Spectrum analysis*. Longmans, Breen and Co.

40 Trowbridge, C. C. (1907). Physical nature of meteor trains. *The Astrophysical Journal, 26*, 95–126.

41 Trowbridge, C. C. (1924). Meteor trains. *Proceedings of the National Academy of Sciences of the United States of America, 10*, 24–41.

42 Hughes, R. F. (1959). Meteor trains. *Smithsonian Contributions to Astrophysics, 3*, 79–94.

43 Beech, M. (1987). On the trail of meteor trains. *Quarterly Journal of the Royal Astronomical Society, 28*, 445–455.

44 For the complete video sequence see: http://www.imo.net/observations/methods/video-observation/examples/

45 Olivier, C. P. (1942). Long enduring meteor trains. *Proceedings of the American Philosophical Society, 85*, 93–135.

46 Jenniskens, P., et al. (2000). FeO "orange arc" emission detected in optical spectrum of Leonid persistent train. *Earth, Moon, and Planets, 82–83*, 429–438.

47 Larsen, K. (2006). Shooting stars: The women directors of the meteor section of the British astronomical association. *The Antiquarian Astronomer, 3*, 75–82.

48 Gilbert, G. N. (1977). Growth and decline of a scientific speciality: The case of radar meteor research. *EOS, transactions of the American Geophysical Union, 58*(5), 273–277.

49 Beech, M. (1998). Meteor astronomy: A mature science? *Earth, Moon, and Planets, 43*, 187–194.

50 Weber, M. (2005). Some apparatuses of meteor astronomy from the pre-electronic era. *WGN, The Journal of IMO, 33*, 111–114.

51 Hoffleit, D. (1988). Yale contributions to meteoric astronomy. *Vistas in Astronomy, 32*, 117–143.
52 Elkin, W. L. (1900). The velocity of meteors deduced from photographs at the Yale observatory. *Astrophysical Journal, 12*, 4–7.
53 Collinson, E. H. (1929). An automatic meteor camera. *Journal of the British Astronomical Association, 39*, 150–153.
54 Opik, E. J. (1953). A vibrating camera for meteor photography. *Irish Astronomical Journal, 2*, 193–202.
55 Whipple, F. (1955). Meteors. *Publications of the Astronomical Society of the Pacific, 67*, 367–386.
56 Hey, J. S. (1973). *The evolution of radio astronomy.* Science History Publications.
57 McKinley, D. W. R. (1961). *Meteor science and engineering.* McGraw-Hill.
58 Porter, J. G. (1944). An analysis of British meteor data. Part 2: Analysis. *Monthly Notices of the Royal Astronomical Society, 104*, 257–272.
59 Whipple, F. (1972). The incentive of a bold hypothesis: Hyperbolic meteors and comets. *Annals of the New York Academy of Science, 198*, 219–224.
60 Jacchia, L. G. (1955). The physical theory of meteors VIII. Fragmentation as a cause of the faint meteor anomaly. *The Astrophysical Journal, 121*, 521–527.
61 Lovell, A. C. B. (1956). *The Observatory, 76*, 2.
62 Opik, E. (1956). The hyperbolic hare. *Irish Astronomical Journal, 4*, 59.
63 Opik, E. (1969). The failures. *Irish Astronomical Journal, 9*, 156–160.
64 Wiegert, P. (2014). Hyperbolic meteors: Interstellar or generated locally by the gravitational slingshot effect? *Icarus, 242*, 112–121.
65 Halley, E. (1714). An account of several extraordinary meteors or lights in the sky. *Philosophical Transactions of the Royal Society, 24*, 159–164.
66 Hughes, D. W. (1990). James Joule and meteors. *Vistas in Astronomy, 33*, 143–148.
67 Herschel, A. S. (1863). The mass of meteoroids. *Proceedings of the British Meteorological Society, 2*, 22.
68 Linderman, F., & Dobson, G. (1923). A theory of meteors, and the density and temperature of the outer atmosphere to which it leads. *Proceedings of the Royal Society, A102*, 411–437.
69 Lovell, A. C. B. (1954). *Meteor astronomy.* Clarendon Press.
70 Opik, E. J. (1958). *Physics of meteor flight in the atmosphere.* Interscience Publishers.

6

The Astronomical Register

The appearance of *The Astronomical Register* in January of 1863 represents a pivotal-point in the development of British amateur astronomy. It aimed, and indeed, largely succeeded, to break the establishment stranglehold on the conveyance of astronomical knowledge and practices. Rather than being a journal established by and for a formal astronomical society, specifically, in Great Britain, the Royal Astronomical Society, the *Register* was intended for the amateur enthusiast. Indeed, the *Register* provided this new and blossoming body of observers with a forum through which they could discuss their observations, air their frustrations, and share their experiences. From the very outset the *Register's* aim was to nurture the growth of amateur astronomy, build a sense of community among amateur astronomers, and, in principle, to pay its own way—perhaps even make a small profit. The appearance of the *Register* was set at a time when there was a growing public interest in astronomy, and at a time when many other hobby-interests (photography, horticulture, cycling, microscopy, and geology, amongst many others) were being catered to by an increasing number of cheap and readily available publications. Indeed, the latter half of the 19th century saw a veritable explosion in the number of periodicals intended for a general, rather than an Oxbridge-educated, readership. Most of these new publications dealt with social, historic, and general-interest topics, but the *Register* set out to shape its contents around an engaged and active readership. Indeed, the reader of the *Register* was assumed to be both interested in astronomy as a subject, and actively involved in making actual astronomical observations. In this latter respect, the *Register* provides a useful window through which the activities of the mid to late Victorian amateur astronomer can be gauged. Specifically,

© The Author(s), under exclusive license to Springer Nature
Switzerland AG 2023
M. Beech, *William Frederick Denning*, Springer Biographies,
https://doi.org/10.1007/978-3-031-44443-2_6

here, the personal for-sale and want advertisements, along with those advertisements placed by commercial dealers, will be reviewed in an attempt to determine what kind of instruments amateur astronomers were using, and/or wanting, and who it was that was supplying the market. Likewise, we shall look at who it was that wrote to and for the *Register*, and what the main areas of interests were. Here, also, we shall find Denning beginning to establish his name and reputation as a key figure within the ranks of amateur astronomers.

6.1   Rise of the *Register*

The first journal devoted to what can be called scientific matters was that published by the Royal Society (RS) of London, with its *Philosophical Transactions* first seeing print in 1665. The first dedicated astronomy periodical, *Memoirs of the Royal Astronomical Society*, however, did not appear until 1825. The Royal Astronomical Society (RAS), which was founded in 1820 (and given Royal Charter in 1831), was to become (and remains) the preeminent astronomical body in the United Kingdom. While many of the founding members of the RAS were also fellows of the Royal Society of London, the core of its membership was derived from among the ranks of university-educated professionals, and the well-to-do (that is, independently wealthy) amateurs. That observing was a primary goal of the RAS is reflected in its motto, *Quicquid nitet notandum* (whatever shines should be observed), and this sits in contrast to that of the Royal Society, *Nullius inverba* (take no one's word for it). The motto of the RAS is inherently active in outlook, while that of the RS is more philosophically reflective. The RAS introduced its (still running) *Monthly Notices* (MNRAS), with the specific aim of publishing peer-reviewed technical articles, short observing reports, along with society news, in 1827. Both the *Memoirs* and the *Monthly Notices* appeared at a time when scientists were beginning to form new bodies of fellowship, with highly specific aims, concerning both research and reach.[1] To this end society journals enabled an outlet through which observations and research could be disseminated, and it enabled members to keep in contact with each other, and be informed of society activities. The articles appearing in the journals of these new national societies, however, tended to cater to the specialist, and they were not specifically aimed at the amateur, or the reader simply wanting to be abreast of latest thinking and research topics. To a certain extent, other

---

[1] The appellation scientist was first coined by William Whewell in 1833. Prior to this 'invention' the title of natural philosopher was the more commonplace name for someone conducting experimental research. For all this, the title of astronomer has a long and ancient heritage.

and earlier periodicals had catered to these latter interests, but the coverage was typically piecemeal. Publications such as *The Gentleman's Magazine* (first published in 1731), *The Saturday Magazine* (first published in 1832), and *The English Mechanic* (first published in 1865) all carried short review and commentary articles on astronomy related topics.

Following the appearance of the MNRAS in 1827, the Royal Geographical Society began to publish its own monthly journal, in 1831, and this step was emulated by the Statistical Society of London, in 1838. Following its formation in 1831, the British Association for the Advancement of Science (BAAS), began to issue annual reports, although these again tended to be directed towards the specialist reader. During the 1840s and 50s, however, numerous specialist journals, not specifically allied to a formal society, began to appear. These included, for example, the publication of the *Annals and Magazine of Natural History* (1841), *The Zoologist* (1843), *The Archaeology Journal* (1844), *The Photographic Journal* (1853) and the *Quarterly Journal of the Microscopically Society* (1853). These latter journals catered directly to the amateur enthusiasts interested in natural history, and in the appearance of new-technologies (especially photography). The appearance of the *Astronomical Register* in 1863, therefore, follows the trend of catering to non-specialist readers—the erstwhile amateur interested in some highly specific area of study.

The *Astronomical Register* was founded by Sandford Gorton (1823—1879) [1, 2]. Gorton was an Oxford-educated, independently wealthy astronomer who constructed a private observatory in London (Fig. 6.1). Gorton published papers in the MNRAS on cometary observations, transits of Titan, lunar occultations, and the appearance of Jupiter's disk, and was elected a Fellow of the RAS in June 1860. It was presumably through his association with the RAS that Gorton realized the need (or niche) for an amateur-based journal. Indeed, in his opening editorial, Gorton writes that the *Register*, "introduces a sort of astronomical Notes and Queries, a medium of communication for amateurs". His hope was to establish a monthly journal that could serve both as a national outlet for the reporting of observations by amateur astronomers, and which could, through its letters section foster discussion, education, and community. To these ends, the *Register* was largely successful, but it was not until issue number 37 (January, 1866) that Gorton was to jubilantly write, "the *Astronomical Register* has been a decided success".

The initial production of the *Register* was almost entirely due to Gorton's efforts, not only did he largely write the general articles that appeared during the first year, but he also set and produced the journal pages on his own printing press. The initial subscription rate was set at 2-shillings per quarter,

Fig. 6.1 Sandford Gorton (in thoughtful pose). Image courtesy of the Royal Astronomical Society

or 4-pence per single copy. After issue number 6, however, the single copy price was raised to 6-pence, with an editorial plea that, "it is to subscribers, moreover, that we must look for the necessary aid to enable us to carry on our paper". Further increases were announced in issue number 12, the subscription rate rising to 3-shillings per quarter, and individual copies being priced at 1-shilling. Such teething problems are not uncommon for any new journal, and Gorton was determined to struggle on, noting that, "it would be truly mortifying to be obliged to discontinue it". In his desire to make the *Register* a going concern, Gorton's tenacity eventually paid off, but it took 8 years. Writing in the January issue (number 97) for 1871, Gorton enthusiastically writes that, "a slight profit upon our publication" has been made, although this was tempered by the comment that, "even now, however, the original loss of the earlier years is not entirely covered". The financial turn-around in the *Register's* fortunes is attributed by Gorton to its growing list of subscribers, but he comments that if even more subscribers could be found, then the number of pages per issue might be increased, and, "we should be able to carry out many excellent suggestions which have been frequently made". It is not possible to be certain about the monthly print run of the *Register*, but even at peak subscribership it likely did not exceed several hundred copies.

After the first year of nursing the *Register* through its production schedule, the publishing was sent-out and performed commercially, with Gorton remaining as editor until his death in 1872. The Reverend John Jackson took over editorial control upon Gorton's death, although he made no significant changes to the *Register's* format. Indeed, when Jackson took over, no mention of an editorial change was even announced. The final issue of the *Register* was published in December 1886, after a lifespan of 24 years. No warning of its imminent demise was announced within its pages, and the final printed words were simple and precise: "*Finis. Valete*". The demise of the *Register* may have been rapid and unexpected, but it does not appear that it was due to financial losses. Indeed, a final notice to subscribers simply states, "the editor purposes to discontinue the publication". At the time of discontinuation, the Rev. Jackson would have been in his early 60's, and was presumably looking towards adopting a quieter life style. It is a little surprising, however, that no new editor was found for the *Register*, but in a financial and readership sense, it was certainly facing more competition from several new astronomy publications, and weekly science magazines. An anonymous editorial comment in the December 1886 issue of the *Journal of the Liverpool Astronomical Society*, records that with the demise of the *Register*, "some of our members will experience a feeling akin to that endured at the prospect of losing an old familiar friend to who they owe a large dept of gratitude". Furthermore, the editorial piece praises Sandford Gorton as, "no ordinary man … in keeping his scientific venture afloat and extending its influence far and wide".

The format of the *Register* was set in place from the very first issue. Typically, each number began with an account of the most recent meeting of the RAS—Gorton, himself transcribing the goings-on. At later times reports from the Liverpool Astronomical Society (founded in 1881) were included, and also *Dun Echt Circulars* from Lord Crawford's Observatory in Scotland. Indeed, while the *Register* was not officially allied to the RAS, the society meetings were a hot bed of latest astronomical news and happenings, and thereby of general interest to anyone following astronomical matters. Not only this, Gorton probably realized that many of the initial subscribers to the *Register* would likely come from the rank and file of the RAS (see Sect. 6.2), and this was an audience that he could *tap* in person. The RAS summary was generally followed by one or two short, general articles on some astronomy related topic: a new discovery, recent research, a famous observatory, or a famous astronomer. Next was a list of *Astronomical Occurrences* for the specific month ahead. This latter list would include information about planet locations, satellite ephemerides, comet and asteroid positions, along with details concerning double stars, star clusters and nebulae that were well placed for viewing. The monthly calendar provides a good indication of those objects that amateurs at that time were particularly interested in seeing, and specifically wanted to observe. Indeed, much of the work described by readers of the *Register*, through their letters and/or short reports, was concerned with drawings and written descriptions of what they had seen through the eyepiece, and/or measured with a micrometer (such measurements, for example, being applied to double star separations). Many letters and notes were associated with observing craters and features upon the Moon's disk, while other accounts looked at features appearing on planetary disks (for example Jupiter's red spot), or presented transit time data for Jovian-planet satellites. Still others reported on sunspots and solar flares, or the details of some bright meteor. Perhaps unsurprisingly the amateur astronomers of the mid-to late 19th century were interested in observing, counting or drawing those same objects that are sought-out by amateur astronomers to this very day. One of the unique features of the *Register* was its inclusion of a general letters section, where readers could raise issues or ask questions. Operating somewhat like the modern internet chatroom, the letters section occasionally stirred controversy, and more than once Gorton, who otherwise wielded a light editorial pen, had to remind correspondents to refrain from making personal attacks, and to keep the tone of their letters civil. Indeed, Gorton felt that it was important that everyone should have their say in any debate, and in his January-issue addresses, for many years, he defended the importance of the Editor's neutrality. Note only did every correspondence section carry the reminder that the Editor did not, "hold [himself] answerable for any opinions expressed by [the] correspondents", but in the March 1865 issue (number 28), an 8-point *Rules of Correspondence* was printed—this included the requirement that all letters to the editor must be signed in full, although the eventual article might only carry the writer's initials, or *nom de plume*.[2] For all this, one of the most important outcomes of the letters section was that it established the idea of a community—a community of amateur enthusiasts that could freely express their thoughts, bicker over technical details, air their frustrations, and report on their observations.

---

[2] Many of the of the pseudonyms were classical in nature and included such appellations as Clericus, Senex, Argus, Juno, Nauticus, and Cyclops, others were more intriguing and include, Oculi Ambo, A Grumbler, A Beginner, and An Enquirer.

6.2   The Subscribers

For the historian one of the highly useful features of the *Astronomical Register* is that each issue contained a name and address list of new subscribers. This primarily enabled other subscribers to identify fellow astronomers (local or otherwise) that they might wish to contact or work with, and it promoted the action of building a sense of national community—the latter aspect becoming more important with the founding of the Observing Astronomical Society (see Chap. 7). Usefully, however, the subscribers list enables an analysis of who was reading the *Astronomical Register* and wherefrom. Figure 6.2 shows the location of subscribers listed in the *Register* during the first two years of publication. In total 133 subscribers are identified during the first year of publication, with a further 54 being added in the second year. During the first two years (January, 1863 to January, 1865) of its production, 4 subscriptions were attached to businesses and institutions of learning (The London Institution, The Royal Institution, The Nottingham Mechanical Institute, and Negretti and Zambra),[3] and 5 were attached to overseas subscribers (from Canada, France, Italy, Uruguay, and South Africa). Of the 187 personal subscriptions 139 (74.3%) are listed under the title of esquire, 32 (17.1%) were identified as clergy (through the title Reverend), and 5 (2.7%) are identified as women.[4] The remaining 11 (5.9%) subscribers are identified under various military titles (Admiral, Captain, Major and Colonel), as a Lord, an Archdecon, and one Professor (this being Charles Piazzi Smyththen Astronomer Royal for Scotland). Of the 187 subscribers 46 (24.6%) identified themselves as being Fellows of the Royal Astronomical Society (FRAS), 5 (2.7%) as being Fellows of the Royal Society, and 2 (1.1%) as being Fellows of the Royal Geographical Society. The majority of subscribers, appear, therfore, to be located outside of any formal scientific society. In terms of physical location, 33.3% had addresses listed in the greater London region (NB. not all London address are indicated in Fig. 6.2). Accordingly, some 64% of subscribers lived outside of the capital region, with 7% of these subscribers being located in the Liverpool/Manchester region. About 5% of subscribers were located in or near Nottingham, with a further 3% being located in

---

[3] Negretti and Zambra was a London-based optical and scientific instrument makers that operated from 1850 to c.2000. The company ran a studio that specialized in travel and areal (that is, by balloon) photography. Other companies and artisans eventually took-out subscriptions, including: J. H. Dallmeyer, John Browning, T. Cooke and Sons, J. Slugg, Yeates and Son, George Calver, and Horne and Thornthwaite.

[4] These pioneers were Mrs. Hannah nèe Gwilt Jackson, Madame La Comtesse Baldelli, Mrs. Weldon, Mrs. Janet Taylor, and Miss W. R. Hall. While women certainly contributed to the *Register's* pages, they were not eligible for election to RAS Fellowship until 1915.

the vercinity of Birmingham. As can be seen in Fig. 6.2, the majority of subscribers are located along a diagonal that runs from Brighton on the southeast coast of England through to Liverpool in the north west. Accordingly, it appears that the reach of the *Astronomical Register* extended well beyond that of the nations capital, and this informs us that substantive pockets of active enthusiasts existed throught much of provincial England, including Scotland and Ireland, in the early to mid 1860s. Additionally, a good fraction of the initial subscribers to the *Register* were drawn from amongst the ranks of professional astronomers; Gorton, writing in the January 1865 issue (number 25), noted that the subscriber's list contans, "many of the most distinguished names known to the scientific world". These included Warren De La Rue (at the time President of the RAS), Willam Huggins, John (and later Alexander) Herschel, J. Norman Lockyer, James Glaisher, and Piazzi Smyth. Indeed, it seems likely that many of the these astronomers, and FRAS subscribers to the *Register* would have been signed-up by Gorton at RAS meetings. In later years other well-known grand-amateurs [3] and professional astronomers, along with science popularisors, subscribed to the *Register*—including, James Nasmyth, William Lassell, James Glaisher, Richard Proctor, the Reverend C. Pritchard (President of the RAS: 1866-68) and Reverend T. Webb (author of the classic text, *Celestial Objects for Common Telescopes*—published in 1859).

6.3   The Rise of the Amateur

When the *Astronomical Register* suffered its sudden closure, in December 1886, amateur astronomy was on a very different footing to that at its first appearance in 1863. The *Register* had indeed, begun the process of promoting a sense of a national community among amateur astronomers, and this was further strengthend by the formation of the British Astronomical Association (BAA) in 1890. This society, still thriving to this day, began publishing its own journal, along very similar lines to those of the *Astronomical Register*, in October 1890 under the editorship of Edward Walter Maunder. The BAA itself can arguable be seen as a more structured (and more successful) outgrowth of the short lived Observing Astronomical Society (OAS), which flourished just briefly from 1869 to 1873 (see Chap. 7). The origins of the OAS are clearly an outgrowth of the *Register*, and as early as issue number 28 (April 1865), a letter form "G. J. W" suggests, "it would be a good thing if the many persons in the country provided with instruments would agree to unite in some work". Tapping into the Victorian ideal of application, advancement, and the imposition of order, G. J. W. was in effect

Fig. 6.2 *Astronomical Register* subscriber location map for January 1863 to January 1865

espousing the idea of citizen science. This same ideal was echoed by the RAS, and accordingly a notice "to amateur observers" was printed in issue number 35 (November, 1865) of the *Register*. The notice was an invitation from the President of the RAS, "to the amateur astronomers of Great Britain and Ireland, to send a short account of their means of observation". The call was to amateurs with "fixed observatories", and the aim was to determine what, "methodoical arrangement may be adopted by those of our readers who are practical observers". Subscribers to the *Register* rallied to the call, and issue number 39 (March, 1866) carried an analysis of the information provided by 48 observatories (these data will be discussed in Sect. 6.4.3). Clearly, it would appear, there was some considerable interest among well-equiped amateur astronomers in establishing a net-work of communication, and in establishing group observational programs. To this end, one of the earliest projects involving only amateur astronomers is that outlined in a letter to the *Register* by Denning (dated May 9, 1868, in Bristol). This project, by "five gentlmen", concerned the continuous monitoring of the Sun's disk, from March 7th to April 21st, in hopes of detecting the supposed planet Vulcan. Denning only mentions three of the observers by name in his letter, with these observers having "powerful telescopes"; two of these observers (James Cook, of Preston, and T.P. Barkas, of Newcastle-on-Tyne) are known to have taken out subscriptions to the *Register*, and both were later included in the list of OAS founding members—Denning, himself, becoming the society's treasurer and secretary. I would suggest that the link that initially brought these geographically dispersed observers together was the *Register* and its practice of printing subscriber addresses. Indeed, Denning closes his letter with the comment that he would be happy to hear from any other observers that might be interested in assiting in future observing campaigns. Issue number 79 of the *Register* (July, 1869) carried a formal announcement of the establishment of the OAS, with, "the purpose of securing concerted observations of interesting astronomical phenomena, and recoding the results obtained". Like the *Register's* subscribers, the initial members [4] of the OAS were widely dispersed, with observers being found within the quadrilateral running from Brighton in Sussex, to Aberdeen in Scotland, to Millbrook in Ireland, and Truro in Cornwall (indeed, a distribution very similar to that seen in Fig. 6.2). The majority of these initial members were located in the midlands (Birmingham, Staffordshire, Derby) and industrial northern England (Manchester, Liverpool, York, Sunderland). Of the 35 initial OAS members only 4 (1%) had address that locate them in London. The OAS was very much a movement of provincial amateaur astronomers.

The initial enthusaism of OAS members seemingly ran high, and over the next several years Denning provided regular reports to the *Register* on OAS work. Multiple search campaings for Vulcan were run, an extensive campaign to view the disk of Venus, at all its phases, was carried out, coordinated meteor studies were conducted, and telescopic surveys of the Moon's surface were organized in the hope of identifing new or changing features. It appears, however, that the OAS membership could not maintain its initial momentum, and while there was no official announcement of its demise, this effectively came about with the close of 1873. At least, at this time no further reports were published in the *Register*. In some sense, the OAS dissolved into the individual carriers of its more prominent founding members. Denning, for example, went on to garner international renown as a sweeper of comets and observer of meteor showers, while others, such as David Gill, moved into professional ranks, with yet others taking-on prominent roles in the BAA when it eventually formed in 1890.

Not only was amateur astronomy in a state of ascendency in the late 1860's, so too was the business of science communication. *The English Mechanic*, for example, first saw publication in 1865, and this magazine regularly carried astronomy related articles. Likewise the journal *Nature*, founded and edited by Joseph Norman Lockyer, first appeared in 1869, and this journal had a distinct astronomical bias. Furthermore, the *Observatory* magazine, founded in 1877, and edited by William Cristie (then Chief of the Royal Observatory Greenwich), specifically took-on the task of reporting the news and discussions arising during RAS meetings. Although not officially tied the RAS, the *Observatory* has been edited by many prominent RAS members, and along with original articles and book reviews, it still reports on RAS meetings to this day.

6.4 For Sale and Wanted

It is reasonably clear from the subscribers lists, as well as its letters section, that the *Register* catered to a diverse audience of astronomers, with the readership spanning the spectrum from the meanest tyro to that of the first-rank professional. The editorial tone, however, was clearly directed towards the hobbyest amateur. These general readers were not assumed to be independently wealthy, but rather enthusiasts working with limited funds, under modest observatory conditions, and against competing time constraints. The question, therefore, becomes, what sort of instruments were these enthusiasts using, and who was manufacturing and supplying them. In an attempt to answer these questions, a review of the 'for sale and wanted' advertisments placed by subscribers within the *Register*, as well as those advertisements taken-out by manufacturers and artisans, has been made. The data from this review, it is argued below, provides us a glimpse into the world, wants, aims, influences, and desires of the mid to late Victorian amateur astronomer.

6.4.1   Subscriber Advertisments

A study of all 24 volumes of the *Register* reveales a total of 138 items for sale, and 44 items wanted advertisements being placed within its pages. Figure 6.3 shows the cumulative sum of these advertiesments from issue 1 to issue 288. The manner in which each data set plots in Fig. 6.3 tells an interesting story. Clearly there were more adverts concerning items for sale than advertisments for items wanted, and the curves indicate that there were numerous intervals, some many years in duration, when no subscriber advertisements were carried in the *Register* at all. The initial policy with respect to placing an advertisement in the *Register* was that there was no cost incurred provided that one was a subscriber, and that the notice was no longer than 3 lines in length (amounting to about 30 words or less). This generous insertion policy probably accounts for the rapid increase in the cumulative number of subscriber advertisments during the first six years of publication. The response procedure for advertisements was that the name, address, and price wanted for the articles on sale were obtained through the Editor, the practice of actually including contact information for the seller and prices wanted only began much later, and after advertisement charges were introduced (point A in Fig. 6.3). It seems likely that the added burden placed upon Gorton, in dealing with each advertiser, and any potential buyers, was taken-on in order to gauge readership interests, drum-up potential subscribers, and to gauge the reach of the *Register*. Indeed, the time spent must have been quite onerous, since there were many pleas to let the Editor know about successful sales in order that notices (and associated correspondence) could be closed. From issue number 8 (October 1863), and in a sense still feeling his way forward, Gorton introduced a new policy (seemingly in response to readers letters), that more detailed advertisments, up to 72 words long, but not exceeding six lines of text, could be placed for the price of one shilling. Longer advertisments were possible, however, at a cost of 2-pence per 12 words. Issue number 72 (December, 1868) included yet another editorial change, indicating that all personal advertisements would now carry an insertian charge (point A in Fig. 6.3). Acknowledging that there was a large number of notices to deal with, Gorton explained that the charge for an advertisement would be, "one shilling per quarter, for twenty words or under". The funds would be payable in advance and must also be accompanied by a quarterly subscription to the *Register* (at that time still set at three shillings per quarter). One suspects that by this stage Gorton was trying to minimize his involvement in the exchange of letters concerning items for sale, and attempting to juggle the problem of limited space against a need to generate income from the *Register*. The response to these new insertion charges for personal advertisments was rapid, and in terms of revenue generation counter productive. Indeed, not a single personal advertisments was carried in the *Register* for the next 28 issues (a time interval of 2 years and 4 months). The first plateau in Fig. 6.3 corresponds to the slump induced by the introduction of advertising charges. It was in issue number 97 (January, 1871), however, that Gorton announced that the *Register* had starting to pay its way—mainly through subscription revenue (point B in Fig. 6.3). From issue number 90 to issue number 111 the number of personal advertisments began to climb again, although from issue 111 (March, 1872), the number of advertisments dropped to zero, this time for 73 issues—a time interval of 6 years and 1 month. The editorial change in 1872 (following Gorton's death—point C in Fig. 6.3) made for little variation in advertising policy, although there was a modest increase in personal advertisments placed in the time interval covering issues 227 to 250. There were no personal advertisments placed in any of the last 48 issues (265 to 289) of the *Register*.

The items that were listed for sale, or placed as wanted, range from the mundain to the quite marvelous. The majority of advertisements were simple and to the point, the very first advertisement (in issue number 1) being fairly typical and reading: "Astronomical telescope, 8 feet focal length, 5 1/2 inches

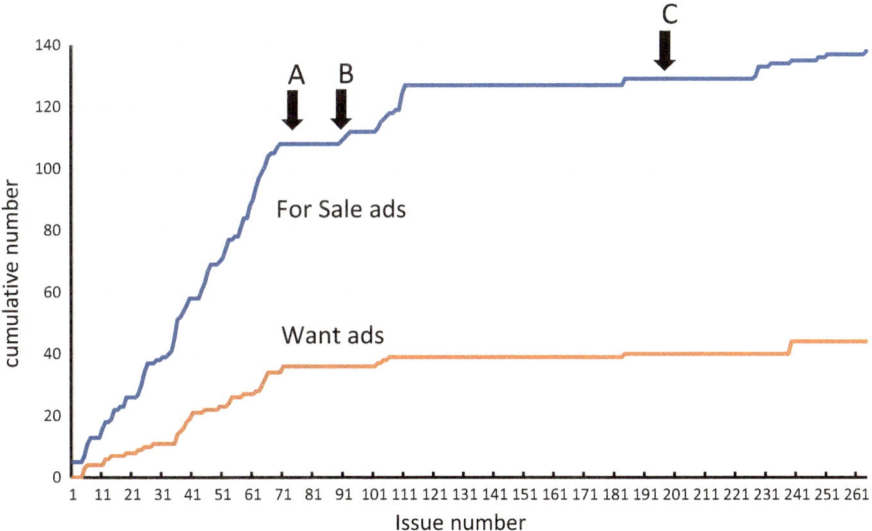

Fig. 6.3 Cumulative sum of advertisements placed by subscribers in the *Register*. The x-axis indicates the issue number, but note that no advertisements were placed in the last 2 volumes (48 issues) and the data plot ends at issue 264; Vol. 22, 1884. Points A, B and C are discussed in the text

aperture, mounted on equatorial stand". Other advertisements offered spectacular objects for sale, including a brass Orrery and Tellurium designed
by Dr. William Pearson (purchased in 1853), and, "the 10-inch equatorial telescope, exhibited in the International Exhibition, London, last year
[1871]". This latter item was made by T Cooke & Sons, of London and
York, and the asking price was £1,100 cash.[5] Yet other advertisements offered
entire observatories (dome, and telescopes) for sale; still others offered clocks,
micrometers, mirror grinding apparatus, and books, star charts and photographic equipment. Along with the personal advertisements there were a
few notices placed by auction houses, and by the executors of estate sales.
Among the auction house offerings was the sale of astronomical instruments from Hartwell Housethis auction following upon the death of Dr.
John Lee.[6] Also included was a notice of the (infamous) auction of observatory items by Sir James South.[7] Estate sales included those of Captain
R. W. H. Hardy, R.N., and Chas J. Corbett F.R.A.S. A good number
of books and manuscripts were offered for sale, including such specialist
works as *Mensurae Micrometricae* by F. G. W. Struve, *Fasciculus Astronomicus*
by Francis Wollaston, and J. W. Lubbock's *Treatise on the Tides*. These
latter texts would likely be of more interest to professional, rather than
amateur astronomers, and this underscores the distinct, twin-audience readership of the *Register*. More typical of amateur interests, however, other
books being offered for sale included: Admiral William Smyth's two-volume,
Cycle of Celestial Objects (first published in 1844), John Hind's *Illustrated
London Astronomy* (first published in 1853), and Edmund Denison's (later
Lord Grimthorpe) *Astronomy Without Mathematics* (first published in 1865).
Volume 2 of Smyth's *Cycle* was particularly sought after within the items-
wanted advertisements (see below). This volume was generally called the
Bedford Catalogue (after the location of Hartwell House, where the obser-
vations were made), and contained Smyth's observations of 1604 double stars
and nebulae.

---

[5] A notice in issue #116 of the *Register* reveals that the telescope was sold at auction, and that, "the
purchaser was Mr. Henley, the telegraph engineer". William Henley pioneered the development of
submarine telegraph cables. The telescope purchase price was £750.

[6] The reverend John Lee constructed a private observatory located at Hartwell House in Buckinghamshire. He was well known for his encouragement and sponsorship of the sciences and was
President of the RAS (1861–1863).

[7] South had a very public row and long legal battle (from 1834 to 1838) with instrument maker
Edward Troughton concerning the construction and commissioning of a new equatorial telescope and
mount. For all this, South was one of the founding members of the Astronomical Society of London
in 1820. He was a renowned observer of double stars, and it was under his Presidency (1829–1831)
that the Astronomical Society received its Royal Charter.

Table 6.1 Analysis of items wanted

Item wanted	Number
Telescope	11
Mountings/stands	6
Sidereal clocks	4
Objective lens/mirrors	2
Spectrum apparatus	1
Celestial globe	1
Wedgwood plaques of astronomers	1
Back issues of the *Register*	3
Back issues of the MNRAS	2
Bedford Catalog[a]	4
Books[b]	7
Lecture illustration instruments	1

Notes [a] The *Bedford Catalog* constituted volume 2 of W. Smyth's *Cycle of Celestial Objects*. [b] Various titles including: Hieronym Schröder's *Selenotopagraphische Fragmente*, Julius Schiller's *Coelum Stellatum Christianum*, William Galbraith's, *Mathematical and Astronomical Tables*, Bartholemew Prescot's *Universe*, and James Bedford's, *The Comet- New Theories of the Universe*

Table 6.1 provides a break-down of want add requirements. The majority (29%), as one might expect, concerned the desire to purchase some specific telescope, with requests for telescope mountings and drives coming in at a relatively distant second place (amounting to 14% of the total). Other items sought after were, sidereal clocks, lenses and mirrors, globes. and observatory equipment. Also among the wanted adds. were a good number of requests for back issues of the *Register*, and the *Monthly Notices of the Royal Astronomical Society*, along with requests for various specialist texts, and astronomy memorabilia.

6.4.2   The Telescopes

Most of the advertisements concerning items for sale simply indicated the type of telescope on offer, refractor or reflector, along with any related eyepieces, filters and mounting (if any). Figure 6.4 provides a plot of aperture size (objective diameter) versus focal length (when given) of those items listed for sale. The majority of instruments on offer were refractors, although a good number of reflecting telescopes were also advertised. These latter telescopes generally being of large aperture—indeed, typically of sizes that would be of interest to semi-professional or grand amateur astronomers. While it does not seem appropriate to perform a detailed statistical analysis, it is clear, by eye, from Fig. 6.4 that the most common telescope aperture was in the 3 to 4-inch diameter range. Telescopes of this size are very much instruments of the amateur; capable of providing details beyond that of the pocket telescope, and allowing for introductory research work (such as, solar studies, Moon observations, double star measurements, scanning for new comets, and planetary disk appearance). Although published after the demise of the *Register*, Captain Willam Noble published a series of articles in the journal *Knoweldge*[8] (later collected in book form and published in 1895), under the title *Hours with a 3-inch Telescope*, picking out this aperture (and a focal length of 42 inches) for its ease of use and gerneral availiablity. Indeed, it is noted in Gorton's obituary that he used a 3 and 3/8th—inch objective refractor (made by Alexander Ross), mounted on a stand built by Thomas Cooke, with a clockdrive by John Dallmeyer. Indeed, it would seem that the telescope advertisments in the *Register*, both for sale and wanted, reflect the desires and requirements of both the serious amateur and the neophyte. For all their ease of use, however, to the serious amateur, a 3 to 4-inch aperture telescope is a transitional instrument, and while such an instrument will enable the observer to learn their trade and hone their observing skills, it is only with larger aperture instruments that genuine research work might really begin. Indeed, the larger telescope apertures indicated in Fig. 6.4, in the 10 to 12-inch diameter range, would only be of interest to the serious amateur or professional observer—that is, those observers already conversant with observing techniques, and involved in some form of organised astronomical work. This very point being emphasised by William F. Denning in his book *Telescopic Work for Starlight Evenings* (first published in 1891), where he comments that 3 to 4 inch aperture telescopes are good for hobby purposes, but for those intent on performing original work, an 8 to 10-inch aperture reflecting telescope is required. Indeed, the majority of Denning's observations and drawings of the disk of Jupiter were made with the aid of a 10-inch aperture With-Browning reflecting telescope. Of the 11 telescope-wanted advertisments placed in the *Register*, 9 specified aperture ranges between 2 and 6-inches, with two looking for 9 and 11-inch aperture telescopes.

The prices being asked for those items advertised for sale (when an asking price was actually mentiond) varied considerably. In the case of refractors with 3 to 4-inch diameter objectives, however, the prices ranged from £12 to £50. These prices typically included several eyepeices and a stand or tripod.

---

[8] Founded by Richard Proctor (an early subscriber to the *Register*) in 1881, *Knowledge* became well known for its extensive correspondence column. Proctor's daughter Mary (1862–1957) assisted in the production of *Knowledge*, and also became a well known and respected lecturer and popularizor of astronomy. She was elected a member of the BAA in 1897, and a Fellow of the RAS in 1916.

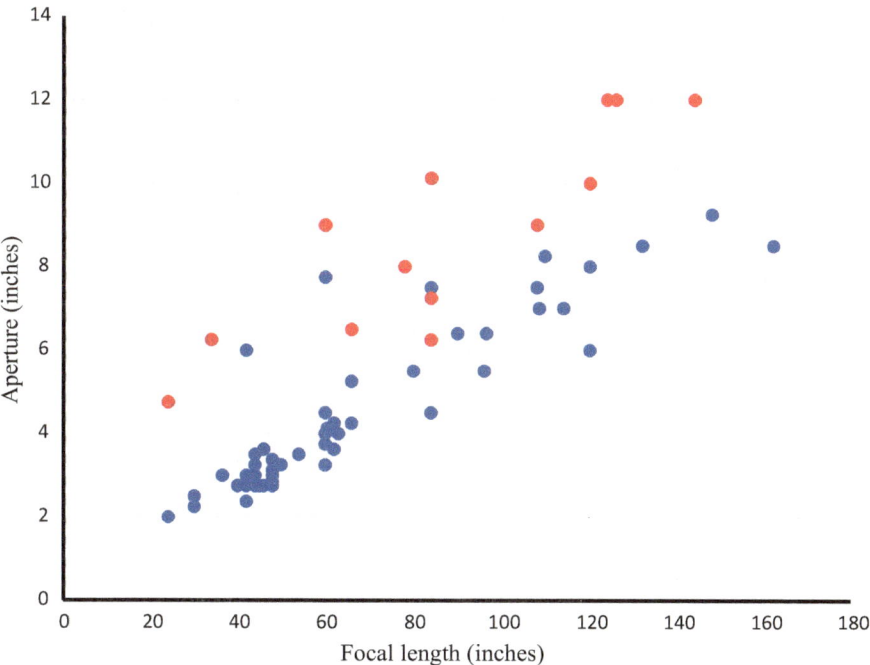

Fig. 6.4 Aperture versus focal length of telescopes listed as being for sale in the *Astronomical Register*. Blue dots indicate a refracting telescope, while red dots correspond to reflectors

In terms of the best deals spotted, one advertiser, offered a 2 ¾-inch aperture refractor, with box, stand and (terrestrial) eyepeice, by Samuel and Benjamin Solomons, for the bargain price of just 4 Guineas. These second-hand prices can be gauged against the cost of a new instrument (without stand), as advertised in the *Register*, by Josiah Slugg of Manchester (Fig. 6.5). Only one reflector had an asking price set against it—this being some £12 for a 7½-inch aperture telescope, including eypieces and mount. Moving to larger apertures, the prices expected increased dramatically; one notice asking 120 Guineas for a 5½-inch aperture refractor (by Thomas Cooke & Sons), another asking £250 for a 5¼-inch refractor (lens by John Dolland), with eyepieces and a stand. Yet another notice asked £250 for an 8-inch objective (lens by William Wray) telescope, including eyepieces, a micrometer and a mount. The sale of instruments from Hartwell House realised a price of £100 for a 4½-inch aperture Cooke & Sons refractor with eyepieces and equatorial stand (this same instrument, which "cost upwards of £300" being exhibited in the Great Exhibiton in 1862). Also at the Hartwell House sale, a 5-inch aperture (unmounted) reflecting telescope by William Wray, with 4 eye-pieces, realised a price of £16 10s. It is not known what Denning payed for his 10-inch With-Browning telescope, but a 10¼—inch, 9-foot speculum telescope, with an equatorial stand, was advertised by John Browning in his 1876 catalog for £181—such a price-tag would have been a considerable outlay of money for Denning considering his generally straightened circumstances.

The annual income of a manual laborer would have been about £50 per annum in the late 1800s, while that of a professional worker (clerk or civil servant) would have amounted to a few hundred pounds per year. To these practitioners, the price of even a small 3 to 4-inch, second-hand telescope (or even a new telescope by Josiah Slugg—Fig. 6.5) with eyepieces and a stand would have required a substantial part of a year's wages to purchase. New or even second-hand telescope ensembles were priced well beyond the financial reach of a laboreer. The expense of buying a finished telescope could be off-set, of course, by home-construction, and towards this market, several manufactures of lenses and speculum mirrors (Fig. 6.6) advertised their wares in the *Register*. Amongst the items wanted advertisements, one subscriber was looking to purchase a, "Metallic speculum, unground, 12 inches in diameter", indicating that in at least one case the desire was to home-construct an impressively-large telescope. Indeed, towards the close of the 19th century, some remarkable instruments were built by what have appropriately been labeled as "working-class scientists" [3]—a new breed of motivated, and skilled amateurs looking (in spite of straitened circumstances) to equip themselves with telescopes having apertures beyond the reach of many professional

SLUGG'S ASTRONOMICAL TELESCOPES,
AND PARALLACTIC STANDS.

The reputation which these telescopes have now gained renders it neddless to say that their cheapness is not the offspring of inferiority. The Superiority of their Definition may be relied upon, and will give satisfaction.

PRICES OF TELESCOPES, WITHOUT STANDS.

2 inches Diameter,	36in. focus		......£2	10s. 0d.
2¼ ,,	,,	36 ,,	 3	10 0
2⅞ ,,	,,	48 ,,	 5	0 0
3¾ ,,	,,	60 ,,	12	10 0
4¼ ,,	,,	70 ,,	25	0 0
6¼ ,,	,,	90 ,,	85	0 0

The Object-glasses are all of the best quality, and may be had without the tubes.—Full particulars on receipt of one stamp.

JOSIAH T. SLUGG,
214, STRETFORD ROAD, MANCHESTER.

Fig. 6.5. Advertisement placed by J. T. Slugg in issue 12 the *Astronomical Register* (December 1863). Notice the assurance by the advertiser that the low prices are not an indication of inferior workmanship

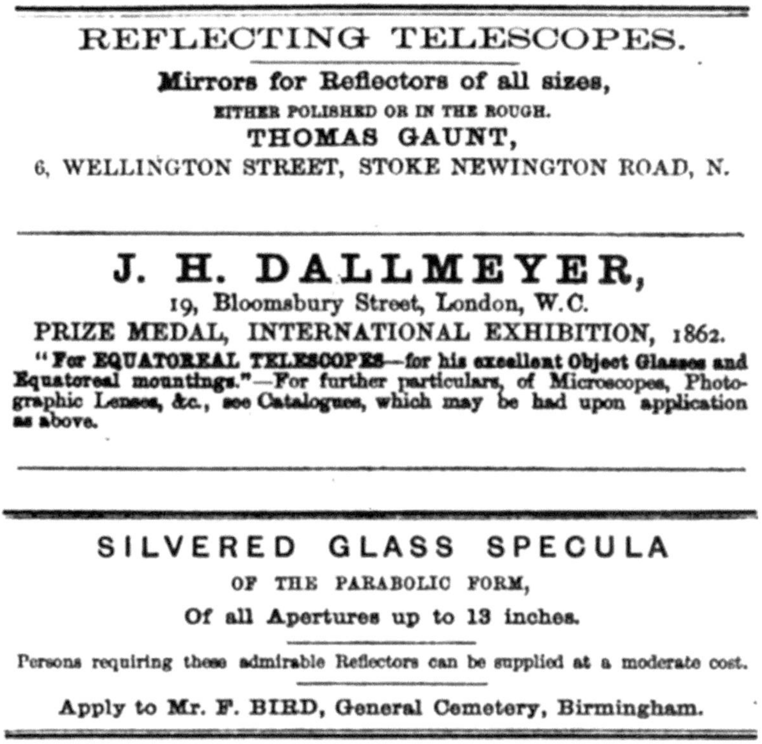

REFLECTING TELESCOPES.

Mirrors for Reflectors of all sizes,

EITHER POLISHED OR IN THE ROUGH.

THOMAS GAUNT,

6, WELLINGTON STREET, STOKE NEWINGTON ROAD, N.

J. H. DALLMEYER,

19, Bloomsbury Street, London, W.C.

PRIZE MEDAL, INTERNATIONAL EXHIBITION, 1862.

"For EQUATOREAL TELESCOPES—for his excellent Object Glasses and Equatoreal mountings."—For further particulars, of Microscopes, Photographic Lenses, &c., see Catalogues, which may be had upon application as above.

SILVERED GLASS SPECULA

OF THE PARABOLIC FORM,

Of all Apertures up to 13 inches.

Persons requiring these admirable Reflectors can be supplied at a moderate cost.

Apply to Mr. F. BIRD, General Cemetery, Birmingham.

Fig. 6.6 Advertisments placed (top) by Thomas Gaunt, (middle) John Dallmeyer and (bottom) F. Bird in issue 12 the *Astronomical Register* (December 1863). Gaunt's advertisement is very straighforward, while Dallmeyer emphasises his International Exhibition prize medal. The advertment by Bird indicates that only moderate costs for his "admirable reflectors" are charged

astronomers. One such example of this is described in the August 1870 issue of the *Register*, and concerns an exhibit by Mr. T. W. Bush at the Workmen's International Exhibition, held in the Agricultural Hall, Islington. A baker by trade, Mr. Bush, from Nottingham, exhibited a hand-built 13-inch aperture, silvered glass reflector of 10-foot focal length, with one reviewer noting that, "the mounting on the whole is simple and massive, and looks as if intended for work".

6.4.3 Commercial Advertisers

Of the 138 personal items-for-sale advertisements placed in the *Register*, 71 (51% of the total) indicate the manufactuer of the instrument or the components on offer—a further 36 gave no spcific manufacture information.

Table 6.2 indicates the specific companies and artisans that were referenced. Among the companies and individual artisans listed in Table 6.2, we find the usual suspects, but also a few surprises. The top four optician/telescope makers mentioned are Thomas Cooke and Sons, of London and York, Edward Troughton and William Simms, of Chalton, S. E. London, John Dolland, of London, and Alvan Clark, of Massachusetts, USA.

Table 6.2 indicates that most amateur astronomers (with an instrument to sell) had purchased their telescopes from one of two British suppliers: Thomas Cooke and Sons, and Troughton and Simms. After these two, and in a distant 3rd place, was American telescope maker Alvan Clark. The remaining telescopes listed in Table 6.2 were constructed by some 10 additional manufacturers, some well known and renowned (such as William Wray, and George With), with others being more obscure. In terms of objective lenses and eyepieces, those manufactured by George Dolland and Andrew Ross dominated. Interestingly, in terms of trade networks, the instruments and articles being offered for sale in the *Register* were drawn from a wide geographical reticulum, ranging from manufacturers situated throughout the United Kingdom, to manufacturers in Germany, Holland, and the United States of America.

A summary of the information received by Gorton at the *Register*, in response to the RAS initivie to catalogue amateur observatories, was printed in its November 1865 issue. In all there were 48 respondents describing observatories with telescopes with apertures ranging in size from just 2-inches, to 14½-inches. Spread amongst 41 refractors and 7 reflectors, the most common aperture size was 4¼-inches. Only 14 of the observatories contained telescopes driven by clockwork, and just 19 observatories were equiped with a sidereal running clock. With respect to refracting telescopes, the top six manufactures (of the 17 listed) 12 (25% of the total) were attributed to Thomas Cooke and Sons, with 3 apiece being attributed to Alvan Clark, Troughton and Simms, James T. Goddard, Josiah Slugg, and Thomas Slater. This ordering, with Thomas Cooke and Sons dominating, is almost identical to that found in Table 6.2. Amongst the reflecting telescopes 2 were attributed to George With, with 1 each being attributed to F. Bird (see Fig. 6.5), the Rev. Henry Key, and (?) Lawton.[9] In terms of the pure number of telescopes being brought, used, and then potentially sold in the *Register*, it would appear that the British amateur market was dominated by instruments constructed by Thomas Cooke and Sons.

---

[9] I have not been able to trace a telescope maker / optician under the name Lawton. It may be a typographic error, however, with the maker intended being astronomer and optician Henry Lawson.

Table 6.2 Trade names mentioned with respect to personal advertisements placed in the *Register*

Company	Comments	Location	No.
Cooke & Sons	Telescopes/mountings	London and York	20
Troughton and Simms	Telescopes/micrometer/ mountings	London	10
Dolland	Eyepieces/speculum mirrors	London	7
Alvan Clark	Objective lens/telescopes	USA	3
Ross	Objective lens/eyepieces	London	2
With	Speculum mirrors	Hereford	2
Slugg	Telescopes	Manchester	2
Horne and Thornthwate	Telescope/photographic apparatus	London	2
Browning	Micrometer/spectroscope	London	2
Slater	Telescope/mounting	London	2
Merz	Telescopes	Holland	2
Negretti and Zambra[a]	Barometers	London	2
Wray	Telescopes	London	2
S & B. Solomons	Telescope	London	1
Jones	Telescope	London	1
(E. J.) Dent[b]	Dipleidoscope[c]	London	1
Pistor and Martins	Telescope	Germany	1
Moreland	Setting circles	?—?	1
Hague	Transit instrument	Bath	1
Frauenhofer	Objective lens	Germany	1
Tulley	Telescope	London	1
Goddard	Telescope	London	1
Cary	24-inch Globes	London	1
Berge	Micrometer	London	1
Silver & Co.	Electrical machine[d]	London	1
Taylor	Mounting	Birmingham	1
Horstmann	Astronomical clock	Bath	1

Column 1 indicates the company or artisan mentioned, column 2 describes the items that they are tyically associated with. Column 3 indicates the location and column 4 shows the number of advertisements mentioning the indicated company or artisan
Notes [a] Negretti and Zambra were listed among the initial subscribers to the *Register*. [b] Edward Dent was a renowned marine chronometer and watchmaker. [c] A dipleidoscope is a prism instrument used to determine the time of true noon. [d] This advertisement concerned a rotating, 12-inch glass plate for the generation of static electricity. Silver and Company was founded by S. W. Silver in 1864, and was largely involved in the production of submarine cables for telegraphy

The business of T. Cooke and Sons was founded by Thomas Cooke in 1837. A self taught optician and instrument maker, Thomas Cooke became well known for the quality of his work, and he eventually opened his own factory in Bishophill, York in 1855; a London office was opened in 1863. It was at the Buckingham Works in Bishophill that Cooke applied innovative assembly-line principles to the production of optical instruments. Being situated outside of London, where the instrument trade had traditionally functioned from within the confines of small specialist workshops, Cooke was able to produce quality telescopes relatively quickly, and in a price range accessible to the amateur astronomer. Indeed, producing a broad range of instruments T. Cooke and Sons developed a strong national and international reputation, and went on to supply many observatories around the world with high-quality telescopes, mounts and drive mechanisms. Thomas died in 1868, and a short obituary notice by Gorton appeared in the November (number 71) issue of the *Register*, where it was noted that, "his success may be judged of the number of instruments bearing his name which are to be found in the hands of observers of every class". To this is added, "we owe it very much that astronomers have been able to supply themselves at home". The company continued to thrive and expand after Thomas's death, but eventually came under the control of Vickers engineering company in 1915. In 1922, T. Cooke and Sons merged with Troughton and Simms. The partnership of (Edward) Troughton and (William) Simms formed in 1826, although the Troughton family had been involved in instrument making since the late 18th century. The company opened a new factory in Charlton (SE. London) in 1860, and from there produced astronomical instruments primarily for the British market.

While the *Astronomical Register* was produced on the basis that it had to pay its own way, it never attracted, or seemingly relied upon, a strong revenue stream from commercial advertisers (we do note, however, that many companies did take out subscriptions). Table 6.3 provides a list of those companies and artisans that did, at one time or another, advertise within the pages of the *Register*. The cost of advertising was outlined in issue 29 (May, 1865) of the *Register*, where it was indicated that a one page (no more than 40 lines) advertisement could be had for 15-shillings. Half, quarter and one-sixth page advertisments could be inserted for 8, 4 and 3-shillings respectively. Most advertisers (recall Fig. 6.6, and see Fig. 6.7) adopted the one-sixth page option. The majority of advertisers were located in London, although advertisers from Manchester, Birmingham, and Dublin were also present. Interestingly, while Cooke and Sons regularly advertised (Fig. 6.7) in the *Register*, Troughton and Simms never advertised within its pages.

Table 6.3 Companies and individual artisans that took-out payed advertising in the *Astronomical Register*

Company	Description	Location
T. Cooks & Sons	Opticians, telescope, mountings, mathematical, and meteorological instrument makers	York and London
Smith, Beck and Beck	Microscopes, telescopes, scientific instruments	London
F. Horne and W. Thornthwaite	Telescopes, microscopes and meteorological instruments	London
J. Browning	Reflecting Telescopes	London
T & H. Doublet	Manufacturing opticians	London
G. Yeates & Son	Telescope and mount	Dublin
T. Gaunt	Mirrors for reflectors	London
J. H. Dallmeyer	Lenses and telescopes	London
J. T. Slugg	Astronomical telescopes and parallactic stands	Manchester
L. P. Casella	Optical and scientific instrument maker	London
E. G. Wood	Optician and manufacturer of philosophical apparatus	London
F. Bird	Speculum mirrors	Birmingham
A. J. Frost	Transit instruments	London

In addition to telescopes and optical aparatus, the *Register* also contained advertisments from various publishing houses and book sellers. These are listed in Table 6.4. Longmans, Green and Co. regularly placed advertisements in the *Register*, and so too did Henry Frowde, of the Clarendon Press Warehouse, London. It is perhaps a little ironic that the only time that the *Observatory* magazine advertised in the *Register* was in the latters final issue in 1886.

6.4.4   The Liverpool Astronomical Society

With the demise of the Observing Astronomical Society (OAS) in 1873 (see Chap. 7), the British amateur astronomer was left with no national association or society to encourage correspondence, companionship, and/or direct observing programs. This absence of organization was to remain in place until the early 1880s. The first inklings that amateur astronomers were, once again, interested in seeing a new national society emerge were expressed in the letters section of the *English Mechanic and World of Science* (EMWS) in early 1880. Writing in the 23 January, 1880 issue of the EMWS, and almost lost amidst a discussion concerning the observation and discovery of

T. COOKE & SONS,
OPTICIANS,
MATHEMATICAL AND METEOROLOGICAL
Instrument Makers,
31, SOUTHAMPTON STREET, STRAND,
LONDON, W. C.

Manufactory—Buckingham Works, York.

YEATES & SON'S
EQUATOREALLY MOUNTED TELESCOPE,
For Students and Amateur Astronomers.

This Instrument is particularly adapted for those who have no convenient place to erect an Observatory, as it possesses portability and great steadiness, combined with perfect freedom of motion in all directions.

YEATES & SON,
2, GRAFTON STREET, DUBLIN.
Opticians and Mathematical Instrument Makers to the University.

Fig. 6.7 Advertisements placed by Thomas Cooke & Sons, and George Yeates and Son within the *Astronomical Register*. While Cooke and Sons promote the large range of their scientific instruments, Yeates and Son indicate that they are catering to the amateur astronomer without acces to an observatory

Table 6.4 Publishers and book sellers that advertised within the *Astronomical Register*

Publisher/magazine	Items	Location
Longmans, Green, & Co.	Astronomy books	London
A. Brothers	Star charts/moon photographs	Manchester
J. D. Potter[a]	Star charts/astronomy books	London
William Wesley	*Telescopic Pictures of the Moon*, painted by Henry Harrison	London
Henry Frowde, Clarendon Press	Astronomy books	London
The Observatory	Magazine (launched 1877)	London
The Illustrated Science Monthly	Magazine (launched 1884)	London

Notes [a] J. D. Potter was the publisher of the *Register*

"red stars", the reverand Thomas Espin, at the Wallasey Rectory in Birkenhead, commented that it was a shame that, in spite of there being many well-equipped observers, there was no organization of effort amongst amateur astronomers. Indeed, he writes, "I have a dim recollection of hearing of an

Amateur Astronomical Society years ago, but I don't know what became of it. Probably it came to grief through want of members". Espin, no doubt, was dimly recalling the OAS, but goes on to remark that, "now every year fresh recruits enter the field, the number of telescopes increases vastly, and I think there would be no lack of support now". Espin's letter brought an immediate response from W. Goodacre, who in the 30 Janaury, 1880 issue of the EMWS noted that Espin had obliged the magazine's readers in calling attention, "to a want which has no doubt been long felt by the ever increasing number of amateur astronomers, whose willing energies are wasted for the want of some proper organization". Goodacre concludes his letter with the comment, "what has become of the Astronomical Observing Society, the record of whose observing were communicated by Mr. W. F. Denning, F.R.A.S.? Would it be possible to revive the same?". Over ensuing weeks, more letters followed. Mr. F. C. Dennett, writing in the 30 Janaury, 1880 issue of the EMWS commented that a society was very much needed, "the isolation so much felt now amongst a great number of those possessing instruments of comparatively small aperture has often blighted a would-be worker". What was needed, Dennett suggested, was a society that would form "a link" between observers—his final remarks being, "will some gentleman of influence take up the task?". A letter from *N'importe* in the 6 February, 1880 issue of the EMWS, further noted that, while he was a long-time observer, he regretted the want of some formal directions on what to observer, and when. To this end it was sugested that *FRAS* (the pseudonym of Captain William Noble), or the Reverend Thomas Webb, or (American astronomer) S. W. Burnham might consider providing information to assist amateurs in their work. The 13 February, 1880 issue of the EMWS saw two further letters on the topic. Under the sudonym of *Plumb-Bob*, one writer suggested that, "could not a new society be founded …. [under the name of] The Junior Astronomical Society, or … The Corresponding Astronomical Society". To this is added the comment that the society would circulate information among its members on what had been observed, and what might usefully be observed. The second writer, *A Sleeping Fellow of the R.A.S*, echoed the want of direction with respect to their observing practices, and called for the organization of a metropolitan society (that is a national, London-based society) to which provincial astronomers could send in their reports. It is known that Denning read (and contributed to) the EMWS, but he was not drawn into the debate concerning the formation of a new national society. In the wake of all the commentaries presented in the EMWS, however, a new and influential amateur society did eventually appear.

The Liverpool Astronomical Society (LAS) was established in 1881. The Society's first President was Richard Johnson FRAS, and the first Secretary was William H. Davies. The "laws" of The Society's activities were originally described in a series of Abstracts of Proceedings, starting in 1882, with the first issue of the formally titled Journal of the LAS (JLAS) appearing in 1883. the laws of the LAS were adopted on 21 July, 1883, and the objectives were outlined as being, "to foster a liking for astronomy, and to encourage astronomical observation, especially amongst the possessors of small telescopes". Furthermore, as articulated by Davies, in a letter written to the *Observatory* magazine, and published in its September 1883 issue, it was clear that there was a need for a, "kind of halfway resting-place between the amateur public and the Royal Astronomical Society". In this manner, while the LAS was founded by local enthusiasts, it soon became a society with national, and even international membership. Importantly (being the first such society in England) the LAS allowed, on an equal footing, female membership. Prominent among the latter members were historian Agnes Clerke, and solar observer Elizabeth Brown.

Regular meetings were held in Liverpool, and observing sections relating to specialist interests were also established. While not being drawn into the debate over the formation of a new society, Denning was, none-the-less, a major player in the Society's early years, being a regular contributor to its journal (see below), and acting as the director to its Meteor Section. Indeed, Denning was elected an honary member of the LAS in 1882, worked as a journal editor in 1886, and served as its President from 1887 to 1888. For all this evolvement, however, it appears that Denning never actually attended a meeting in Liverpool.

The LAS membership numbers grew rapidly. LAS Secretary William Davies indicated that in 1883 the Society had about 70 members, but this had increased to over 400 members by 1886. Indeed, in the latter year, the then President, T. G. E. Elger (see Chap. 7), enthused about the national character of the LAS, and he called for the establishment of branches outside of Liverpool. The call was presient, but problematic, and by 1890 the Society was in both organizational disarray and financial debt. A crisis was precipitated at this time by the near simultaneous resignation of its Treasurer (William Davies), the sudden death of its President (the Reverend S. J. Perry[10]), and a Secretary (W. Ellison) that was incapacitated by illness. These dramatic events resulted in the near demise of the Society, but it began to

---

[10] Stephen Joseph Perry was a Fellow of the Royal Society, and taught physics and mathematics at Stonyhurst College. His death occurred during an expedition to observe the 22 December 1889 solar eclipse in French Guiana.

regroup over ensuing years, and it is still operational to this very day. Perhaps what saved the LAS from total collapse, however, was the emergence and founding of the British Astronomical Association in October of 1890. This Association, located in London, effectively took-on the role of a national coordinating forum for amateur astronomers.

One of the most important actives established by the LAS was the publication of a journal (JLAS). This publication helped to establish a community amongst its geographically separated members, and it satisfied the Society's aim of promoting an interest in astronomy. During the first several years the JLAS was financed through membership subscriptions and donations to the Society, but after October 1884, the Journal began to carry trade advertisements, and also publish for-sale and wanted notices. Since the LAS was a society formed by amateur astronomers, and because its Journal catered directly to amateur interests, it is interesting to ask who was advertising in its pages, and how, if at all, the advertisements in the JLAS differed from those appearing in the *Register*. To this end, a study has been made of the trade advertisments that were placed in the Journal in the interval October 1884 to December 1889 (when the Journal ceased publication). The results of this survey are presented in Table 6.5.

Over the 5 year interval in our study, advertisements from 34 distinct businesses and individuals were found. Immediately, this is a much larger advertising base than that found for the *Register* (indeed, it represents a 70 percent increase in the number of advertisers). Furthermore, the range of products being advertised was larger and much more diverse. In this respect, one finds an advertisement for Mrs. A. James's "Herbal Ointments", and an advertisement for the "Facile" – a dwarf penny-farthing-style safety-bicycle made by Ellis and Co. in London. In terms of location 41 percent of the advertisers were based in London, with 26 being located in Liverpool (indicative of a strong local presence). The remaining advertisers, however, were widely spread across the British Isles, being located in Birmingham, Newcastle-upon-Tyne, Hereford, Ramsgate, Chelmsford, Cirencester, Edinburgh, and Dublin. By far the greater number of advertisers were allied to the publishing trade, with some 44 percent of all advertisers (15 in all) falling under this banner. A diverse range of books were advertised, ranging from astronomical topics, to health care and medicine, maps of the Moon, and cardboard disks for solar observations. LAS Secretary (1885-6) and journal editor, J. W. Appleton advertised his religious text that explored human genius, and Adam and Charles Black of Edinburgh placed advertisements for the 2nd edition of Agnes Clerke's book *A Popular History of Astronomy—* Clerke being one of the more promenent, and widely read, LAS members.

Table 6.5 List of businesses that advertised in the Journal of the Liverpool Astronomical Society between October 1884 and December 1889. Column 1 indicates the business name; column 2 shows the business location, and colum 3 provides brief details of the business

Advertiser	Location	Comments
George Calver	Chelmsford	Telescope/instrument maker
James Wheeler	Liverpool	Optician—photography—lantern slides
J. J. Atkinson	Liverpool	Photographic supplies
W. Watson & Sons	London	Telescopes ("agent ot Mr. Grubb of Dublin")
M. Aronsberg & Co.	Liverpool	Optician and mathematical instrument maker
W. F. Archer & Sons	Liverpool	Telescopes and lantern slides
Hodges, Figgis & Co.	Dublin	Publisher/book seller
Edward Stanford	London	Publisher/book seller
Parry and Hale	Liverpool	Book seller
W. Stevenson	Liverpool	Second hand telescopes
William Wesley	London	Publisher/book seller
Charles Griffen & Co.	London	Publisher
Horne, Thornthwaite and Wood	London	Moon maps
J. W. Appleton, FRAS	Liverpool	Book advertisement: *Genius—what it is, and what it has accomplised.*
Mrs. A. James	London	Herbal ointments
Caplatzi's Science Depot	London	Telescopes, micrometers, and stands
James Blevin	Liverpool	Publisher
Ellis & Co.	London	The "Facile" safety bicycle
Tricycling Journal	London	Edited by Tom Moore
The Camera	London	Edited by T. C. Hepworth
The Observatory	London	Founded by William Christie in 1877
J. Lancaster & Son	Birmigham	Optician, camera supplies
Pitman	London	Publisher—self help, hygien and medicine
E. Brown	Cirencester	Cardboard disks for sun spot positions
Longmans, Green & Co.	London	Publisher/book seller
J. Linscott	Ramsgate	Telescopes–brought tools from G. A. With
G. A. With	Hereford	Silvered glass specula
Adam and Charles Black	Edinburgh	Publisher: advertising *A Popular History of Astronomy* by Agnes Clerke

(continued)

Table 6.5 (continued)

Advertiser	Location	Comments
Parker's	London	Publisher
Renshaw	London	Publisher—medical essays
Mr. I. H. Isaacs	Liverpool	Private language and mathematics lessons
Mawson & Swan	Newcastle	Photographic equipment
W. Wray	London	Optician and telescope maker
W. Bevlin & Son	Liverpool	Printer and publisher

Among the individuals who placed advertisments in the journal were, I. H. Isaacs (one of the editors of the JLAS) who offered coaching in languagers and mathematics, and James Gill (LAS President 1893–1894) who advertised, in 1884, the commensement of astronomy classes, to be held in Liverpool, upon the completion of which one could obtain a Certificate of Competence. The suggestion that such Certificates might be formally awarded by the Society was raised by Mr. F. Woulfe Brenan at the 12 March, 1888 meeting of the LAS, where he suggested, "the society should institute periodical examinations in astronomy". At the same meeting (which was held at the Royal Institution, in London) Denning vocalised his support, noting that, "many persons felt a shade of contempt for the assumed scientific dignitaries of men who strung a lot of letters after their names, but who had to pass no examination whatsoever for proficiency in the science they professed". At the same meeting, the minutes note that the Douglas, Isle of Man, branch of the LAS had formed an educational program. Denning had, in fact, raised the possibility of establishing formal astronomy programs as early as 1883, when in the 8 March issue of *Nature*, he noted, "it seems a thing to be deplored that in this country there are no establishments where astronomy is made a special subject for teaching". In this same article he also bemoaned the lot of the amateur astronomer, commenting, "it is a fact to be regretted that many promising amateurs have had to relinquish, prematurely, all astronomical work on account of circumstances". Indeed, in more dire and expressive tones he continued, "Life is real, life is earnest; the telescope must be neglected for the ploughshare, and the solitary though happy hours of vigil must be given over to Morpheus[11]". No formal moves to establish the granting of Certificates by the Council of the LAS appear to have been made, however, and indeed, by 1889 the society was dealing with the 'real and earnest' fall-out concerning its financial problems.

---

[11] Morpheus being the ancient Greek God of sleep and dreams—not the Captain of the Nebuchadnezzar in the Matrix movie franchise.

Perhaps surprisingly, given the nature and presumed readership of the journal, only 21 percent (that is 7 companies and individuals) specifically, and repeatedly, advertised the sale of astronomical instruments in the JLAS. These were:

- George Calver (London)
- G. A. With (Hereford)
- J. Linscott (Ramsgate)
- W. Wray (London)
- W. Watson and Sons (London)
- W. F. Archer and Sons (Liverpool)
- Caplatzi's Science Depot (London)

None of these businesses placed advertisements in the *Register*. Telescope and mirror makers Calver, With and Wray are well known, while others are more obscure. It is perhaps suprising that Cooke and Sons did not advertise in the JLAS, but it is also the case that they would have been well-known to all amateur astronomers. John Linscott of Ramsgate emphasised in his advertisments that he had purchaced George With's instruments, while With was simply advertising the remainder of his stock. W. Watson and Sons of London (Fig. 6.8) emphasised that they were, "Opticians to Her Majesty's Government", and the sole London agent for telescopes constructed by "Mr. Grubb of Dublin" (this being the famed telescope maker Sir [after 1887] Howard Grubb). Linscott is known to have advertised in the *Observatory* and the *English Mechanic and World of Science*, but little else is known of his activities. William Watson was an optical instrument maker, who established a workshop, in High Holborn, in 1837. Following his death in 1881, however, Watson's sons took over the business, which seems to have thrived, and in 1900 Watson and Sons brought-out the optical instruments business of John Browning. The firm of W. F. Archer was established in 1848, and focussed on the production of optical instruments. In the 1880's after his three sons had joined the business, Archer and Sons began to manufacture oil and lime-light optical lanterns. The company received medals of recommendation from the Liverpool International Exhibition in 1886, and from the Photographic Exibition in 1888. In the 1890s the company began to make and sell various lines of camera equipment and magic lantern projectors. Anthony Caplatzi, and his variously called Science Depot or Emporium, was a mathematical and philosophical instrument maker located near the British Museum in London. His Science Depot was established in 1862 and existed untill the time of his death circa 1900. Perhaps the most memorable anecdote about the

Science Depot is that it was mentioned in *The Island of Doctor Moreau* (first published in 1896), by H. G. Wells. Quizzing the story's interlocuter, Edward Pendrick, Moreau's assistant Montgomery asks, "is Caplatzi still flourishing? What a shop that was!". Since Wells had lodgings in the vicinity of the British Museum, it is presumed that he knew of, and probably even frequented, the Depot.

Five of the advertisers (15 percent) listed in Table 6.5 billed themselves as opticians, and suppliers of photographic equipment and lantern slides—this clearly reflecting the growing interest in photography and in virtual-travel through slide images. Along with the advertisements from book publishers three were from magazines and journals. The *Tricycling Journal* advertised in the JLAS, and so too did *The Camera*. The fomer was edited by Tom Moore, and carried stories on cycling events, and adventures, along with articles on photography, astronomy and music. *The Camera* was edited by Thomas Hepworth, who was a lecturer on photography at Birkbeck Institution. Indeed, Hepworth was a well-known public speaker, and wrote numerous books on photography and on inks. The third magazine to advertise in the

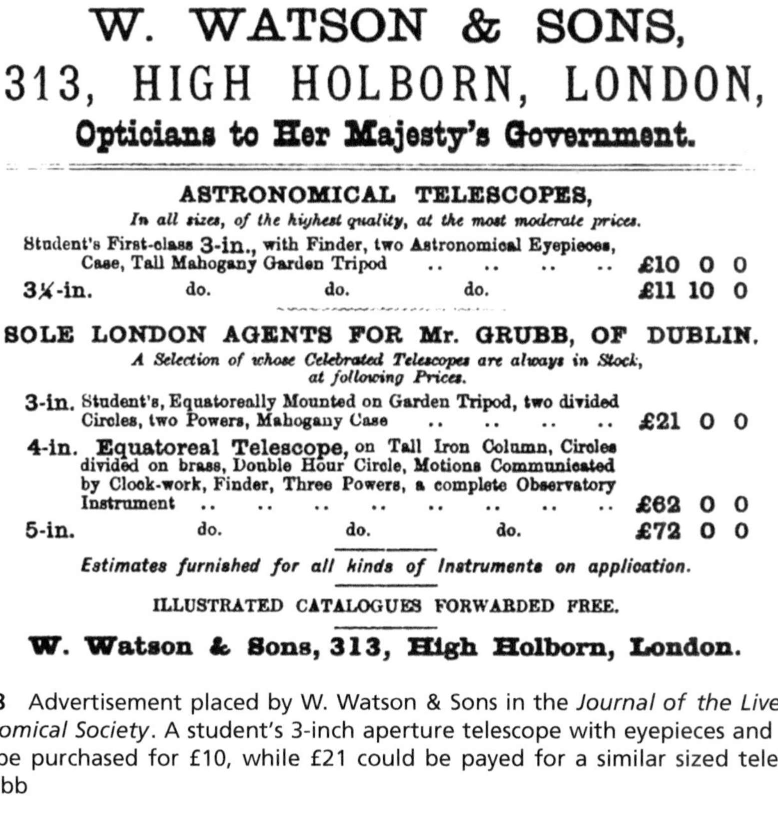

Fig. 6.8 Advertisement placed by W. Watson & Sons in the *Journal of the Liverpool Astronomical Society*. A student's 3-inch aperture telescope with eyepieces and stand could be purchased for £10, while £21 could be payed for a similar sized telescope by Grubb

JLAS was the *Observatory*, with the text of its advertisment indicating that it carried full reports of the "Royal and Liverpool Astronomical Societies". Furthermore, the *Observatory* advertisment indicated that it carried correspondence by numerous eminent astronomers (Denning being one of the authors indicated).

A total of 27 personal for sale notices were identified in our survey of the JLAS. When specifically mentioned these telescopes were indicated as being manufactured by Calver, Cooke and Sons, Browning, With and Wray. One advertisements offered a 16-foot high observatory dome equipped with a 6-inch refractor by renowned instrument maker and optician Charles Tulley—the asking price was £75. An analysis of the telescope apertures indicated in the for sale notices indicates that 48% (13 in all) were in the 4 of 4½-inch aperture range. Some 5 instruments (18.5%) were being offered in the 8 and 8½ inch aperture range, a further 6 instruments were offered in the 1 to 3¼ aperture range. This range in telescope apertures is consistent with the findings from the personal advertisements published in the *Register*, and further supports the idea that most amateurs were using modest-sized telescopes, with the 4-inch aperturte telescope dominanting,

The *Register* and the JLAS catered to essentailly the same audience of amateur astronomers, and they had similar editorial policies. Both journals had to pay there own way, and while the *Register* was austensibly a commercial concern, relying on subscriptions and advertising, the JLAS was subsidized by Society membership fees and donations. It would seem, however, that the editor and original owner of the *Register*, Sandford Gorton, strove to cover costs more through subscriptions than by advertising revinue. In contrast, the JLAS took advantage of an expanded range of advertisers, all catering to the ever-growing population of middle-class, amateur enthusiasts—be their interests located within astronomy, photography, or cycling. Furthermore, while both the *Register* and JLAS drew upon advertising revienue from telescope makers (largely) located in London, the JLAS was able to exploit locally generated (that is from Liverpool) advertising revinue.

6.5   Denning's Contributions to the *Register*

The first issue of the *Astronomical Register* appeared in 1863, when Denning was just 15 years old. The first indication that Denning had begun communications with Gorton are found in the January 1868 issue of the *Register*, where in the Brief Communications section a reply to a query by "W.F.D" is given. Although the actual question is not printed, Gorton indicates that the information being sought on "elements for asteroids" can be found within the *Nautical Almanac*. The query is perhaps a little surprising, since Denning paid little attention to observing asteroids, but at this time he may have been interested in testing the light gathering power and charateristics of his newly acquired 4½—inch refractor. Denning (now 20 years old) is listed among the new subscribers in the February 1868 *Register*, and his first communication, "observations of Jupiter's satellites" is printed in the April issue. More communications to the *Register*, on a vast number of topics (Table 6.6), began to follow, and Denning soon distinguished himself as a dedicated observer. In total Denning was to submit 82 correspondence letters and observing notes to the *Astronomical Register* between 1868 and 1886. Typically the communications were some 300 to 500 words in length (some up to seveal thousand words), matter of fact in thier content, simply listing what was being observed, how it was being observed, and what had been seen on specific dates. Quite often the communications would also include a sentence, or two, calling upon readers to join-in future observing campaigns—especially the communications concerning OAS work (see Chap. 7). Indeed, from the very outset it would appear Denning was keen to promote organized work among amateur astronomers. The first 17 of Denning's astronomy related publications appeared in the *Astronomical Register*, over the time interval from April 1868 to December 1870, with the first article beyond its pages appearing in the 1 December, 1871 issue of the journal *Nature*. Indeed, the *Nature* article, on observations of the planet Venus, was the only article that Denning published in 1871, and this communication appeared after a year-long gap in his *Astronomical Register* correspondence. As will be outlined in the following chapter, 1871 was apparently a difficult year for Denning, and it effectively corresponds to the year in which the Observing Astronomical Society (OAS) was to slide in to a slow demise (see Chap. 7).

The breakdown of topics and number of communications in Table 6.6 reveals that Denning's primary interests were related to meteor observations (31 letters, or 38% of the total contribution), planet Jupiter (16 letters, or 20% of the total contribution), and comets (8 letters, or 10% of the total contribution). These same topics, as seen in Chaps. 2 and 3, would dominate Denning's later research interests. Indeed, it is clear from his contributions to the *Register*, that Denning was never strongly concerned with stellar astronomy (binary star measurements, and/or variable star observations), but was primarily interested in observing meteors. Table 6.6 further indicates that Denning was particularly interested in recording the activity of the August Perseid and November Leonid meteor showers, although he also

Table 6.6 Breakdown of observing topics that Denning commented upon in the *Astronomical Register* (1868–1886)

Sun	Sunspots	2
Planets	Vulcan	4
	Mercury	2
	Venus	3
	Mars	2
	Jupiter—planet markings	3
	Jupiter—red spot	9
	Jupiter—satellites	4
	Saturn	3
	Uranus	1
Moon	Visibility when new	2
Comets	various	8
Meteors	Fireballs (various)	5
	Persieds	4
	Leonids	5
	Andromedids	2
	Misc. showers	15
Orion	Trapezium variability	2
Occulation	Regulus (α Leonis)	1
Aurora	Displays of	2
Telescopes	Magnifying powers of	1
Queries	General points	2
OAS	Summaries of work	4

Columns 1 and 2 identify the field and topic, while column 3 indicates the total number of communications. In the last line of the table, OAS corresponds to the Observing Astronomical Society

presented observations relating to the Andromedid, the Lyrid, the Orionid, the Geminid and the Quadrantid meteor showers.

A measure of Dennings contributions to the *Astronomical Register* can be gleaned from the general index to the first 20 volumes (1863–1882). In Table 6.7 I have gathered together the top 5 contributors with respect to letters, contributions and commentaries. Of the top 5 contributors, Denning ranks 3rd behind meteorologist/selenologist William Radcliffe Birt, and author Richard Anthony Proctor (1837–1888). Birt was one of the very first subscribers to the *Astronomical Register*, and he wrote for many years a monthly column in the *Register* on observing the Moon. Birt was also known to Denning as one of the founding membesr of the OAS. Richard Proctor became better known in his later years as a prominent popularizer of astronomy, although he also contributed many original research articles to the *Monthly Notices of the Royal Astronomical Society*—producing, in 1867, one of the earliest maps of Martian surface albedo features. Fourth in our list, Captain William Noble (1826–1904) additionally went-on to become better known as a founding member, and first President, of the British Astronomical Association.

Of the names listed in Table 6.7, that of George James Walker (1812–1886) is the most obscure. A brief, but largely lacking, obituary was given by Reverend Jackson in the June 1886 *Register*, where we learn that Walker was gazetted to the 13th Light Dragoons and married the daughter Anna of Elizabeth and Daniel Corrie in May of 1837. Daniel Corrie was the first Bishop of Madras, India, where Walker was then stationed. The War Office in London for 20 January 1834 indicates that cornet G. J. Walker had advanced, by purchase, to the title of Lieutenant. This same title he retired on 8 November 1842, and moved to Teignmouth, Devon. His first son, Daniel Corrie Walker, was born on 12 February 1838, with 8 other siblings being born between 1841 and 1858. Upon retirement to Teignmouth, Walker developed his interest in classical and oriental studies, and became a member of the Plymouth Brethren (a non-conformist, non-denominational Christian group that advocated the supreme authority of the Bible). Jackson was to write, "he was a practical and learned astronomer, taking the keenest interest in every-thing relating to the subject".

Table 6.7 Indexed articles in volumes 1 to 20 (1863–1882) of the *Astronomical Register*

Name	Citations	Comments
W. R. Birt	82	Meteorologist/selenologist
	129	Regular *AR* column on observing the Moon
		OAS founding member
		Elected FRAS 1859
R. A. Proctor	72	Founding editor of *Knowledge* magazine
		Author/science (astronomy) popularizor
		Elected FRAS 1866
		Honory Secretary RAS 1872
W. F. Denning	66	Tresurerer OAS
		Elected FRAS 1877
		Honorary member LAS 1882
W. Noble	60	Army Captain
		1st President of BAA (1890–2)
		Elected FRAS 1855
G. J. Walker	56	Amateur astronomer—Teignmouth, Devon
		Classical and Oriental Studies
		Lietenent 13th Regiment of Light Dragoons

Only the top 5 contributors are listed

In 1863, the start date for the *Astronomical Register*, Birt, Proctor, Noble and Walker were 59, 26, 37, and 51 years of age respectively. Denning was 15 years old, and still 5 years away from submitting his first communication to the *Register*. Indeed, of our top 5 contributors to the *Astronomical Register* (between 1863–1882), Denning was very much the rising star, the young and enthusiastic up-and-comer, who would shape and ultimately define amateur astronomy over the closing decades of the 19th century.

Denning's ascendency to public recognition during the 1870s, as a name recorded within popular books, scientific journals and newspapers, is illustrated in Fig. 6.9. This diagram shows a Google Books Ngram[12] result for the search *n*-grams W. F. Denning, W. R. Birt, R. A. Proctor, and Captain W. Noble. The Ngram curves are not strictly definiative, since there are numerous biases in their production, but they are indicative. The prodigious output and popular reach of Richard Proctor is evident in Fig. 6.9, with references to his name starting to rise after he became a Fellow of the Royal Astronomical Society (RAS) in 1866, and peaking during the mid-1870s, the time at which he became Honorary Secretary of the RAS. William Noble was elected a Fellow of the RAS in 1855, but references to his name only pick-up circa 1890, at which time he was involved in the formation of the British Astronomical Association. William Birt was elected a Fellow of the RAS in 1859, and began his extensive correspondence in the *Astronomical Register* (recall Table 6.7) starting in 1863. The frequency usage of the name W. F. Denning rises steadily through the 1870s and 1880s, peaking circa 1900, the time at which he was awarded the Gold Medal of the RAS. The second peak circa 1920 relates to Denning's discovery of Nova Cygni (recall Chap. 3). George James Walker (the fifth entry in Table 6.7) is not included in Fig. 6.9 since the Ngram frequency for the appearance of his name is essentially zero.

6.6   Chapter Summary

The *Astronomical Register* was established in January of 1863 in order to become "a medium of communication for amateurs and others". Its founding editor, Stanford Gorton, intended that it should pay its own way, and cater to the growing audience of amateur enthusiasts interested in the sciences,

---

[12] It is estimated that over 130 million distinct book titles have been published worldwide. Over 40 million of these books have been scanned by Google Books. Ngram is an online search engine offered by Google Books to search the frequency usage of a specified word string (*n*-gram) within its scanned volumes. The results from an Ngram search are displayed as a smoothed line of frequency usage versus year. The data plots from Ngram are not corrected for book-selection biases, but they are generally taken to be representative of trends.

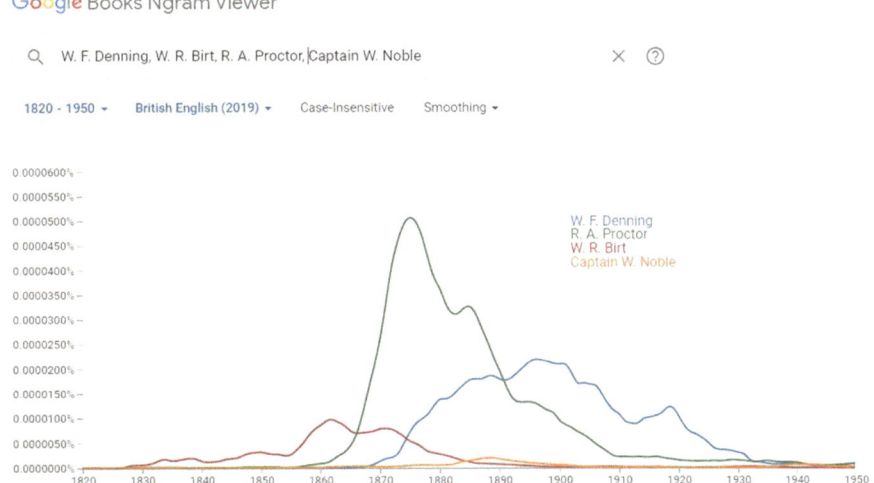

Fig. 6.9 Google Books Ngram analysis on the search words W. F. Denning (blue), W. R. Birt (red), R. A. Proctor (green), and Captain W. Noble (ochre). The time interval covered is that between 1820 and 1950. See text for details

and astronomy in particular. Its aims were to educate, inform, and promote discourse between amateur astronomers, building thereby, a sense of national community. Based upon the subscriber information printed in the *Register*, it is found that it had a widely dispersed readership (Fig. 6.2) from across England, Scotland and to some extent Ireland, and the world beyond. During its lifetime, the *Register* printed the reports of the short-lived (from 1869 to 1873) Observing Astronomical Society, and it catered to its subscribers by printing personal advertisements concerning items wanted and items for sale. Also, to a space-limited degree, it carried advertisements, relating to amateur astronomy needs, placed by manufacturers and individual artisans. An anaylsis of the personal wants and needs advertisments placed within the *Register* (Tables 6.1 and 6.2) reveals a broad range of astronomical ephemera being traded: from telescopes, lenses, mirrors, spectrum analysis equipment, books and celestial globes. An analysis of advertisment information concerning telescope apertures (Fig. 6.4) reveals that the most commonly used (and sought-after) telescope was that with a 3 to 4-inch objective lens. Items manufactured by T. Cooke and Sons (followed by those manufactured by Troughton and Simms) were predominantly mentioned in personal advertiesments, although a wide-range of telescope makers were identified (Tables 6.2 and 6.3). Most commercial advertisers were located in the greater London area, but other addresses including York, Birmingham, Manchster, and Dublin are additionally found.

The *Astronomical Register* appeared at a time when amateur astronomers were first beginning to find each other and form loosely-constructed communities outside that of the professionally-run national societies. Through its then innovative inclusion of a letters section the *Register* enabled amateur astronomers to have a public voice. It also enabled communication between geographically diverse enthusiasts to take place, helping to foster, thereby, a series of collaborative observing projects by amateur astronomers. When the *Astronomical Register* suffered its sudden demise in late 1886, amateur astronomy was on a very different footing to that at its first appearance in 1863. The *Register* had indeed, begun the process of forming a sense of an amateur astronomical community, and this was further strengthened by the formation of the Liverpool Astronomical Society (LAS) in 1881, and the British Astronomical Association (BAA) in 1890. Both of these societies, still thriving to this day, began publishing their own journals, along similar lines to the *Astronomical Register*: The monthly Journal of the LAS was edited by Herbert Sadler,[13] while that of the BAA was edited by Edward Walter Maunder.[14] Not only was amateur astronomy in a state of ascendency in the late 1880's, but so too was the activity of scientific communication, the two growing synergistically, side-by-side. Furthermore, Denning, as one of the major contributors to the *Register*, emerged thought its pages as an enthusiastic observer, a strong promotor of amateur astronomy, and a tireless organizer of coordinated research campaigns. Through the *Register* Denning's name became linked, by its multiple appearances, with well-known and established amateur figures (specifically, Birt and Proctor), thereby lifting his public profile. Indeed, Denning used the *Register* to pronounce upon his works and interests, to build a leadership portfolio, and establish himself as a nationally respected figure amongst amateur astronomers.

Bibliography

1 Johnson, P. (1990). The astronomical register 1863–86. *Journal of the British Astronomical Association, 100*, 62–66.

---

[13] Sadler studied Hebrew at Cambridge University, but followed no professional career. His astronomical interests were concerned with double stars, star catalogs, and selenography. He was elected a Fellow of the RAS in 1876.

[14] Maunder worked at the Royal Observatory in Greenwich, and along with his wife, Annie Scott Dill Russell, a "lady computer" at the Observatory, discovered the "butterfly diagram" associated with the Sun's 11-year sunspot cycle. Maunder was a founding member of the BAA and was elected a Fellow of the RAS in 1875.

2 Lightman, B. (2018). The mid-Victorian period and the astronomical register (1863–1866): A medium of communication for amateurs and others. *Public Understanding of Science, 27*(5), 629–636.

3 Chapman, A. (2017). *The victorian amateur astronomer: independent astronomical research in Britain 1820-1920*. Gracewing.

4 Baum, R. (2012). Gems from the astronomical register. *Journal of the British Astronomical Associations, 122*, 125–126.

7

The Observing Astronomical Society

While the *Astronomical Register* began the process of building a community of amateur astronomers, pooled from across Britain, allowing them a forum to present their thoughts, questions, and observational notes, the Observing Astronomical Society (OAS) offered the more serious amateur a chance to participate in observational projects. In concept it was to be a society of practical observers, equipped with good astronomical instruments, intent upon performing formalized, diligent research work. This was not, as such, to be a club for the casual observer, but rather a movement of observers who would dedicate time to specific observational projects.

The OAS was far from being the first amateur astronomical society to form in England—credit for the fist such move being given to the Leeds Astronomical Society (LeAS), which formed in 1859. Indeed, the LeAS organized in the wake of the week-long meeting of the British Association for the Advancement of Science held, that year, in Leeds. The founding members, in fact, approached Sir John Herschel for advice, and afforded him an honorary presidency.[1] The LeAS had no specific ambition of becoming a national society, it was purely a group of like-minded amateur astronomers interested in pursuing practical observational work and promoting an interest in astronomy within their local community. For all this, the Society thrived. Indeed, it went-on to garner a remarkable array of patrons, with John Herschel being joined by Sir George Biddell Airy (the Astronomer Royal),

---

[1] Herschel also presented the society with a 3½ -inch aperture refracting telescope made by Ross in London. This telescope is still in the society's possession.

© The Author(s), under exclusive license to Springer Nature Switzerland AG 2023
M. Beech, *William Frederick Denning*, Springer Biographies,
https://doi.org/10.1007/978-3-031-44443-2_7

John Wrottesley (President of the Royal Society), and numerous regional and national dignitaries. In spite of this impressive support, the Society entered a phase of relative dormancy during the 1870s and 80s, only to be revived in a more resilient and active form in 1892—largely through the efforts of local residents David Booth,[2] and Henry Townshend. The Society began to published a journal of society activities, observational notes, and articles in 1895, and this publication continued, somewhat under the radar of non-society readers, until 1921.

That an attempt might be made to form a national body for amateur observers was, by the late nineteenth century, inevitable. For all this, the circumstances surrounding the formation of the OAS remain obscure. The success of the *Astronomical Register* had certainly shown that a large number of amateur astronomers existed within the British Isles, and it was only natural that a sub-set of the more seriously minded amateurs would attempt to form a network through which they could coordinate their activities. The formation of the OAS was announced in multiple publications in July of 1869. The editor of the *Astronomical Register* wished the society success, and at some earlier stage must have agreed to publish reports from the Society (see Fig. 7.1). Indeed, many of the initial members of the OAS were subscribers to the *Register*, and had already published articles, letters and observing notes within its pages. The *Register's* practice of publishing subscriber address presumably helped in the communication between those observers who set-out to form the OAS. It seems reasonably clear that Denning, who would have then been just 21 years old, was the driving force behind the creation of the society, and indeed, the society's announcement notice in the July 30 (1869) issue of *The Mechanics Magazine* actually states, "Mr. William F. Denning, …, has succeeded in forming a society of gentlemen processing astronomical instruments for securing concerted observations of interesting astronomical objects". In this account, it is the indication that Denning had "succeeded" that is of note, and again suggestive that he was the primary instigator behind the establishment of the OAS.

The formal structure of the Society was already established by the time that it began to court (that is after July 1869) new members. The President was to be the Reverend Robert Eli Hooppell, while the Treasurer and Secretary was to be Denning. In addition, there was a committee of 5 members—these being:

- T. P. Barkas (Newcastle-on-Tyne)

---

[2] Booth was the first Director of the BAA's Meteor Section (1890–1892)—see Chap. 5.

A NEW OBSERVING SOCIETY.

We have always great pleasure in directing the attention of our readers to any project which is likely to aid the spread of practical astronomy ; and the Observing Society, of which the following is a prospectus, will doubtless, if energetically carried out, be of great service for this purpose. Many— if not all—who have at present joined it, are among our subscribers, and it will be seen that the observations are to be published in our pages. We can only add our cordial wishes for its success.

PROPOSED "OBSERVING" ASTRONOMICAL SOCIETY.—It is proposed to form a Society of gentlemen possessing astronomical instruments, for the purpose of securing concerted observation of interesting astronomical phenomena, and recording the results obtained. The mode of proceeding proposed is as follows :—The Secretary will forward, monthly, to each member, a list of the astronomical phenomena expected during the ensuing month. Those members who observe any, or all, of the phenomena will send in reports of the same to the Secretary by the 12th day of the following month. The Secretary will then select the most valuable and interesting observations from the reports he may receive, and have them published in the next number of the *Astronomical Register*. A slip from the *Register* containing the report will be sent to each member.

It is proposed that the following be the Officers of the Society :—*President*, the Rev. R. E. HOOPPELL, M.A., LL.D., F.R.A.S., Winterbottom Nautical College, South Shields ; *Treasurer and Secretary*, WILLIAM F. DENNING, Esq., Ashley Road, Bristol ; *Committee :* T. P. BARKAS, Esq., Grainger Street, Newcastle-on-Tyne ; JAMES COOK, Esq., Ribbleton Lane, Preston ; A. W. BLACKLOCK, Esq., Waterloo Place, Manchester ; H. MICHELL WHITLEY, Esq.. Penarth, Truro ; A. P. HOLDEN, Esq., 107 Hoxton Street, London. The President and Treasurer and Secretary to be ex officio members of the Committee. It is likewise proposed that the subscription be payable yearly on the 15th of June; that the amount be Five Shillings for the first year ; that in subsequent years it be, if possible, less ; that all intending members who send in their subscriptions by the 1st of July of the present year be deemed actual members ; that after the 1st of July a gentleman wishing to join the Society be recommended by a member, to whom he must be personally known, and through whom he must send his application. The Secretary will be happy to give any further information that may be desired. Subscriptions may be forwarded to him either by Post Office Order or in stamps.

Fig. 7.1 Notice announcing the formation of the Observing Astronomical Society. From the July 1869 issue of the *Astronomical Register*

- James Cook (Preston)
- W. Blacklock (Manchester)
- H. Michell Whitley (Truro)
- P. Holden (London).

It is immediately clear from the list of committee members that the majority where from outside of London, indeed the greater number were located in northern industrial England. As indicated in Fig. 7.1, the President, and Treasurer and Secretary were to be ex-officio members of the Committee. In Denning's case an ex-officio status is understandable, as instigating founder of the society, and given his observing experience and publication record. The Presidency of Reverend Hooppell was most likely one of a willing dignitary—indeed, the records indicate that he submitted no observational data to society reports (see below). For all the establishment of a formal committee structure, it would appear that the vast majority of work, that is, running the society, overseeing correspondence, and preparing observing reports, fell upon Denning's young shoulders.

The Reverend Robert Eli Hooppell (Fig. 7.2) was born in Rotherhithe, Surrey on 30 January, 1833. Educated at St. Olave's Grammar School in Southwark, he went-up to St. John's College, Cambridge in 1851. Hooppell graduated 40th wrangler[3] in the mathematical tripos in 1855, and obtained a first-class in the moral science tripos in 1856. He was ordained a deacon in 1857 (becoming a priest in 1859), and thereafter, he proceeded to an M.A. in 1858, and Doctor of Law (LLD) in 1865. Between 1855 and 1861, Hooppell was mathematics master at Beaumaris Grammar School on the isle of Anglesea, serving as English Chaplin at Menai Bridge from 1859 to 1861, after which he moved to South Shields to take-on the Headmastership of Winterbottom Nautical College. *The Illustrated London News* for 23 January 1869 covered the opening of the new college building, and noted that, "Dr. Hooppell has spared no pains or expense to provide means in particular for the teaching experimentally of the physical sciences, and his evening lectures have become deservedly popular". In 1875 Hooppell was awarded the degree of Doctor of Civil Law by the University of Durham, and, stepping down as Headmaster of the Nautical College in that same year, he was instituted to the rectory of Byers Green in county Durham. Hooppell passed away on 23 August 1895.

Hooppell was elected a Fellow of the Royal Astronomical Society in 1865, but submitted no research papers or observational reports to its journals. Connected to his teachings at the Winterbottom Nautical College, Hooppell published *Tabular Forms for Facilitating the Calculation of Certain Nautical Problems—A practical Introduction to Navigation and Nautical Astronomy* in 1871. Copies of this text were available through the publisher J. D. Potter[4]—the same publisher of the *Astronomical Register*. Hooppell became a Fellow of the Society if Antiquaries of Newcastle-upon-Tyne in 1877, and he contributed papers to the society's Proceedings concerning Roman names

---

[3] First wrangler and Smiths Prize winner was James Savage, who unfortunately drowned in that same year. Second wrangler was Leonard Henry Courtney (1st Baron Courtnay of Penwith), who became a radical politician advocating for proportional representation, and was a prominent supporter of women's suffrage. He practiced as a lawyer, and taught political economy at University College London.

[4] In their 1913 catalog Hooppell's text was still available for 3 shillings and 6 pence, indicating that it had a long print run.

Photo Elliott and Fry, Baker Street.

THE LATE REV. DR. R. E. HOOPPELL.

Fig. 7.2 The Reverend Robert Eli Hooppell (1833–1895)

and ruins in the South Shields region.[5] He additionally submitted many reports, particularly on meteorology, to the *Transactions of the Tyneside Naturalists Field Club*. In terms of being a public figure, Hooppell most noticeably agitated for the repeal of the Infectious Diseases Acts, first introduced by the British Government in 1864. These acts were concerned with the rampant increase in sexually transmitted diseases amongst men in the British armed forces, and the Acts allowed, specifically, for women (particularly prostitutes) to be forcibly incarcerated if suspected of being infected. Hooppell spoke out about the Acts and their denial of women's rights (along with reformer Charles Bell Taylor) at the Social Science Congress[6] held in Sheffield in 1869, and again at the British Association meeting held in Edinburgh in 1871. The acts were eventually repealed in 1886.

---

[5] Additional archaeological papers were published by Hooppell *in Archaeologia Aeliana*, the *Illustrated Archaeologist*, and the *British Archeological Journal*.

[6] A congress was held each year by the Social Science Association, founded by Lord Brougham, from the time of its formation in 1857 through to 1884.

Fig. 7.3 The Winterbottom Nautical College, South Shields. Opened on 12 January 1869, and built through the bequest of £20,700 by Dr. Thomas Masterman Winterbottom (1766–1859). In 1984 the Nautical College was merged with the Hebburn Technical College, to become the South Tyneside College. Image from *The Illustrated London News*

It is not obviously clear why Hooppell was sought out to be the President of the OAS—nor is it clear why he accepted the position—it was certainly not an active role on his part. None-the-less, he did have a background in mathematics and navigation, and an interest in astronomy, but no established track-record of observational work. At the time of the formation of the OAS, however, he was the esteemed headmaster of the Winterbottom Nautical College (Fig. 7.3).

The Two Worlds (a journal devoted to spiritualism, occult science, ethics, religion and reform) for September 4th, 1891 carried a lengthy obituary of Novocastrian Thomas Pallister Barkas (Fig. 7.4). Here we lean that he was born, raised, and died in Newcastle-on-Tyne. Interested in natural history from an early age, Thomas, upon leaving school at the age of 14, went on to join the family construction firm in 1833. Upon stepping down from the business in 1843, Thomas returned to his interests in natural history and began lecturing on scientific topics. Indeed, over the next 30 years it is estimated that Thomas delivered thousands of lectures and talks. In 1845 he embarked on a new venture and opened a bookshop, which was eventually turned into an Art Gallery and News Room (complete with a telegraph feed) in 1870. Barkas was elected an alderman of the City of Newcastle in 1866, and served on the Public Libraries Committee. Starting in 1868 Barkas became increasingly interested in the field of geology, and was elected a Fellow of the Geological Society in 1869. His major work in this area, *Illustrated Guide to the Fish, Amphibian, Reptilian, and Supposed Mammalian Remains of the Northumberland Carboniferous Strata* (London: W. M. Hutchings), saw publication in 1873. This work, dedicated to "the working miners of Northumberland" was based upon an extensive collection of carboniferous fossils gleaned from local collieries. His interests were catholic, however, and he lectured on mesmerism and spiritualism,[7] as well as astronomy—the account given in *The Two Worlds* obituary indicates that, "his lectures on astronomy will be long remembered in many a local colliery village into which he carried the latest news of comet and star". In terms of work related to the OAS, Barkas was involved in several campaigns, using a 4 ¼-inch refractor, looking to detect the supposed planet Vulcan (see below).

Arthur Woolsey Blacklock (1840–1934) was born in Brighton, Sussex. Nothing is known of his formative years, but he became a medical doctor, gaining his MD from Aberdeen University in 1872. He established, with Henry Brady, a medical practice in Gateshead in 1875, but by mutual consent the partnership ended in 1881. The *Irish Medical Directory* for 1878 indicates that Blacklock was Assistant Medical Officer at the Durham County Lunatic Asylum. At the time of serving on the OAS committee Blacklock was resident in Manchester. By 1905, however, he had established a new medical practice in Ipswich. He is buried, along with his second wife, Annie Letitia Page (married in 1906), in Ipswich Cemetery. While never a Fellow of the Royal Astronomical Society or member of the British Astronomical Society, Blacklock was a regular contributor to *The English Mechanic* magazine, and

---

[7] He published an account of his studies in this area in, *Outlines of 10 years' Investigations into the Phenomenon of Modern Spiritualism* (Frederick Pitman, London) in 1862.

Fig. 7.4 Thomas Pallister Barkas (1819–1891)

the *Astronomical Register*, but submitted no known reports to the OAS. He is known, however, to have constructed several telescopes, including a Newtonian reflector of aperture 5¾-inches. He was one of the founding members of the Ipswich branch of the Chaldæan Society for Astronomy in 1921. This latter society being formed in London c.1916 with the aim of promoting naked-eye astronomy (recall Chap. 3).

According to the *Bibliotheca Cornubiensis* (compiled by G. C. Boase and W. P. Courtney. Longmans, Green, Reader, and Dyer, London. 1878), Henry Mitchell Whitley was born in Truro, Devon, on May 3rd, 1845. He attended Truro Grammar School from 1857 to 1862, and after moving to London in 1872, he became an Associate of the Institute of Civil Engineers in 1874. From 1875 to 1877 he was Assistant Engineer to the North Eastern railway line, Newcastle-on-Tyne. Between 1866 and 1872 Whitley contributed a number of articles to the *Journal of the Royal Institute of Cornwall*, covering topics from churches, rainfall and marine ecology. From 1870 onwards, he contributed articles to *The Engineer* (on locomotive practices), *The Surveyor* (on street and city planning), the *Astronomical Register*, and observational reports to the OAS (see below). Whitley was elected a member of the Geological Society in 1874. It is not recorded when Henry Whitley passed away, but according to an index search of the Journal of the Royal Institute of Cornwall his last contribution was read in 1906.

As we move to the final two members of the OAS council, virtually nothing is known of their personal histories. James Cook lived in Preston, and was elected a Fellow of the Royal Astronomical Society on 14 January,

1881. He contributed observations to the OAS but is otherwise obscure. As for Albert P. Holden, all that is known is an address on Hoxton Street, in the borough of Hackney, London. Holden submitted numerous reports to the OAS, and was active in the Hackney Scientific Assocaition,[8] founded in 1867 by William Radcliffe Birt.[9] Holden presented papers on the conjunctions of Mercury, and the dimensions of the binary star system Algol to the Association, and he also prepared a catalog of variable stars for its members.

A listing of initial OAS members (as of 1 November 1869), some 28 persons in all, has been published by Richard Baum[10] [1], and, as noted by Baum, many of the members went-on to become prominent members of the British Astronomical Association, with several joining the professional scientific ranks—notably, David Gill (1843–1914), and Oliver Lodge (1851–1940). Not all of the initial members listed were active in submitting reports, indeed a survey of OAS reports published in the *Astronomical Register* indicates that only 21 of the initial 35 members (that is, including committee members) contributed observational notes.

The subscription fee for new OAS members was set at 5-shillings per annum (with the possibility of reduced fees being set in following years), and this membership fee could be paid in postage stamps.[11] It was also required that any gentleman wishing to become a member of the OAS would need to be recommended by an existing member who knew them personally. It is nowhere stated that women could not join the OAS, but no female observers are evident in the initial list of members, nor in any subsequent observational reports. The notice published in the July 30 issue of *The Mechanics Magazine* indicates that the society had 26 members. Notices placed in the *Register*, however, reveal that by November 1869 the membership had risen to 28, and that by August 1870 it was 46.

As OAS secretary, Denning submitted the first report on the society's activities in the November 1869 issue of the *Astronomical Register*. Further reports followed at regular intervals during 1870, but thereafter the frequency of reports published dropped dramatically. Table 7.1 shows the month/year distribution of OAS reports published in the *Register*. It seems clear from

---

[8] In 1871 this became the Metropolitan Scientific Association.

[9] Working with John Herschel, Birt conducted research on atmospheric and meteorological phenomenon, and became a well-known selenographer. Birt was also listed among the founding members of the OAS.

[10] This list of initial; OAS members being a fortuitous find affixed within the binding of a second-hand copy of volume 9 (1869) of the *Astronomical Register*.

[11] A short letter or card could be sent for just a few pennies in the late 1800's, and with 5-shillings one could purchase some 60 1-pence stamps. Payment in stamps was no-doubt a practical move and reflective of the intensive exchange of letters concerning OAS projects and correspondence.

Table 7.1 OAS reports published in the *Astronomical Register*

	1869	1870	1871	1872	1873
January		Report 2		AP	
February			VE		
March	VU	Report 3			RAS review—VE
April		Report 4—VU			
May		Report 5			
June		Report 6			
July	Formation	LE	Report 9		
August		Report 7—RE		Report 12	Report 14—VE
September					
October			Report 10		
November	Report 1	Report 8			
December			Report 11—VE	Report 13	

Key: VE = Venus markings campaign. VU = Vulcan search campaign. LE = lunar eclipse on July 12th. AP = Denning's *Astronomical Phenomena in* 1872. RE = Re-election of committee announcement

the report distribution that the society, and the enthusiasm of its members, flourished during 1870, but trailed-off thereafter. The last report submitted by Denning on behalf of the OAS appeared in the August 1873 issue of the *Register*, and this particular report concerned one of the society's more concerted and successful campaigns. While Denning submitted numerous letters and articles to the *Register* throughout 1873, none indicated an association with the OAS, and indeed, while there was no formal announcement of the OAS folding it effectively ceased to function after the 13th report published in December of 1872. What I have labelled as the 14th report (Table 7.1) published in August of 1873, while based upon observations of planet Venus (see below) obtained by OAS members, no OAS participation or membership is actually acknowledged, and nor does Denning use his title of Secretary and Treasurer. Just a few months earlier, however, at the March 1873 meeting of the Royal Astronomical Society, Denning reviewed the same observational material and directly acknowledge the contribution of OAS members.

A breakdown of active OAS observers, along with their areas of study, is provided in table 7.2. The Sun, Moon and planets were by far the most common objects observed, with relatively few observations being made of stars or star systems. The most popular objects to study, with respect to individual members, were planet Jupiter, meteors, the Sun, planet Venus, and the Moon's disk.

The most prolific OAS observers (in the sense of being mentioned in reports) were Denning and,

- Thomas William Backhouse
- Thomas Gwyn Empy Elger
- Henry Ormesher
- John Birmingham
- Henry Mitchell Whitley
- Reverend Samuel Jenkins Johnson.

Of these observers, only the reverend S. J. Johnson was not among the initial members of the OAS. Thomas W. Backhouse (1842–1920) studied at University College, London, and was elected a Fellow of the Royal Astronomical Society in 1873. Interested in astronomy from a young age, he built an observatory, equipped with a 4½-inch reflector, at his home of West Hendon House. Indeed, being independently wealthy Backhouse actually employed an assistant to work at his observatory. In 1911 he published a *Catalogue of 9842 Stars, Or, All Stars Very Conspicuous to the Naked Eye, for the Epoch of 1900*. In addition to his astronomical interests, Backhouse was long interested in meteorology, and served as Vice-President to the Royal Meteorological Society twice in 1918 and 1919. Thomas G. E. Elger (1836–1897) was well known and celebrated for his selenography work, and was elected a Fellow of the Royal Astronomical Society in 1871. A lunar crater has been named after Elger, and he is remembered for his highly praised work *The Moon: A full description and map of its principal physical features* (George Phillip & Son, Liverpool: 1895). Elger was a founding member of the British Astronomical Association, the Director of the BAA Lunar Section from 1891 to 1896, and President of the Liverpool Astronomical Society during 1888. John Birmingham (1816–1884) was born in County Galway, Ireland, and was recognized as a talented astronomer, geologist, and poet. He was an only child, and a life-long bachelor. In terms of astronomy, he is best known for his discovery (or first sighting) of the recurrent novae T Coronae Borealis in 1866, and for his study of red stars, producing a catalogue of 658 such objects in 1877—the latter work earning him the Cunningham Medal of the Royal Irish Academy in 1884. The lunar crater Birmingham was named in his honor.

Table 7.2   Areas of study undertaken by OAS members

Subject	Observer	Location	Issue #
Sun	T. W. Backhouse	Sunderland	85, 88, 89, 90, 92,
	W. F. Denning	Bristol	103,
	T. G. E. Elger	Bedford	106, 120
	A. P. Holden	London	89, 95, 103
	Rev. F. Howlett	Hurst Green	85, 92, 95, 120
	Rev. S. J. Johnson	Crediton	88, 89, 95
	H. Ormesher	Manchester	88
	G. J. Walker	Teignmouth	89, 92
	H. M. Whitley	Truro	87, 89, 95
			88
			88, 92
Mercury	J. Birmingham	Tuam (IRL)	89
	Rev. S. J. Johnson	Crediton	120
Venus	J. Birmingham	Tuam (IRL)	89, 106
	A. P. Holden	London	89
	H. W. Hollis	Derby	106, 108
	H. Ormesher	Manchester	88, 95, 106
	Thos. Petty	Oxford	85
	T. W. Webb	Hardwicke	108
	H. M. Whitley	Truro	88
	F. Worthington	Altrincham	108
Venus study (d)	W. F. Denning	Bristol	99, 106, 128, 129
Lunar topology	J. Birmingham	Taum (IRL)	82, 87, 88, 92
	W. R. Birt	London	87, 88
	J. Cook	Preston	82, 87
	E. Crossley	Halifax	87
	T. G. E. Elger	Bedford	85, 87
	J. Gledhill	Halifax	87
	H. Ormesher	Manchester	90
	H. Pratt	Brighton	87
	H. M. Whitley	Truro	82, 90, 92, 95
Lunar eclipse	W. F. Denning	Bristol	95
(July 12, 1870)	Rev. S. J. Johnson	Crediton	95
	O. J. Lodge	Hanley	95
	E. Neison (1)	London	95
	Rev. R. Prowde	Northallerton	95
Lunar occultation	Rev. S. J. Johnson	Crediton	90
	J. C. Lambert	Sleaford	95
	H. Ormesher	Manchester	82
	G. J. Walker	Teignmouth	90, 92
Mars	A. P. Holden	London	103

(continued)

Table 7.2 (continued)

Subject	Observer	Location	Issue #
Jupiter	T. W. Backhouse	Sunderland	85, 87, 120
	J. Birmingham	Tuam (IRL)	88
	J. Cook	Preston	87
	W. F. Denning	Bristol	85
	G. E. Elger	Bedford	85, 88
	R. Forward	Southsea	87
	D. Gill	Aberdeen	89
	A. P. Holden	London	85, 103
	H. W. Hollis	Keele	120
	E. Neison (1)	London	103
	H. Ormesher	Manchester	82, 85, 87
	E. Salter	Manchester	85, 87
	T. Whitehouse	W. Bromwich	85
	H. W. Whitley	Truro	88
Jovian Moons	G. J. Walker	Teignmouth	88
Saturn	H. M. Whitley	Truro	92, 95
Uranus	H. M. Whitley	Truro	90
Aurora	W. Andrews	Coventry	89
	W. F. Denning	Bristol	120
	T. G. E. Elger	Bedford	87, 89
	Rev. S. J. Johnson	Crediton	87, 88, 90
	H. Ormesher	Manchester	85, 88
	H. M. Whitley	Truro	85, 88, 90
Parahelia	T. P. Barkas	Newcastle	85
Meteors	T. W. Backhouse	Sunderland	82, 85, 89
	W. F. Denning	Bristol	82, 106, 120
	A. P. Holden	London	90
	Rev. S. J. Johnson	Crediton	87, 88, 90, 106, 120
	J. C. Lambert	Sleaford	95
	E. Neison (1)	London	120
	H. Ormesher	Manchester	82
	Rev. R. Prowde	Northallerton	82
	E. Salter	Manchester	87
	G. H. Walker	Teignmouth	92
	H. M. Whitley	Truro	92
Comets (a)	J. Birmingham	Tuam (IRL)	106
	C. Hill	Bristol	82, 106
	G. J. Walker	Teignmouth	92
Double stars	J. Birmingham	Taum (IRL)	88
	D. Gill	Aberdeen	89
	A. P. Holden	London	90
Trapezium star cluster in Orion	J. Browning	London	116
	E. Salter	Manchester	87, 88
	T. Owen	Manchester	116
	R. J. Ryle	Burton-on-Trent	116

(continued)

Table 7.2 (continued)

Subject	Observer	Location	Issue #
Stars (individual)	T. W. Backhouse	Sunderland	120
T Cor. Borealis	J. Birmingham	Taum (IRL)	92
New red star (b)	H. Ormesher	Manchester	82
Mu Ceti			
Pleiades (c)	A. P. Holden	London	120
Vulcan (d)	W. F. Denning	Bristol	88, 89, 99, 108

Key: (a) Comet observations: C. Hill and G. J. Walker specifically indicate that they are observing Winnecke's comet. John Birmingham, however, indicates that he is observing comet's I and II. Technically, comets I and II for 1871 correspond to C/1871 G1 Winnecke and C/1871 L1 Temple respectively. G. F. Chambers discussed the confusion of comet naming in a paper to the *Astronomical Register*, **10**, 91–92, 1872
(b) Detection of an uncatalogued red star in the constellation of Cygnus
(c) Concerning the visibility of the nebulosity in the cluster
(d) See separate discussion on the various campaigns and discussion
    (1) This is the penname of Edmund Neville Nevill

Samuel Jenkins Johnson (1845–1905) was born in Atherton, Lancashire and became interested in astronomy at an early age. He went up to Oxford, and graduated from St. Johns College in 1867. He was ordained deacon in 1868 and priest in 1869. Reverend Johnson was elected a Fellow of the Royal Astronomical Society in 1872, and submitted many papers to the *Monthly Notices* on eclipse phenomena—publishing *Eclipses Past and Future* (James Parker and Co., Oxford & London) in 1874, and *Historical and Future Eclipses* (James Parker and Co., Oxford & London) in 1896. Henry Ormesher presently remains obscure, and we have found no accounts of his life and/or works other than his contributions to the OAS and the *Astronomical Register*.

Perhaps the most successful observing campaign conducted by OAS members was that concerning a study of planet Venus. Writing in the February 1871 issue of the *Astronomical Register* (issue # 99), Denning explained that the OAS committee had decided to undertake a series of systematic observations of Venus during one complete revolution (that is one synodic period of 584 days). Indeed, Denning noted that the committee had decided to form a Venus observing sub-committee, and asked to hear from interested observers. The proposal, which was also published in the *Times of London* newspaper, and the 31 March issue of the *English Mechanic and World of Science*, called for observations of Venus to commence of March 1871 and run through to October 1872. The announcement of the Venus observing project in the *English Mechanic* is particularly interesting in that it provides a name list of 34 observers who had already agreed to contribute

(Table 7.3). Of these observers 13 are known OAS members, and 9 were Fellows of the Royal Astronomical Society. It is an impressive list of volunteers, and many were well equipped with large aperture telescopes. This was not just some make-do-project, but a project that had attracted some serious, and well experienced observers.

Geographically the observer distribution, as revealed by Table 6.3, spanned the domains of England, Wales, Scotland, and Ireland, and the distribution of telescope sizes ranged from that of a 3-inch aperture refractor, to a 13-inch reflector. Some 25% of the observers listed in Table 7.3 were Fellows of the RAS, although only 8% where both OAS members and FRAS. These latter numbers begin to confirm the notion that the majority of OAS members were indeed, amateur astronomers working in an organized fashion in what spare moments they could find. Figure 7.5 is a histogram plot of the telescope apertures listed in Table 7.3. From this figure, it is seen that most of the observers (66%—or 23 out of the 35 total data set) were using telescopes in the 3–6-inch aperture range, while only 4 (11%) were using apertures of 10-inches and larger. Most of the OAS observers, therefore, were using modest aperture instruments.

Denning circulated an observing methodology notice to those observers expressing an interest in studying Venus, and this same note was reproduced some 20 years later in G. F. Chambers, *A Handbook of Descriptive and Practical Astronomy: II instrument and practical astronomy* (4th ed. Clarendon Press, Oxford, 1890). Denning, with typical zeal and encouragement, explained, "every fine day should be turned to account, and, when any markings or special features are visible, watch them as long as possible and note any changes either in shape or position". With respect to sketching, Denning indicated that, "simple *fidelity of reproduction*" is what is required, giving, "the appearances seen just as they impress the eye". It is also suggested that all observations should be made during the daytime, since this provides for higher seeing[12] elevations that should, thereby, be less liable to atmospheric distortion.

The first reports on the Venus observing campaign appeared in *The English Mechanic and World of Science*, for 20 October 1871 (Volume 14), and in the *Astronomical Register* for December 1871 (issue #108). These reports cover observations made between June and early August, and it is noted that

---

[12] Seeing is the term used to account for atmospheric scintillation—the lower the altitude at which observations are being made, so the greater the depth of atmosphere an object (star or planet) is seen through, and the worse the seeing will be. Seeing is technically at a minimum in the observer's zenith.

Table 7.3 List of observers engaged in the OAS Venus campaign

Name	Location	Telescope aperture (inch.)	OAS	FRAS
W. I. Banks	London	3.75		
L. Berry	Rotherham	3.25		
J. Birmingham	Taum, Ireland	4.5	Yes	
A. W. Blacklock	Godalming	6.0	Yes	
D. H. Blair	Barnsley	7.5		
J. H. Brabazon	Clonmel, Ireland	6.5		
J. R. Bridson	Bolton le Moor	6.25	Yes	
G. F. Chambers	Bickley	4.0		Yes
J. Cook	Preston	10.0	Yes	
C. J. Corlett	Thames Ditton	8.0		Yes
A. Cunningham	Devises	8.25		
W. F. Denning	Bristol	10.0	Yes	(1)
S. Drew	Sheffield	5.25		
T. G. E. Elger	Bedford	4.0	Yes	Yes
W. J. Ferguson	Glasgow	–		
R. J. Gould	Reading	5.0		
A. P. Holden	London	3.0		
R. E. Hooppell (2)	North Shields	7.0	Yes	Yes
H. W. Hollis	Newcastle	12.0 and 6.0	Yes	Yes
E. B. Knobel	Burton-on-Trent	8.5	Yes	
E. G. Loder	Cambridge	7.5		
J. C. Lambert	Sleaford	3.0		
R. Langdon	Collumpton	6.0		
E. Neison (3)	London	6.0		
H. Ormesher	Manchester	5.25	Yes	
T. Petty	Oxford	–	Yes	
G. Proctor	Kesway	4.5		
G. M. Seabroke (*)	Newcastle	8.5		Yes
T. W. Webb	Hay	9.25		Yes
W. West	Brixton	3.0		
A. M. Whitley	Truro	6.5	Yes	
W. G. Williams	Conway	4.25		Yes
T. Wilson	Skipton	4.24	Yes	
W. Windsor	Brighton	4.0		
H. Wortham	Royston	6.0		Yes
F. Worthington (*)	Altrincham	13.0		

Columns 1 and 2 provide the name and location of the observer, while column 3 indicates the telescope aperture being employed. Column 4 indicates if the observer was among the initial OAS membership list found by Baum [1]. Column 5 indicates if the observer was a Fellow of the Royal Astronomical Society. The data is from the 31 March, 1871 issue of the *English Mechanic and World of Science*, except for the entries marked (*) which are from the 20 October 1871 issue of the *English Mechanic and World of Science*. **Key**: (1) At the time Denning was not FRAS, his election not proceeding until 1877. (2) President of the OAS. (3) This is the penname of Edmund Neville Nevill

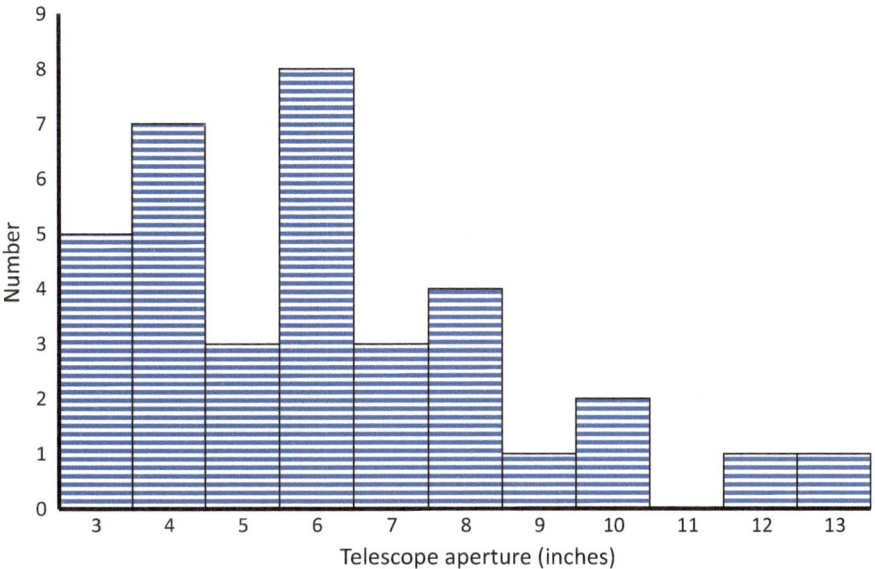

Fig. 7.5 Distribution of telescope aperture sizes used in the OAS Venus observing campaign

several observers had recorded dark cloud-like appearances on several occasions. Most observers, however, had nothing out of the ordinary to report. Several sketches by G. M. Seabroke, F. Worthington and H. Ormesher were reproduced in the *English Mechanics* account (see Fig. 7.6), but only a few drawings were ever made on the same date by different observers. However, it was suggested that the 11 June observations by G. M. Seabroke and F. Worthington suggested independent records of the same cloud feature (images 8 and 10 in Fig. 7.6). Denning also produced a summary of observations in the 7 December issue of *Nature*, indicating that several observers had reported seeing bright spots, which it was suggested might be surface craters. Recall at this time, and indeed well into the twentieth century, it was not known that the surface of Venus was entirely hidden from view at optical wavelengths.[13] Denning also reported in his *Nature* article that the dark, cloud-like features seen on Venus were similar to those recorded on the surface of Mars.

In spite of the success of the Venus observing campaign, no further reports on OAS activity occurred. Indeed, no mention of activity is found for the

---

[13] That Venus had an atmosphere had been realized since the mid-eighteenth century. That the surface was not visible at optical wavelengths was only slowly realized, however. Furthermore, it was not until 1920 that dark markings were first photographed in the atmosphere—these images, however, were only prominent at wavelengths in the ultraviolet region of the electromagnetic spectrum.

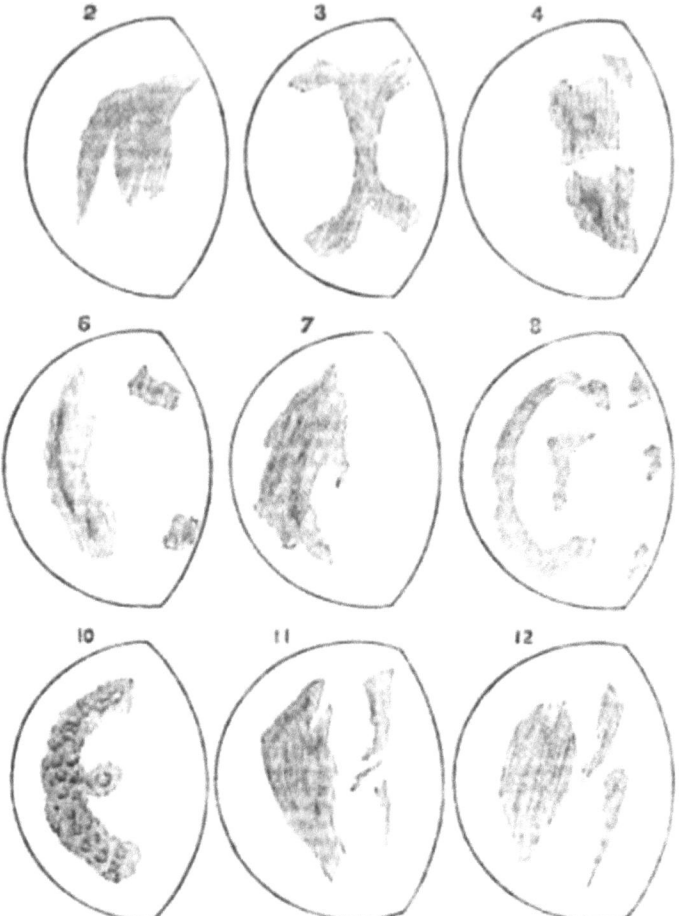

Fig. 7.6 Suspected markings on Venus as recorded by G. M. Seabroke (2, 3, 4, and 8), F. Worthington (10), and Henry Ormesher (11 and 12). Images 8 and 10 were recorded on June 11, 1871, by different, and well separated observers. These two sketches show remarkably similar markings, suggestive of a true visual phenomenon being present. From the 20 October 1871 issue of *The English Mechanic and World of Science*

entire duration of 1872. As seen in Table 7.1, this particular year saw very little reporting from the OAS, and indeed, it is not until August 1873 (issue # 128) that the *Astronomical Register* carries any account of the Venus observing campaign. Interestingly, in this account Denning simply provides his name, with no mention of the OAS being made. This final account provides observations from H. W. Hollis, Rev. T. W. Webb. J. Birmingham, T. G. E.

Elger, and Edmund Neison[14], all of whom indicate that they had observed no markings of any kind. Indeed, Denning concludes that, "it is evident that many skilled observers have quite failed to satisfy themselves as to the existence of the markings on Venus". Denning, further suggests that those apparent positive sightings can be explained by poor image resolution and observer "fancy". Such conclusions did not provide for any positive end-game, and while two letters appeared in the September 1873 (issue #129) of the *Register*—one agreeing with Denning's conclusions, and one arguing in a contrary manner—no further mention is made either of the Venus observing campaign, or indeed, the OAS as a functioning body.

Denning reviewed the history of Cytherean markings in his *Telescopic Work for Starlight Evenings* (in 1891), but made no reference to the OAS survey—suggestive of his reluctance to acknowledge the Society's work at that later time. He argued, however, that "delicate", "faint grey areas, without definite boundaries" [2], might be seen, and attributed these to being either surface markings, or regions of "different reflective power" in localized regions of its atmosphere. Indeed, while it was generally assumed that Venus supported a dense, highly reflective atmosphere, this was not universal the case. Amateur astronomer and artist John Brett[15] was to write in the January 1877 issue of the *Monthly Notices* [3], for example, that the surface of Venus was a highly burnished (mirror-like) substrate overlain by a dense envelope of glass. And, in this respect, he demonstrated to the assembled RAS fellows a model for Venus consisting of "a glass bulb filled with Mercury" with which he argued that the observed bright ring phenomena observed during the 1874 transit of Venus, could be reproduced. Indeed, Brett suggested that an observer with a large telescope might look at the center of Venus during the next transit (in 1882) for a reflection of the Earth. Perhaps needless to say, Brett's ideas concerning the structure of Venus gained no traction amongst other astronomers.

The search for hypothetical planet Vulcan preoccupied many observers, situated around the globe, during the closing decades of the nineteenth century. Had it actually existed, any finder would have certainly gained world-wide notoriety, and in principle the search was open to anyone who

---

[14] Edmund Neison (1849–1940), whose real name was Edmund Neville Nevill, was educated at Harrow School and New College, Oxford. He specialized in selenography, and from 1882 to 1911 he was Government Astronomer at the Natal Observatory in Durban. Nevill was elected FRAS in 1873, and FRS in 1908.

[15] Born in 1831 to an army veterinarian, Brett was a keen astronomer throughout his entire life. Attending the Royal Academy Schools in 1853, Brett was strongly influenced by the pre-Raphaelite movement, and became well known for his highly detailed landscape paintings. He was elected a Fellow of the RAS in 1871 and published articles relating to astronomy in *Nature,* the *Monthly Notices*, and the *Observatory*.

could observe the Sun's disk with even a modest aperture telescope. It was, in many ways, a problem that suited the amateur astronomer perfectly. Furthermore, it was a project that suited an organized, group observational approach. While many professional observers took great pains and efforts to place themselves along solar eclipse tracks, in the hopes of finding the fleeting planet at the time of totality, the amateur could in principle observer the Sun at any convenient moment during the day with the hope of catching the planet in transit. The Sun, however, can be deceptive, or more specifically, sunspots can fool even the most experienced of observers. A fine example of one such false positive detection is found within William Herschel's notebook on solar observations [4]. While not actually scanning for Vulcan (it had not been invoked at the time), Herschel records for 12 March, 1800, "there is a beautiful planetary black spot in the southern half of the Sun. It is perfectly round and is not a sunspot". Thinking that he had found his second planet, Herschel called in witnesses to his discovery. In came visiting guests Sophia Baldwin and Miss Wilson, and then "Mrs. Herschel", and all confirmed the observation. Then other visitors to Observatory House, Mr. & Mrs. Weyland, confirmed the sighting, and finally Scottish physician, natural philosopher, and tutor at Rugby College, James Lind. All joined in the excitement, and Herschel concluded his account with the comments, "it resembled a transit of Mercury over the Sun". One can only imagine the sense of wonderment felt in the Herschel household that evening, but sadly any joy of discovery was to be crushed the following day. For 13 March, Herschel records, "I saw the same round spot. It is as fine a deception as can be imagined. It is perfectly round and well defined". Rallying to the discovery, however, Herschel conjectured that the spot might be the caldera of a conical volcano extending above the luminous solar clouds.[16]

Planet Vulcan had been penned into existence by Urbain Jean Joseph Le Verrier, in 1845, in an attempt to explain the anomalous perihelion advancement in the orbit of planet Mercury [5]. Indeed, to Le Verrier, Vulcan was a straightforward mathematical requirement. Given that the observed perihelion advancement of Mercury's orbit was true, so Newtonian theory absolutely required a perturbing body to produce the necessary effect[17] (Fig. 7.7). In the light of such certainty, Edmond Lescarbault (infamously) thought that he had caught sight of the new planet crossing the Sun's disk on

---

[16] Herschel had earlier suggested that the Sun was a solid, planet-like body, surrounded by a highly luminous atmosphere. Sunspots, he reasoned, were gaps in the upper cloud deck through which the dark undersurface could be seen. Indeed, he even suggested that the Sun's 'surface' might be habitable.

[17] Technically, it need not be one object, but could be a group of orbiting objects, or an inner asteroid belt.

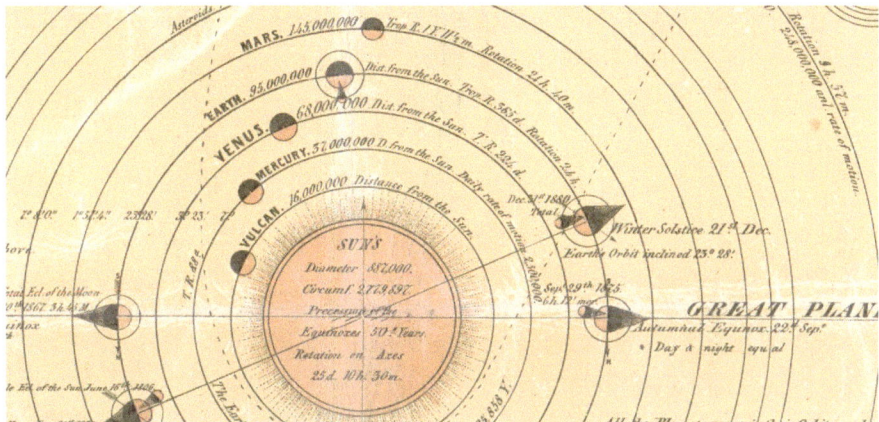

Fig. 7.7 Portion of a solar system map produced for schools by Hall Colby (Rochester, NY) in 1846. Vulcan is indicated as having an orbital radius of 16 million miles. Note that no planetary rotation rates are given for either Venus or Mercury (since, indeed, they were not then known). Image courtesy of the US Library of Congress

29 March, 1859. In spite of less than clear observing circumstances, Le Verrier praised Lescarbault, and used his account to predict an orbit for Vulcan—the hypothetical planet having a mass and size about the same as that of Mercury, with a mean heliocentric distance of 0.14 AU, and an orbital period of 19 days and 17 h. The search to confirm Lescarbault's apparent detection was soon taken-up by other astronomers.

Denning's first attempt at an organize search for Vulcan is described in a letter, dated 9 May, 1868, published in the *Astronomical Register*. In this letter he informs the readers that he, along with 5 other observers, had systematically monitored the Sun's disk from 7 March through to 21 April. While many sunspot groups were noted, there were no indications of a suspect Vulcan in transit. Denning mentions the names of only three of his companion observers, these being James Cook, Arthur Blacklock, and F. Wheatley. Of these three observers, the first two were to be listed, a year later, as OAS founding committee members. While the first campaign was technically unsuccessful (as we now know it would have to be), Denning indicated that a further survey was scheduled for July, and placed a call for interested parties to contact him. This summer campaign seems to have not taken place, and we next find Denning organizing a search from 14 March (1869) through to 14 April. In this case a total of 16 observers are credited with taking part, the name list now including Philip Vallance (of Storrington), and Chas Hill (of Bristol) who were to be listed as founding OAS members. Again, no positive transit detections were reported, and it appears that the weather was generally poor across England during the campaign. With the OAS being formed in July of 1869, the next announcement of a Vulcan survey proceeded under its auspices. This third campaign ran from 20 March (1870) to 10 April, and 25 observers pooled their time and resources. This time the weather appears to have been generally favorable, and many sunspot groups were recorded—but no transiting Vulcan. Ever enthusiastic, however, Denning ends his report on the campaign with the intention that he plans to expand the observer distribution to include "more distant nations". In this manner a world-wide distribution of astronomers was being envisioned, with the Sun being monitored continuously. This latter campaign failed to be realized, however, and indeed, OAS members performed no further systematic surveys of the Sun for Vulcan transits after its 1870 effort. This being said, writing in the *Astronomical Register* for February 1871, Denning announced that a Vulcan transit survey for 20 March through to 10 April had been planned, but no report on this campaign appears to have been published. Denning returned to the discussion of Vulcan in a letter to the *Astronomical Register* dated 2 November 1871. In this letter he suggests that observers might look for the wayward planet during the 12 December 1871 solar eclipse—although the track of totality was located over Indonesia and Northern Australia. Further to his letter, however, Denning added a postscript concerning two letters that he had received, one from Robert Wilson in Manchester, and the other from William Waite in London, concerning possible transit observations of Vulcan. That by Wilson was recorded on 1 August 1858, and that by Waite was made sometime between June 1860 and June 1863—the latter observer clearly being very vague as to its supposed occurrence time. The inclusion of Waite's account produced a strong response from one *Astronomical Register* reader. Indeed, in the January 1872 issue of the *Register* a letter from "A SUBSCRIBER" railed, "Might I venture to hint to the Honorary Secretary [Denning] that quality rather than quantity is what is desirable in such communications as those with which he favours us". Referring to this "hint", another reader, "R.E.J", responded in the March issue of the *Register* that Denning was right to bring Waite's account forward. "A SUBSCRIBER" was not impressed by this response, and dismissed Waite's observation as "mental pablum". In the June issue of the *Register*, Denning criticized "A SUBSCRIBER" for hiding behind a pseudonym, but argued that any and all observations relating to Vulcan, no matter how vague, had some merit. In the same issue, "R.E.J" was less gentlemanly, and chastised "A SUBSCRIBER" for their "puerile sentiments". It was at this juncture that Gorton made a rare move, and added an Editor's Note to the effect that, "We cannot insert any more letters upon this controversy". The matter thus ended, but the episode does indicate that passions had been aroused amongst amateur observers.

While Denning organized no further searches for Vulcan after the 1870 campaign, he continued to maintain an interest in the hypothetical planet. Indeed, writing in volume 4 of *Science for All* (published for the year 1888), Denning noted that, "the whole question is at the present time in a most doubtful state". To this he added that, "nothing definite is to be gleaned from past observations", and that, "mathematicians have laboured unsuccessfully to reduce them [the past observations] to a tangible form". Denning's final comments reflect his Baconian outlook, and he argues that only more observations could resolve the situation, noting specifically that the procurement of daily solar photographs will be particularly important. Interest in Vulcan continued well into the first decade of the twentieth century, but by the early 1890's Denning had moved on, becoming more interested in other observing projects. Indeed, in his *Telescopic Work for Starlight Evenings* (published in 1891), Denning writes of Vulcan that, "absolute proof is lacking, and every year the idea is losing strength in the absence of any confirmation of a reliable kind". Furthermore, and with some finality, Denning added, "the ghost of Vulcan may be said to have been laid, and we may regard it as proven that no major planet revolves [interior to Mercury]".

Given the successful founding of the OAS, and the enthusiasm of its reports and member contributions throughout 1870, the question arises as to why did it fail so soon. Indeed, Table 7.1 reveals that the society was already in decline by the close of 1871. The reasons for the Society's demise are no doubt multi-leveled, and sadly there is no clear paper-trail to guide us. It may be said that the society did not fold through a lack of interest, since many of its members continued to report individual observations (in the *Astronomical Register* and elsewhere), and to produce astronomical works, long after the society had ceased to function. Perhaps the choice of the two major OAS observing projects, markings on Venus and the search for Vulcan, could be criticized as being ill advised given that their goals were somewhat vague—it not being clear what measure of success or failure (that is observation-based conclusions) might be leveled against them. When does a null-result turn into a final result? In the case of Vulcan, searches only stopped once Einstein negated the need for any such body in terms of his general theory of relativity extension of Newtonian dynamics. Likewise, when does one stop looking for markings in the upper cloud deck of Venus in order to conclude that markings never occur (at least at optical wavelengths)? Perhaps it was through a lack of cohesion and strong leadership that the society folded, although, as seen, there were numerous attempts to coordinate observations by the membership—these included monitoring the trapezium nebula in Orion for variability, and monitoring the Moon's disk for changes in surface features. Likewise, it does not seem that OAS members lost interest in being part of a national astronomical community since many became leading members in the Liverpool Astronomical Society, when it formed in 1881, and the British Astronomical Association, when it formed in 1890. Ultimately, perhaps, the time was just not right, in the early 1870's, for a strong and cohesive national society to form, and then remain strong, it requiring a few more decades of development in the group-psych of amateur astronomers for the right combination of circumstances to come about.

Bibliography

1 Baum, R. (2012). *Journal of the British Astronomical Association, 122*, 125–126.
2 Denning, W. F. (1891). *Telescopic work for starlight evenings* (pp. 150–151). Taylor and Francis.
3 Brett, J. (1877). The pecular reflection hypothesis and its bearing on the transit of Venus. *Monthly Notices of the Royal Astronomical Society, 37*, 126–127.
4 Hoyt, D., & Schatten, K. (1992). Sir William Herschel's notebooks: Abstracts of solar observations. *The Astrophysical Journal Suppliment Series, 78*, 301–340.
5 Baum, R., & Sheehan, W. (1997). *In search of planet Vulcan—The ghost in Newton's clockwork universe*. Plenum Trade.

Epilogue

Edwin Hubble suggested in the very first chapter of his ground-breaking book *The Realm of the Nebulae* (first published in 1936) that, "the history of astronomy is the history of receding boundaries". The same can be meaningfully said of biographies. Indeed, no biography is ever complete, and no biography is totally unbiased. They teeter on the effervescent divide between the known, and the yet to be discovered, and they are flavoured by time and distance. In one way, or another, each biographer sees a storied life in different ways, and they wriggle and shift their perspective in order to bring out specific details. Furthermore, when the subject is long departed, time and entropy will blur the image, and erase important records. William Frederick Denning, in spite of all that has been written in this text, remains aloof. As a human being we see him only abstractly, gaining but a susurration of a storied life long-lived. Yes, much can be gleaned from Denning's writings and his scientific palmarès, but this is not the human being, rather, it is the selected thoughts, and actions of a human being preserved on paper. For all this, what emerges from the dusted pages is a fascinating snapshot of a remarkable man and grand amateur observer. The tapestry that constitutes the legacy of William Denning is by no means finished, but at least, it is our hope, some structure and color has been added to its form.

The profile that has emerged in the preceding pages places Denning among the preeminent amateur practitioners of late nineteenth century astronomy. Of humble origins, he was unfailing in his enthusiasm, and he found a quiet beauty in the world both above and around him. As Hector Macpherson was to remark of Denning in 1905, "who among us would not be proud to imitate his single hearted and enthusiastic devotion to the discovery of truth"

© The Editor(s) (if applicable) and The Author(s), under exclusive license to Springer Nature Switzerland AG 2023
M. Beech, *William Frederick Denning*, Springer Biographies,
https://doi.org/10.1007/978-3-031-44443-2

[1]. Indeed, to Denning the search for truth was paramount. His search, however, was not always tempered by a critical rationalism. He was a Baconian scientist in the truest sense of its meaning, and he accepted nothing other than that which he saw with his own eyes. It was, however, a lack of critical reasoning with respect to his interpretation of his observations that saw Denning formulate, and then staunchly defend the stationary radiant concept. It is unfortunate that Denning is mainly remembered today for this erroneous idea, his many other important contributions now being mostly forgotten. Indeed, being unreasonably conservative to change, and defending one's reductions is all part of a healthy scientific outlook, and Denning had, at least initially, very good reasons to believe that stationary radiants were real— he was not the only observer who reported finding them. Yes, the concept was eventually shown to be wrong, but that applies to most ideas in the history of science. If nothing else, Denning's life-long dedication to the pursuit of astronomy should provide inspiration to us all, irrespective of the long-term scientific value attributed to some of his findings. The apparent observation of stationary radiants certainly added confusion to the developing field of meteor astronomy, and Denning's dominance ultimately become stifling. He was nonetheless the most important figure in late-nineteenth century British meteor astronomy. Indeed, his projection as Britain's foremost meteor astronomer was certainly justified initially, but his later curmudgeonly dominance is a little more difficult to understand. Certainly, he had honestly earned the respect of his fellow observers and colleagues, but by the second decade of the twentieth century, the key moment of revolutionary change, he was a more isolated, impoverished, and aged gentleman. The revolution in meteor astronomy practices, brewing since the start of the twentieth century, did indeed, claim Denning as its highest victim, but many others were silently dispatched. It may well be that the all too human desire to create heroes and demons lies at the root of Denning's later (and even contemporary) condemnation. To have a victor one has to have, or even invent, a villain. To the victor the spoils, and Denning was from the old guard. He was one of the many pioneers who had to make all manner of assumptions about what meteor astronomy was all about, and what meteors are telling us about the heavens, and like the other pioneers he got some important things right and others wrong. Such is science.

Thomas Kuhn has suggested that science progresses in fits and starts predicated upon occasional revolutions, or paradigm shifts [2]. Between such revolutions, it is argued, the theoretical framework within a specific field and the observational data are seen to be in agreement—at least by the majority of practitioners. Prior to the onset of a paradigm shift, however, a rift begins to grow between theoretic expectation, observational data, and general interpretation, and, according to Kuhn's idea, this culminates in a shift in practitioner perspective—a new outlook is found, one that realigns group-think. Kuhn's ideas on paradigm shifts are no longer held to provide a good interpretation of the history of science in general, but they do appear to apply to many features in the history of meteor astronomy [3]. The first revolution was that relating to the 1833 and 1866 Leonid storms. These two events, purely observational ones, catapulted meteor astronomy from a minor distraction to a major field of astronomical research—it literally established the field as one beyond the processes of data collecting alone. Specifically, it was with respect to the 1866 Leonid storm (recall Chap. 5) that meteors, as a light phenomenon, were linked to meteoroid streams derived from comets. Meteors were henceforth objects with a very definite place in solar system history and planetary astronomy. During Denning's lifetime, and more broadly from circa 1800 to 1900, meteor astronomy saw its founding and initial consolidation. A major paradigm shift occurred circa 1925, when stationary radaints were finally seen as an erroneous feature. Part of this latter shift was the general acceptance of new radiant-reduction protocols, as promoted and pushed for by Charles Olivier, but it was also forced by an increased theoretical understanding of the make-up and formation of meteoroid streams, and specifically cometary orbits and cometary decay. Another paradigm shift occurred circa 1950 when the idea of predominant hyperbolic velocities was abandoned. This latter revolution was driven by the post World War II development of radar instrumentation. It is interesting to note, however, that while developments in radar instrumentation played a significant role in advancing meteor astronomy, photographic techniques, introduced much earlier circa 1900, initially did not. Indeed, the era in which photographic studies really took-off and began to contributed valuable scientific data was the 1960s. At this time dedicated multi-camera survey networks were established across the United States, Canada, and Europe.[1] Much of the motivation behind these early surveys was to detect meteorite-dropping fireball events, but they were additionally

---

[1] The first dedicated network of cameras for capturing meteors and fireballs was established at the Ondřejov Observatory in the Czech Republic in 1959, and this network has continued in operation through to the present day as part of the geographically wider European Fireball Network. Numerous meteorites have now been recovered as a consequence of running this camera network. The Prairie Meteorite Network was established in the mid-western United states in 1964, and ran until 1975. Observations from this network resulted in the collection of the Lost City meteorite, which fell 3 January 1970. The Meteorite Observation and Recovery Project operated in the Canadian prairie provinces from 1971 to 1985, and saw success in the recovery of the Innisfree meteorite, which fell on 5 February, 1977. Most recently, in the first decade of this century, the Desert Fireball Network has been established in Australia. Additional fireball networks, run by amateur astronomers, have been established at multiple sites across the globe. These latter networks typically employ digital camera systems, and use computer interfacing in order to operate autonomously.

invaluable in producing refined observational data relating to meteoroid ablation, and meteor rates. It was also at this time, the 1960's and 70's that the
advent of space exploration prompted the need for a better understanding
of the Earth's upper atmosphere and the space environment. Figure A.1
provides some idea of the dramatic development of meteoritic astronomy
in the 1970's. The diagram is based upon data extracted from the Astrophysics Data System (ADS) maintained by the Smithsonian Astrophysical
Observatory under a grant from NASA. The ADS is a digital library portal
for research in astronomy and physics, and it provides access to some 15
million records. Using the search word "meteors" in the abstracts field yields
some 3013 articles being published between 1781 and 2020. For the entire
nineteenth century, and much of the 20th, the number of articles published
per year was relatively constant, and rarely exceeded 5 to 10 per year. Post
1975, however, the annual publication rate began to increase dramatically.
The reasons for this increase are no-doubt many and varied, but probably
include such features as an up-tick in the number of journals catering to
astronomical research, an increase in the number of professional researchers
working in the field, an increase in the funding being made available for
professional research programs, and the development of new instrumental
observing techniques (such as advanced radar, low light level TV observations,
and CCD cameras). The formation of the International Meteor Organization
in 1988 was also pivotal in promoting meteor astronomy, and in helping
to establish, and coordinate a new generation of meteor observers [4, 5].
Furthermore, the return of comet 109P/Swift-Tuttle to perihelion in 1992,
and the concomitant increase in the activity in the Perseid meteor shower,
provided inspiration to both observers and theoreticians. This return also
inspired the development of detailed numerical models looking to describe
meteoroid stream formation and evolution [6, 7]. Additionally, it was also
realized in the early 1990s that the possibility of enhanced meteoroid fluxes,
at the times of meteor storms and outbursts, presented an impact threat to
Earth-orbiting satellites [8]. In this latter respect, the Leonids, once again,
played an important role in promoting a general interest in the field of meteor
astronomy. This was especially so with the realization that dramatic activity
should occur around the time of the perihelion return of comet 55P/Temple-
Tuttle in 1998. Indeed, spectacular Leonid storms were recorded in 1999,
2001 and 2002.

   In spite of all that has been written in previous chapters, it is still not
entirely clear why Denning chose astronomy as his main field of study. He
might just as easily, for so it appears, turned to subjects such as animal
behavior, meteorology, or writing on popular science. We are simply left

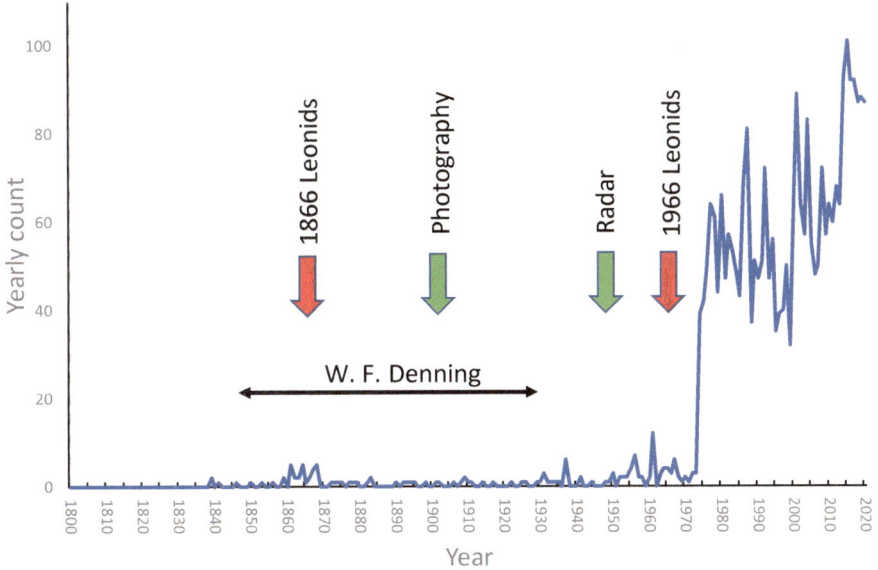

Fig. A.1. The yearly number of refereed research papers relating to meteor astronomy from 1800 to 2020. The data is from the ADS website and the search string "meteors" has been used in the abstract field. Also shown are a few key dates for the 1866 and 1966 Leonid meteor storms, along with arrows showing the approximate times when the new instrumental techniques of photography (c. 1900) and radar (c. 1950) were introduced. The horizontal arrow shows, purely for comparison, the lifespan of W. F. Denning

with Denning's statement that he started observing celestial objects in 1865. Certainly, the field of meteor astronomy was ripe for an observer of Denning's caliber when he first started viewing the heavens. The spectacular Leonid meteor storm of November 1866, and the discovery, by Giovanni Schiaparelli, that meteoroid streams were derived from comets were sufficient reasons to promote meteoric studies to the forefront of astronomical research both at an amateur and a professional level. In contrast to John Donne's poetic decree that no man is an island, however, Denning, in many ways, was an islet adrift from the greater archipelago. An individual, set apart, not in despair, however, but a Robison Crusoe like figure who used what was available and within his reach to build a flourishing domain and make-store of a great legacy. He largely isolated himself from the physical company of others and the direct, face-to face, communion of astronomical societies, but, and in spite of this, through a relentless barrage of letter writing and journalistic goading, he helped establish an active community of amateur astronomers throughout the British Isles. Indeed, we have found Denning to be part of a deep network of gifted amateurs and professional observers spread throughout the British Isles

Fig. A.2. William Frederick Denning. Photograph by J. Webb of Bristol, dated 1904

and beyond. From his isolated domain Denning helped fashion, establish and promote the new scientific field of meteor astronomy, and he helped pave the way for amateur contributions to be taken seriously by the institutionalized elite.

The motivation behind Denning's research studies is reasonably clear. Not only was he interested in furthering his own understanding, and that of science, he was also drawn to the excitement of discovery. He was to comment in 1895, for example, that "while observing, the constant feeling of expectation that a new discovery, or at least an important observation, may be made at any moment, induces a little excitement and keeps one active". Here we find the essential Denning. A man confident that success will follow in the footsteps of application and self-sacrifice, and a man who is prepared to endure discomfort, poverty, and celibacy for the cause of science; that "ceaseless running stream". Indeed, Denning was a man of principle, self-confidence and passion. His passion, however, was that for the opportunity to study the universe in all its many guises. I do not think that Denning had any delusions of academic greatness, although I do believe that he was a great amateur astronomer. He was serious minded, sometimes snappy in his response to other astronomers, but most of all happy when out observing in his garden. In this manner our parting image of Denning should be one of a passionate and dedicated amateur astronomer—a man enthralled by the heavens, and a man who was dedicated to the act of observing. Indeed, Denning's image is not one of a great and revered genius of science, rather it is one of a gentle man who found solace amongst the stars and constellations (Fig. A.2). He found joy under the dark skies of Bristol, while those around him slept soundly in their beds. It is a shadowy image that we see moving at the telescope, an earnest observer silhouetted against the crisp dark sky, and it is with Denning's contented words that we conclude our imagery: "I have supped and imbibed moderately, and even had my weed at the telescope. When I discovered the periodical comet of 1894, on March 26 of that year, I was enjoying my pipe, and it is fortunate for me that the little stranger was not blotted out amid the wreaths of smoke" [9].

Bibliography

1. Macpherson, H. (1905). *Astronomers of today and their work*. Gall and Inglis.
2. Kuhn, T. (1962). *The structure of scientific revolutions*. University of Chicago Press.
3. Beech, M. (1988). Is meteor astronomy a mature science? *Earth, Moon and Planets, 43*, 187–194.
4. Gyssens, M., Knofel, A., Rendtel, J., & Roggemans, P. (1991). The international meteor organization. In *International Astronomical Union Colloquium, Origin and Evolution of interplanetary dust* (vol. 126, pp. 335–338).
5. Roggemans, P. (2006). History: the 25th international meteor conference. *WGN. The journal of IMO, 34*(4), 107–110.
6. Brown, P., & Jones, J. (1998). Simulation of the formation and evolution of the Perseid meteoroid stream. *Icarus, 133*, 36–68.
7. Jenniskens, P. (1998). On the dynamics of meteoroid streams. *Earth, Planets and Space, 50*, 555–567.
8. Beech, M., & Brown, P. (1993). Impact probabilities on artificial satellites for the 1993 Perseid meteoroid stream. *Monthly Notices of the Royal Astronomical Society, 262*, L35–L36.

9. "A self-made English Astronomer", published in the *North British and Ladies Journal*, April 4th, 1904.

Index

© The Editor(s) (if applicable) and The Author(s), under exclusive
license to Springer Nature Switzerland AG 2023
M. Beech, *William Frederick Denning*, Springer Biographies,
https://doi.org/10.1007/978-3-031-44443-2